Genome Mapping and Molecular Breeding in Plants
Volume 6

Series Editor: Chittaranjan Kole

Volumes of the Series
Genome Mapping and Molecular Breeding in Plants

Chittaranjan Kole (Ed.)

Technical Crops

With 35 Illustrations, 2 in Color

 Springer

CHITTARANJAN KOLE
Department of Horticulture
316 Tyson Building
The Pennsylvania State University
University Park, PA 16802
USA

e-mail: cuk10@psu.edu

Library of Congress Control Number: 2006933736

ISBN-13 978-3-540-34537-4 Springer Berlin Heidelberg New York

Springer is a part of Springer Science+Business Media
springer.com

Editor: Dr. Sabine Schreck, Heidelberg, Germany
Desk Editor: Dr. Jutta Lindenborn, Heidelberg, Germany
Cover design: WMX Design GmbH, Heidelberg, Germany
Typesetting and production: LE-TeX Jelonek, Schmidt & Vöckler GbR, Leipzig, Germany
39/3100/YL 5 4 3 2 1 0 – Printed on acid-free paper

Preface to the Series

Genome science has emerged unequivocally as the leading discipline of this new millennium. Progress in molecular biology during the last century has provided critical inputs for building a solid foundation for this discipline. However, it has gained fast momentum particularly in the last two decades with the advent of genetic linkage mapping with RFLP markers in humans in 1980. Since then it has been flourishing at a stupendous pace with the development of newly emerging tools and techniques. All these events are due to the concerted global efforts directed at the delineation of genomes and their improvement.

Genetic linkage maps based on molecular markers are now available for almost all plants of significant academic and economic interest, and the list of plants is growing regularly. A large number of economic genes have been mapped, tagged, cloned, sequenced, or characterized for expression and are being used for genetic tailoring of plants through molecular breeding. An array of markers in the arsenal from RFLP to SNP; tools such as BAC, YAC, ESTs, and microarrays; local physical maps of target genomic regions; and the employment of bioinformatics contributing to all the "-omics" disciplines are making the journey more and more enriching. Most naturally, the plants we commonly grow on our farms, forests, orchards, plantations, and labs have attracted emphatic attention, and deservedly so. The two-way shuttling from phenotype to genotype (or gene) and genotypte (gene) to phenotype has made the canvas much vaster. One could have easily compiled the vital information on genome mapping in economic plants within some 50 pages in the 1980s or within 500 pages in the 1990s. In the middle of the first decade of this century, even 5,000 pages would not suffice! Clearly genome mapping is no longer a mere "promising" branch of the life science; it has emerged as a full-fledged subject in its own right with promising branches of its own. Sequencing of the *Arabidopsis* genome was complete in 2000. The early 21st century witnessed the complete genome sequence of rice. Many more plant genomes are waiting in the wings of the national and international genome initiatives on individual plants or families.

The huge volume of information generated on genome analysis and improvement is dispersed mainly throughout the pages of periodicals in the form of review papers or scientific articles. There is a need for a ready reference for students and scientists alike that could provide more than just a glimpse of the present status of genome analysis and its use for genetic improvement. I personally felt the gap sorely when I failed to suggest any reference works to students and colleagues interested in the subject. This is the primary reason I conceived of a series on genome mapping and molecular breeding in plants.

There is not a single organism on earth that has no economic worth or concern for humanity. Information on genomes of lower organisms is abundant and highly useful from academic and applied points of view. Information on higher animals including humans is vast and useful. However, we first thought to concentrate only on the plants relevant to our daily lives, the agronomic, horticultural and technical crops, and forest trees, in the present series. We will come up soon with commentaries on food and fiber animals, wildlife and companion animals, laboratory animals, fishes and aquatic animals, beneficial and harmful insects, plant- and animal-associated microbes, and primates including humans in our next "genome series" dedicated to animals and microbes. In this series, 82 chapters devoted to plants or their groups have been included. We tried to include most of the plants in which significant progress has been made. We have also included preliminary works on some so-called minor and orphan crops in this series. We would be happy to include reviews on more such crops that deserve immediate national and international attention and support. The extent of coverage in terms of the number of pages, however, has nothing to do with the relative importance of a plant or plant group. Nor does the sequence of the chapters have any correlation to the importance of the plants discussed in the volumes. A simple rule of convenience has been followed.

I feel myself fortunate to have received highly positive responses from nearly 300 scientists of some 30-plus countries who contributed the chapters for this series. Scientists actively involved in analyzing and improving particular genomes contributed each and every chapter. I thank them all profoundly. I made a conscientious effort to assemble the best possible team of authors for certain chapters devoted to the important plants. In general, the lead authors of most chapters organized their teams. I extend my gratitude to them all.

The number of plants of economic relevance is enormous. They are classified from various angles. I have presented them using the most conventional approach. The volumes thus include cereals and millets (Volume I), oilseeds (Volume II), pulse, sugar and tuber crops (Volume III), fruits and nuts (Volume IV), vegetables (Volume V), technical crops including fiber and forage crops, ornamentals, plantation crops, and medicinal and aromatic plants (Volume VI), and forest trees (Volume VII).

A significant amount of information might be duplicated across the closely related species or genera, particularly where results of comparative mapping have been discussed. However, some readers would have liked to have had a chapter on a particular plant or plant group complete in itself. I ask all the readers to bear with me for such redundancy.

Obviously the contents and coverage of different chapters will vary depending on the effort expended and progress achieved. Some plants have received more attention for advanced works. We have included only introductory reviews on fundamental aspects on them since reviews in these areas are available elsewhere. On other plants, including the "orphan" crop plants, a substantial amount of information has been included on the basic aspects. This approach will be reflected in the illustrations as well.

It is mainly my research students and professional colleagues who sparked my interest in conceptualizing and pursuing this series. If this series serves its purpose, then the major credit goes to them. I would never have ventured to take up this huge task of editing without their constant support. Working and interacting with many people, particularly at the Laboratory of Molecular Biology and Biotechnology of the Orissa University of Agriculture and Technology, Bhubaneswar, India as its founder principal investigator; the Indo-Russian Center for Biotechnology, Allahabad, India as its first project coordinator; the then-USSR Academy of Sciences in Moscow; the University of Wisconsin at Madison; and The Pennsylvania State University, among institutions, and at EMBO, EUCARPIA, and Plant and Animal Genome meetings among the scientific gatherings have also inspired me and instilled confidence in my ability to accomplish this job.

I feel very fortunate for the inspiration and encouragement I have received from many dignified scientists from around the world, particularly Prof. Arthur

Kornberg, Prof. Franklin W. Stahl, Dr. Norman E. Borlaug, Dr. David V. Goeddel, Prof. Phillip A. Sharp, Prof. Gunter Blobel, and Prof. Lee Hartwell, who kindly opined on the utility of the series for students, academicians, and industry scientists of this and later generations. I express my deep regards and gratitude to them all for providing inspiration and extending generous comments.

I have been especially blessed by God with an affectionate student community and very cordial research students throughout my teaching career. I am thankful to all of them for their regards and feelings for me. I am grateful to all my teachers and colleagues for the blessings, assistance, and affection they showered on me throughout my career at various levels and places. I am equally indebted to the few critics who helped me to become professionally sounder and morally stronger.

My wife Phullara and our two children Sourav and Devleena have been of great help to me, as always, while I was engaged in editing this series. Phullara has taken pains ("pleasure" she would say) all along to assume most of my domestic responsibilities and to allow me to devote maximum possible time to my professional activities, including editing this series. Sourav and Devleena have always shown maturity and patience in allowing me to remain glued to my PC or "printed papers" ("P3" as they would say). For this series, they assisted me with Internet searches, maintenance of all hard and soft copies, and various timely inputs.

Some figures included by the authors in their chapters were published elsewhere previously. The authors have obtained permission from the concerned publishers or authors to use them again for their chapters and expressed due acknowledgement. However, as an editor I record my acknowledgements to all such publishers and authors for their generosity and good will.

I look forward to your valuable criticisms and feedback for further improvement of the series.

Publishing a book series like this requires diligence, patience, and understanding on the part of the publisher, and I am grateful to the people at Springer for having all these qualities in abundance and for their dedication to seeing this series through to completion. Their professionalism and attention to detail throughout the entire process of bringing this series to the reader made them a genuine pleasure to work with. Any enjoyment the reader may derive from this books is due in no small measure to their efforts.

Pennsylvania,   Chittaranjan Kole
10 January 2006

Preface to the Volume

I believe I owe an explanation to the readers regarding the subtitle of this volume of *Technical Crops* and its contents. There are some crop groups which include a few crop plants, but each of them has immense agricultural importance. These groups comprise fiber crops, forage crops, ornamentals, narcotic crops, plantation crops, and medicinal and aromatic plants. We have included nine chapters on these crop groups devoted to cotton, forage crops, ornamentals, oil palm, coffee, tea, cocoa, rubber, and medicinal and aromatic plants in this volume.

Cotton is the most important fiber crop and progress made on this crop is quite appreciable and up to expectation. The future NSF (USA) Cotton Genome Project will surely facilitate further progress. However, other fiber crops grown in developing countries also require attention and financial support. Jute is an important fiber crop; however, almost no work has been done on this crop on genome mapping and molecular breeding. Sunhemp and sisal are still in the list of "beggar crops." Flax is another fiber crop that could have been included in this volume; however, considering its primary importance as a source of vegetable oil, we placed it in Vol. 2 dedicated to oilseeds. On the basis of the same logic, we have included cotton in this volume, not in Vol. 2, albeit with full appreciation of its importance as one of the leading oil-yielding crops.

Plantations of oil palm, coffee, tea, cocoa, and rubber provide raw materials for a large number of small-scale industries and provide livelihood for innumerable households in developing countries in Asia and Africa. Currently, improvement of some of these crops, particularly oil palm, coffee, and cocoa, is receiving appreciable international support. We believe all these crops deserve more generous support at national and international levels. We could present comprehensive reviews on traditional genetic and breeding efforts along with preliminary works done on molecular areas in these crops in this volume. We are thankful to all the authors of the chapters on these crops. We omitted, however, the work done on tobacco, an important narcotic crop and wish to include it in future editions.

The three groups on forage crops, ornamentals, and medicinal and aromatic plants include a large number of plants in each of them. Only a few plants in these groups have attracted attention for molecular work. The authors of the chapters on these crop groups have also discussed this. We hope each of these groups will require independent volumes in the near future to present commentary on genome mapping and molecular breeding efforts in their member crop plants. I should mention here that we had a chapter on *Capsicums* in Vol. 5 of the series dedicated to vegetables. A section on them has again been included in the chapter "Medicinal and Aromatic Plants" in this volume. I believe readers will find these two reviews to be completely different in approach and content.

Many crop plants in this volume are being confronted with competition from "synthetic products." However, growing consciousness for the healthy and ecofriendly attributes of natural plant products is going to reverse the trend of consumer preference. The time has come to realize the threats posed by synthetic products and to welcome again organic and herbal commodities and embrace natural style living, and to live in and with nature.

Pennsylvania,   Chittaranjan Kole
4 July 2006

Contents

Contributors

François Anthony
Institute of Research for sustainable
Development (IRD)
GeneTrop, BP 64501, 34394
Montpellier Cedex 5, France

Ranjana Bhattacharjee
International Institute of Tropical
Agriculture (IITA), Ibadan, Nigeria
r.bhattacharjee@cgiar.org

Norbert Billotte
CIRAD (CIRAD-CP) TA 80/03
Avenue Agropolis, 34398
Montpellier Cedex, France

Curt Brubaker
Center for Plant Biodiversity Research
CSIRO Plant Industry, GPO Box 1600
Canberra, ACT 2601, Australia

Hongwei Cai
Forage Crop Research Institute (FCRI)
Japan Grassland Agriculture
& Forage Seed Association
388-5, Higashiakada, Nasushiobara
Tochigi 329-2742, Japan
hcai@jfsass.or.jp

Peng Chee
Department of Crop & Soil Science
Coastal Plain Experiment Station
NESPAL Building, University
of Georgia, 2356 Rainwater Road
Tifton, GA 31793-0748, USA

Thomas Debener
Department of Molecular Breeding
Institute of Plant Genetics, Faculty
of Natural Sciences, Hannover
University
Herrenhäuser Str. 2, 30419, Hannover
Germany
debener@genetik.uni-hannover.de

Frederic Dumortier
DAMI, OPRS, New Britain Palm
Oil Ltd., P.O. Box 165, Kimbe
West New Britain Province
Papua New Guinea

Masahiro Fujimori
Yamanashi Prefectural Dairy
Experiment Station, 621-2
Nagasaka-Kamijo, Nagasaka, Hokuto
Yamanashi 408-0021, Japan

Farah Hafeez
Department of Genetics, Downing Site
Cambridge University, Downing Street
Cambridge CB2 2EH, UK

Maiko Inoue
Forage Crop Research Institute (FCRI)
Japan Grassland Agriculture
& Forage Seed Association
388-5, Higashiakada, Nasushiobara
Tochigi 329-2742, Japan

Suman P.S. Khanuja
Central Institute of Medicinal
and Aromatic Plants, P.O. CIMAP
Lucknow 226015, India

P. Lava Kumar
International Crops Research Institute
for the Semi-Arid Tropics (ICRISAT)
Patancheru, India

Philippe Lashermes
Institute of Research for sustainable
Development (IRD)
GeneTrop, BP 64501, 34394
Montpellier Cedex 5
France
philippe.lashermes@mpl.ird.fr

Marcus Linde
Department of Molecular Breeding
Institute of Plant Genetics, Faculty
of Natural Sciences, Hannover

University, Herrenhäuser Str. 2
30419, Hannover, Germany

Sean Mayes
Division of Biosciences
Sutton Boninghton Campus
Nottingham University
Loughborough LE12 5RD, UK
sean.mayes@nottingham.ac.uk

Don MacDonald
Department of Genetics, Downing Site
Cambridge University, Downing Street
Cambridge CB2 2EH, UK

Zuzana Price
Department of Genetics, Downing Site
Cambridge University, Downing Street
Cambridge CB2 2EH, UK

P.M. Priyadarshan
Rubber Research Institute of India
Regional Station, Agartala 799006
Tripura, India
pmpriyadarshan@gmail.com

Ajit K. Shasany
Genetic Resources and Biotechnology
Division, Central Institute of Medicinal
and Aromatic Plants, P.O. CIMAP
Lucknow 226015, India
khanujazy@yahoo.yom

Ashutosh K. Shukla
Genetic Resources and Biotechnology

Division, Central Institute of Medicinal
and Aromatic Plants, P.O. CIMAP
Lucknow 226015, India

Junichi Tanaka
Genetic Resources and Breeding
Team & Tea Genome Research Team
National Institute of Vegetable
and Tea Science, National Agriculture
and Bio-Oriented Research
Organization, Seto-cho 87, Makurazaki
Kagoshima, 898-0087, Japan
tanajun@affrc.go.jp

Fumiya Taniguchi
Genetic Resources and Breeding
Team & Tea Genome Research Team
National Institute of Vegetable
and Tea Science, National Agriculture
and Bio-Oriented Research
Organization, Seto-cho 87, Makurazaki
Kagoshima, 898-0087, Japan

Mauricio Ulloa
USDA-ARS, WICS, Research Unit
Cotton Enhancement Program
17053 N. Shafter Avenue, Shafter
CA 93263, USA
mulloa@pw.ars.usda.gov

Zifu Yan
Plant Research International
Wageningen University
and Research Centre, P.O. Box 16
6700 AA Wageningen, The Netherlands

Abbreviations

AFIS	Advanced fiber information system
AFLP	Amplified fragment length polymorphism
ASGR	Apospory-specific genomic region
AVROS	Algemene Verneiging Rubber Planters Oostkust, Sumatra
BAC	Bacterial artificial chromosome
BAP	Benzylaminopurine
BC	Backcross
BLAST	Basic Local Alignment Search Tool
BPM	Balai Penelitian Medan
BPRO	Breeding populations of restricted origin
CAP	Coordinated Agricultural Project
CAPS	Cleaved amplified polymorphism sequence
cDNA	Complementary DNA
CIM	Composite interval mapping
CIMAP	Central Institute of Medicinal and Aromatic Plants (Lucknow, India)
CIRAD	Centre de Coopération Internationale en Recherché Agronomique pour le Développement
CLCuV	*Cotton leaf curl virus*
cM	centimorgan
CMS	Cytoplasmic male sterility
cpDNA	Chloroplast DNA
CSSV	*Cocoa swollen shoot virus*
CYP80B1	P450 hydroxylase
2,4-D	2,4-Dichlorophenoxyacetic acid
DAF	DNA amplification fingerprinting
DH	Doubled haploid
EST	Expressed sequence tag
F and Fx	Clones from the collections and recombinations of Ford Company in Brazil
FG	Fall growth
FI	Freezing injury
FIS	Family and individual selection
GAM	Generation-wise assortative mating
GA_3	Gibberellic acid
GCA	General combining ability
GG	Gough Gardens
GISH	Genomic in situ hybridization
Gl	Glenshiel
GLS	Gray leaf spot
GT	Gondang Tapen
HVI	High-volume instrument
IAC	Instituto Agronomico de Campinas, São Paulo, Brazil
IAN	Instituto Agronomico do Norte
IBA	Indolebutyric acid
IPP	Isopentenyl pyrophosphate

IRAP	Interretrotransposon amplified polymorphism
IRCA	Institut de Recherches sur le Caoutchouc en Afrique
IRD	Institute of Research for sustainable Development (Paris, France)
IRRDB	International Rubber Research and Development Board
ISSR	Intersimple sequence repeat
ITS	Internal transcribed spacer
kb	kilobase
LD	Linkage disequilibrium
LOD	Logarithm of odds
LR	Likelihood ratio
LSCT	Large-scale clone trial
LTR-RTN	Long-terminal-repeat retrotransposon
MAB	Marker-assisted breeding
MAS	Marker-assisted selection
Mb	Million bases
Mbp	Million base pairs
MPM	Maximum parsimony method
MPOB	Malaysian Palm Oil Board
mtDNA	Mitochondrial DNA
NAA	Naphthaleneacetic acid
NDF	Neutral detergent fiber
NIBGE	National Institute for Biotechnology & Genetic Engineering
NJM	Neighbor-joining method
NMT	N-Methyltransferase
PAUP	Phylogenetic Analysis Using Parsimony
PB	Prang Besar Rubber Estate
PBIG	Prang Besar Isolated Gardens
PCA	Principal component analysis
PCR	Polymerase chain reaction
PHB	Polyhydroxybutyrate
Pil	Pilmoor Proefstation, Indonesia
PK	Protein kinase
PR	Proefstation, Indonesia
QTL	Quantitative trait locus
RAPD	Randomly amplified polymorphic DNA
rDNA	Ribosomal DNA
REF	Rubber elongation factor
REMAP	Retrotransposon microsatellite amplified polymorphism
Rf	Fertility restoration
RFLP	Restriction fragment length polymorphism
RGA	Resistance gene analog
RHM	Radiation hybrid mapping
RIL	Recombinant inbred line
RRIC	Rubber Research Institute of Ceylon
RRII	Rubber Research Institute of India
RRIM	Rubber Research Institute of Malaysia
RRIT	Rubber Research Institute of Thailand
RRIV	Rubber Research Institute of Vietnam
rRNA	Ribosomal RNA
RRS	Reciprocal recurrent selection
SALB	South American leaf blight

SCA	Specific combining ability
SCAR	Sequence-characterized amplified region
SCATC	South China Academy of Tropical Crops
SDRF	Single-dose restriction fragment
SET	Seedling evaluation trial
SNP	Single nucleotide polymorphism
SRF	Seedling root florescence
SRPP	Small rubber particle protein
SSCT	Small-scale clonal trial
SSR	Simple sequence repeat
STMS	Sequence-tagged microsatellite site
STS	Sequence-tagged site
TIA	Terpenoid indole alkaloid
TILLING	Targeted induced local lesions in genomes
Tjir	Tjirandji, Indonesia
UPGMA	Unweighted pair group method with arithmetic mean
WI	Winter injury
WSC	Water-soluble carbohydrate content
WWRH	Radiation hybrid wide-cross whole genome

1 Cotton

Mauricio Ulloa[1], Curt Brubaker[2], and Peng Chee[3]

[1] WICS, Research Unit, Cotton Enhancement Program, ASDA-ARS, 17053 N. Shafter Avenue, Shafter, CA 93263, USA
  e-mail: mulloa@pw.ars.usda.gov
[2] Centre for Plant Biodiversity Research, CSIRO Plant Industry, GPO Box 1600, Canberra, ACT 2601, Australia
[3] Dep Crop & Soil Science, University of Georgia, Coastal Plain Experiment Station, 2356 Rainwater Road, Tifton,
  GA 31793-0748, USA

1.1
Introduction

The most important renewable natural textile fiber worldwide and the world's sixth largest source of vegetable oil is cotton. After losing some ground to synthetic fibers in the past, the demand for cotton has been steadily growing. Between 1998 and 2003, there were five consecutive records for cotton consumption with a net gain of 2.6 million tons of cotton finding its way to textile mills (Valderrama 2004). The other side of this coin is that the competitiveness of the supply market has also increased. Production is also increasing, as is yield per hectare, and the price of cotton relative to that of other textile fibers is declining (Valderrama 2004; Anonymous 2005). The cotton growers who survive through the next decade will be those who can produce the most cotton per unit input and produce cotton whose characteristics attract price premiums. This is the message being delivered to cotton breeders around the world. While perennial concerns regarding pest and disease resistance and yield remain, water-use efficiency/drought tolerance, fiber quality and uniformity are high priorities for the future of cotton improvement. The challenge for breeders is that the high-priority breeding objectives for the next decade involve phenotypically complex traits controlled by many interacting genes. Making real breeding gains will require an unprecedented understanding of the molecular genetics of these complex traits. The purpose of this chapter is to provide the reader with an understanding of recent research advances in cotton molecular DNA technologies and genetic linkage mapping. These are the primary tools breeders will use to reach their objectives in the next decade.

While much attention has been focused on genetic linkage maps and quantitative trait locus (QTL) analyses in the recent literature, the fundamental analytical process, genetic linkage analysis, dates back to Mendel. The sophisticated genetic linkage maps of the past decade are the result of our ability to look at DNA sequence differences directly with molecular markers and to use powerful computers that analyze genetic linkage at scales that were impossible to do by hand. Unable to view genes directly, a geneticist could, for a small number of genes controlling a trait with a discrete and variable phenotype, infer the number of genes involved and their mode of action and interaction with other genes. However, the application of this information to cotton breeding was limited because the number of agronomically useful traits that are phenotypically discrete and variable is low, too low for genomewide sampling of genetic linkages. The critical agronomic traits (e.g., yield, quality, disease resistance) were controlled by complex interactions of many genes distributed across the genome that produced a phenotype that varied continuously between two extremes.

The event that allowed the modern cotton breeder to dissect complex agronomic traits genetically was the discovery that DNA-modifying enzymes could be isolated from living organisms (e.g., restriction enzymes, DNA and RNA polymerases, kinases, ligases) and used to manipulate and characterize the cotton genome. The new generation of cotton breeders, the molecular breeders, could use these molecular tools to locate and characterize the sequences of DNA-encoding proteins and the DNA – protein binding regions that controlled their expression, developmentally and temporally. Even when the inferred gene is yet to be located, sequenced, and characterized, molecular breeders could use these natural DNA modifying enzymes to identify sequence differ-

Genome Mapping and Molecular Breeding in Plants, Volume 6
Technical Crops
C. Kole (Ed.)
© Springer-Verlag Berlin Heidelberg 2007

ences among breeding lines and then apply traditional genetic linkage analyses to identify sequence differences (e.g., randomly amplified polymorphic DNAs, RAPDs, restriction fragment length polymorphisms, RFLPs, amplified fragment length polymorphisms, AFLPs, simple sequence repeats, SSRs, sequence-characterized amplified regions, SCARs, single nucleotide polymorphisms, SNPs) that could be used as surrogates for the inferred genes (marker-assisted selection, MAS). Molecular markers are being continuously developed, which allow cotton geneticists to sample all regions of the cotton genome. When this genome saturation is combined with quantitative genetic analyses, cotton breeders finally will have the tools they need to identify the location of the genes (QTLs) conditioning the expression of complex agronomic traits, such as yield (Ulloa and Meredith 2000; Paterson et al. 2003), water tolerance , and water-use efficiency (Ulloa et al. 2000; Saranga et al. 2001).

The fundamental goal of genetic linkage mapping is, and always has been, gene discovery, or when the gene itself is yet to be located, the identification of a linked surrogate that can be used in its stead. Identifying the genes conditioning a trait, their mode of expression, and how they interact gives the cotton breeder the greatest leverage in manipulating the trait to the desired effect. Traditionally this was a process of combining the most effective alleles, from what was available in the germplasm into a single genotype. The advent of transgenic methodologies and the discovery of which genes are interacting to condition a critical trait allows cotton breeders to look beyond traditional germplasm for useful genes, but the efficacy of transgenic technology is entirely dependent on gene discovery. Over the past decade new gene discovery tools have emerged that do not depend on genetic linkage analyses (e.g., bacterial artificial chromosome, BAC, physical mapping, microarrays, and differential display). While these new tools provide exciting new avenues for gene discovery that circumvent some of the logistical limitations of genetic linkage mapping, genetic linkage mapping will continue to be a critical tool for gene discovery in cotton breeding. So this is a particularly appropriate time to review how genetic linkage mapping can be effectively integrated into future gene discovery methodologies.

To understand the complexities of genetic linkage mapping in cotton, we must first understand the evolution and structure of the cotton genome. Understanding the cotton genome is complex because cotton, the lint used in textiles, is derived from the seed trichomes (hairs) of not one, but four *Gossypium* species: two Old World diploid species, *G. arboreum* L. and *G. herbaceum* L., and two New World allotetraploid species, *G. barbadense* L. and *G. hirsutum* L. Seed trichomes are common in the plant world, and while seed trichomes of the lengths we see in the four *Gossypium* cotton species are certainly uncommon, it is not the length of fiber that makes cotton unique. The singular feature of cotton fibers that sets them apart from other plant seed trichomes and other non-cotton *Gossypium* species is the fact that the cellulose microfibrils are laid down in a spiral arrangement. When the boll (a capsule in botanical terms) opens and the seed trichomes desiccate, they collapse into convoluted ribbons; this is what makes cotton fibers "spinnable." While this chapter will focus primarily on the most widely cultivated of the four species, *G. hirsutum*, the other three cannot be ignored in any discussion of cotton genome mapping. *G. barbadense* contributes a significant proportion of the extra-long staple lint required for a significant niche of the cotton market and, importantly, is typically used as a parent to generate genetic populations and/or segregating families for genetic linkage mapping. *G. arboreum* and *G. herbaceum*, which are still cultivated in Africa and Asia, are important because they are the only extant representatives of the lineage in which the convoluted seed trichome first evolved (Saunders 1961), and are therefore critical experimental models for gene discovery. Also important are the New World D-genome diploid species, which contributed the second genome to the tetraploid cottons and are an important source of genes for lint quality.

1.1.1
Origin and Domestication of Cotton

The cotton genus, *Gossypium* (Malvaceae) comprises 45 diploid and five allotetraploid species (Fryxell 1979, 1992; Percival et al. 1999; Ulloa et al. 2006). *Gossypium* species are found in tropical and subtropical regions of Australia, Africa, Arabia, Asia, and the Americas. Cotton breeders typically group the *Gossypium* species into one of eight diploid genomes (A – G and K) and one allotetraploid genome (AD) (Beasley 1940, 1942; Phillips and Strickland 1966; Edwards and Mirza 1979; reviewed by Endrizzi et al. 1985; Stewart 1995; Percival et al. 1999). These genomic designations are functional groups based on the similarities in chromosome size and structure and the success of interspecific crosses. In general, within genomes hybrids are fertile and their chromosomes recombine during meiosis; however, intergenomic hybrids are generally infertile and there are few stable bivalents at meiosis (Stewart 1995). *G. barbadense* and *G. hirsutum* are AD-genome tetraploids; *G. arboreum* and *G. herbaceum* are A-genome diploids.

The diploid *Gossypium* species have 13 pairs of chromosomes, while the tetraploid species have 26 chromosome pairs. Haploid nuclear DNA amounts (1C) range from 980 to 3,425 Mbp (Edwards et al. 1974; Kadir 1976; Bennett et al. 1982; Endrizzi et al. 1985; Michaelson et al. 1991; Stewart 1995; Grover et al. 2004). The length of the A-genome nuclear DNA is approximately 1,860 Mbp. The D genome is smaller with a haploid length of 980 Mbp. Genome size in the AD tetraploids is roughly additive: 2,835 Mbp.

1.1.2
Evolution of Cotton

Evolution of the Old World Cottons
The Old World A-genome diploid cottons belong to an early *Gossypium* lineage that comprises the 14 African and Arabian species (genomes A, B, E, F). Within this group the evidence suggests that each of the genomes represents a monophyletic group of species (Wendel and Albert 1992; Seelanan et al. 1997). Because wild populations of *G. arboreum* have never been identified, it has been suggested that that *G. arboreum* is simply a recent segregate of domesticated forms of *G. herbaceum* (Hutchinson et al. 1947; Hutchinson 1954, 1959). As discussed later, cytogenetic, genetic, and phylogenetic studies, however, clearly indicate that the two A-genome species diverged prior to the independent domestication of each species (Wendel et al. 1989; Liu et al. 2001).

Evolution of the New World Cottons
The New World tetraploid cottons are allotetraploids that combine one genome derived from an A-genome ancestor and a second genome from a D-genome ancestor (Endrizzi et al. 1985; Galau and Wilkins 1989; Wendel 1989; Brubaker et al. 1999; Liu et al. 2001; Senchina et al. 2003; Wendel and Cronn 2003), two lineages that diverged about 6.0–7.3 million years ago (Senchina et al. 2003). How this occurred remains a mystery. All the known AD tetraploid species are indigenous to the New World, but there is no evi-

dence of any A-genome species in the New World, and there are no D-genome species outside the New World. Nonetheless, 1.3–1.7 million years ago an A-genome plant was pollinated by a D-genome plant and produced an AD diploid hybrid, which, in a single polyploidization event, gave rise to an AD allotetraploid with an A-genome cytoplasm (Wendel 1989; Cronn et al. 1996; Small et al. 1998; Senchina et al. 2003; Wendel and Cronn 2003). After the original AD tetraploid entity arose, it subsequently diverged into three distinct lineages (Seelanan et al. 1997; Small et al. 1998; Cronn et al. 2002). One lineage is represented by *G. mustelinum*, which is found only in a small region of northeast Brazil. The two remaining lineages contain a domesticated species and wild island endemic species: *G. barbadense/G. darwinii* (Galapagos Islands) and *G. hirsutum/G. tomentosum* (Hawaii).

Of the extant A- and D-genome species, *G. arboreum* and *G. herbaceum* are the best diploid models of the tetraploid At subgenome and *G. raimondii* is the best model of the Dt subgenome (Wendel and Cronn 2003). The gross structural arrangement of the *G. arboreum* genome differs from that of the At subgenome by only two translocations, while the *G. herbaceum* genome differs from the At subgenome by three translocations (Brown and Menzel 1950; Gerstel 1953; Menzel and Brown 1954; Endrizzi et al. 1985; Brubaker et al. 1999), but the phylogenetic data, however, clearly indicate that these two species diverged after the polyploidization event and therefore both of the extant A-genome species are phylogenetically equidistant from the At subgenome and equally appropriate living models of the At subgenome. There has been considerable speculation over the years as to which extant D-genome species is the best model of the Dt subgenome (Endrizzi et al. 1985; Wendel 1989; Wendel et al. 1995), but the most recent and convincing data clearly indicate that *G. raimondii* is the closest living relative of the ancestor of the Dt subgenome and therefore its best living model (reviewed by Wendel and Cronn 2003).

1.1.3
Morphology and Taxonomy of the Cotton Species

The Old World Diploid Cotton Species
The two A-genome cotton species, *G. arboreum* and *G. herbaceum*, are shrubs or subshrubs less than 2-m tall with palmately lobed leaves and yellow flowers (also in *G. arboreum* white to red to purple) that are

surrounded by three broadly ovate, laciniate, epicalyx bracts with cordate bases. *Gossypium arboreum* is distributed throughout Asia (the Indian subcontinent, China, and Southeast Asia), but can also be found in southern Arabia, northern and eastern coastal Africa. *Gossypium herbaceum* is primarily African and Arabian in distribution, but under cultivation has spread into the Middle East, India, Iran, and Afghanistan. The morphology of the two species overlaps and misidentification is common, but the two species, in general, can be differentiated on the basis of the number of the epicalyx teeth, epicalyx shape, leaf lobe shape and sinus depth, and boll shape (Hutchinson and Ghose 1937; Hutchinson et al.1947; Fryxell 1979, 1992; Wendel et al. 1989; Stanton et al. 1994). *G. arboreum* typically has elongated (rather than globose) bolls and longer epicalyx bracts with fewer teeth relative to *G. herbaceum*, which, in turn, typically has leaves with shorter, broader (ovate rather than lanceolate) leaf lobes.

Although there is considerable morphological overlap between *G. arboreum* and *G. herbaceum*, genetic and cytogenetic studies clearly indicate that they are two distinct species. When *G. arboreum* and *G. herbaceum* are hybridized the F_1 is vigorous and fertile, but pollen fertility is less than 60%. The progeny of *G. arboreum* and *G. herbaceum* exhibit high levels of F_2 breakdown This is evident in the high frequency of inviable seeds, moribund seedlings, and anomalous morphologies (Silow 1944; Stephens 1950; Gerstel and Sarvella 1956; Phillips 1961). In further generations, there is a clear bimodal segregation toward the parents. Furthermore, where the two species are sympatric (northern Africa, Arabia, northern and western India) each species remains distinct (Silow 1944; Hutchinson et al. 1947). The species differ by a reciprocal translocation (Gerstel 1953; Gerstel and Sarvella 1956), and, genetically, *G. arboreum* and *G. herbaceum* have high proportions of private alleles, 31 and 19%, respectively. The estimated genetic identity between *G. arboreum* and *G. herbaceum* ($I = 0.74$) is typical of the level of genetic differentiation between species, and is much lower than expected for populations within a species.

The immense range of morphological variation in each of the A-genome species has lured taxonomists into identifying a bewildering range of formal and informal subspecific entities (Hutchinson and Ghose 1937; Silow 1944; Hutchinson et al. 1947; Hutchinson 1950); however, the only taxon that is useful is *G. herbaceum* ssp. *africanum* (Fryxell 1992).

This group of southern African populations represents the only known wild forms of *G. herbaceum*. They are components of natural vegetation rather than feral derivatives in ruderal habitats (Hutchinson 1954). The remaining subspecific entities in both species do not represent evolutionary lineages that warrant taxonomic recognition. They are most appropriately treated as morphogeographic races that represent infraspecific genetic coalitions that correspond with nodes of domestication and human-mediated germplasm diffusion pathways (reviewed by Brubaker et al. 1999).

The most primitive agronomic forms of *G. arboreum* are assigned to race *indicum*. Although a few annualized forms may be included here, race *indicum* plants are perennial cultigens grown along the eastern coast of Africa, including Madagascar, and in India (the Rozi cottons). Morphologically, they range from short multibranched shrubs to columnar subshrubs. The lint is sparse and coarse although there is some evidence of improvement in some genotypes. The race *soudanense* cottons (Senaar tree cottons), in contrast, are large perennial shrubs or small trees (stem diameters can reach 7–8 cm) of northwestern Africa and the Sudan. The lint is of moderately higher quality than the race *indicum* cottons. A third group of predominately perennial cottons, race *burmanicum*, occurs in northeastern India. An annualized segregate ecotype of race *burmanicum*, known as race *cernuum*, is found in Assam. The two exclusively races *bengalense* (northern India) and *sinense* (China) are characterized by very early maturity and are grown in areas subject to frost.

The cultivated derivatives of *G. herbaceum* can be assigned to one of four races. Race *acerifolium* from northern Africa and Arabia is the least improved agronomically. These perennial multibranched shrubs have small bolls and the lint is coarse and sparse. The three annualized races *kuljianum* (western China), *persicum* (Iran and Afghanistan, and *wightianum* (western peninsular India) vary from small slender sparsely branched subshrubs with small bolls with scanty low-quality lint to large stout moderately branched shrubs with large bolls and copious high-quality lint.

The New World Diploid Gossypium D-Genome Species

Although the 14 diploid species of the D genome are not cotton species in the strict sense (they do not bear spinnable lint), they merit consideration because of their contribution to the genome of the tetraploid cotton species. The majority of D-genome species are found in Mexico. Five species are adapted to the desert environments of Baja, California (*G. armourianum* Kearney, *G. harknessii* Brandegee, and *G. davidsonii* Kellogg) and northwestern mainland Mexico (*G. turneri* Fryxell and *G. thurberi* Todaro). An additional seven species [*Gossypium* sp. nov.(see later), *G. aridum* (Rose & Standley) Skovsted, *G. lobatum* Gentry, *G. laxum* Phillips, *G. schwendimanii* Fryx. & Koch, *G. gossypioides* (Ulbrich) Standley, and *G. trilobum* (Mociño & Sessé ex DC.) Skovsted] are located in the Pacific coast states of Mexico and, with the exception of the last species, are arborescent in growth habit (Ulloa et al. 2006). *G. raimondii* Ulbrich is endemic to Peru, while *G. klotzschianum* Andersson is found in the Galápagos Islands. The D-genome species (subgenus *Houzingenia*) are classified into two sections: section *Houzingenia* (*G. trilobum*, *G. thurberi*, *G. klotzschianum*, *G. davidsonii*, *G. harknessii*, *G. armourianum*, *G. turneri*) and section Erioxylum (*Gossypium* sp. nov., *G. aridum*, *G. laxum*, *G. lobatum*, *G. gossypioides*, *G. raimondii*, *G. schwendimanii*).

To assess the current status of *Gossypium* genetic resources in Mexico, Ulloa et al. (2006) conducted a series of USA – Mexico collaborative expeditions in areas and states where cultivated and wild *Gossypium* were known to occur, but that had not been surveyed for several decades. They collected seven populations of known species, *G. aridum*, *G. barbadense*, *G. gossypioides*, *G. hirsutum*, *G. laxum*, *G. lobatum*, and *G. schwendimanii*, and one undescribed wild diploid *Gossypium* taxon during the survey. In situ conservation of some of these species is threatened, and as these cottons disappear, the accessions that are preserved ex situ will be the surviving source of genetic diversity that was once present in southern and western Mexico (Ulloa et al. 2006).

Gossypium sp. nov. is a new undescribed taxon reported by Ulloa et al. (2006). At a small settlement (Oxtutla) in eastern Guerrero, a *Gossypium* taxon was encountered that initially, in its defoliated condition, resembled *G. aridum*. Observations of greenhouse plants and subsequent site observations of flowering plants (with some leaves remaining), indicated that this population represents an undescribed taxon belonging to subsection *Erioxylum*. Its distribution range is unknown, and because this species is newly discovered, current documentation by Ulloa et al.

(2006) provides the baseline for its in situ status. More recently, Alvarez et al. (2005) studied the 13 described D-genome species as well as this new undescribed taxon (US72). The sequences of three low-copy nuclear genes also suggest that this new taxon should be recognized at specific level. US72 taxon appears to be sister to *G. laxum* (Alvarez et al. 2005). At present, it is known only from remote steep-sided ravines in Guerrero. In view of the remoteness of the area, it does not appear to be under immediate threat, but it is likely that human activity could threaten its survival in the future.

The New World Tetraploid Cotton Species

The two New World AD tetraploid cotton species *G. barbadense* and *G. hirsutum* are large shrubs to small trees (Fryxell 1979, 1992). *G. barbadense* is indigenous to South America but does extend into Mesoamerica and the Caribbean. The center of morphological diversity for *G. hirsutum* is Mesoamerica, but the indigenous range includes the Caribbean, northern South America, and some Pacific Islands. Distinguishing *G. barbadense* from *G. hirsutum* morphologically is much easier than it is for Old World cotton species. Although there is some morphological overlap where the species are sympatric, misidentification is rarer. *G. barbadense* has large bright yellow flowers (8 cm) and large epicalyx bracts (4–6 cm). *G. hirsutum* has smaller cream or pale yellow flowers (2–5 cm) and epicalyx bracts (2–4.5 cm). The short (2–4 cm) smooth, ovoid to subglobose *G. hirsutum* boll is quite distinct relative to the long (3.5–6 cm), deeply pitted, narrow ovoid *G. barbadense* boll. *G. barbadense* and *G. hirsutum* are not distinguished by any gross cytostructural rearrangements (Gerstel and Sarvella 1956). As was the case for *G. arboreum* × *G. herbaceum* hybrids, the F_1 hybrids are fertile, but there is significant F_2 breakdown and subsequent segregation toward the parental types in later generations (Stephens 1950). Genetic discrimination between the two species is unambiguous. They are fixed or nearly fixed for alternate alleles at 11 isozyme loci and 18 RFLP alleles (Percy and Wendel 1990; Wendel et al. 1992; Brubaker and Wendel 1994), and the two species have different ribosomal DNA sequences (Wendel et al. 1995), and chloroplast genomes (Wendel and Albert 1992).

As was the case with the Old World cottons, the morphological diversity among the New World cottons that emerged from millennia of human selection inspired the description of a large number of formal and informal subspecific entities (reviewed by Brubaker et al. 1999). Of these, only one infraspecific taxon within *G. barbadense* is worthy of note, albeit as a separate species, and Hutchinson's (1951) seven morphogeographic races of *G. hirsutum* are useful in understanding the domestication of this species.

G. barbadense was originally thought to comprise three discrete entities: variety *barbadense*, variety *darwinii*, and variety *brasiliense* (Hutchinson et al. 1947). *Gossypium barbadense* var. *darwinii* is the allotetraploid cotton endemic to the Galápagos Islands. Although morphologically similar to mainland *G. barbadense* populations, the Galápagos Island plants represent a distinct gene pool (Wendel and Percy 1990; Wendel and Albert 1992), and are now recognized as *G. darwinii* (Fryxell 1979, 1992). *Gossypium* var. *brasiliense* is the name applied to the 'kidney-seeded' cottons of the Amazon basin. The diagnostic feature of these cottons, the fusion of all the seeds in each locule of the capsule into a single mass, however, is controlled a single recessive gene (Turcotte and Percy 1990). These cultigens are otherwise genetically indistinguishable from other cultigens of *G. barbadense* (Percy and Wendel 1990) and are no longer recognized as a formal infraspecific taxon (Fryxell 1979, 1992).

G. hirsutum was also subdivided into three varieties at one time: variety *hirsutum*, which included Upland cotton and other early cropping or annualized forms, variety *punctatum*, and variety *marie-galante* (Hutchinson et al. 1947). This original formal taxonomic circumscription was later replaced by an informal system of seven morphogeographical races (Hutchinson 1951). Race *yucatanense* comprises the only truly wild forms of *G. hirsutum* found on the northern coast of the Yucatan Peninsula. Across the indigenous range of *G. hirsutum*, the most agronomically primitive forms are typically assigned to race *punctatum*. They are mostly small shrubs and although they may bear a large number of bolls, the yield and lint quality is low. A large proportion of race *punctatum* accessions also have tufted seeds (i.e., they lack the fuzz layer typical of more advance cultigens). The most agronomically advanced of the indigenous cottons are assigned to race *latifolium*. Race *latifolium* cottons are typically compact shrubs with medium to very large bolls bearing good-quality lint. Of the indigenous *G. hirsutum* races, race *latifolium* is the most similar to the modern elite cultivars, genetically and morphologically (Brubaker and Wendel 1994). Race *marie-galante* contains the strongly ar-

borescent and highly photoperiodic cottons of southern Mesoamerica and northern South America. Race *palmeri* comprises a morphologically distinct group of cottons found in Oaxaca and Guerrero. Although their strongly laciniate leaves and pyramidal shape set them apart from other *G. hirsutum* types, they are genetically embedded within *G. hirsutum* and are no longer recognized as a formal taxonomic entity (Fryxell 1992; Brubaker and Wendel 1993). The remaining two races, *morrilli* and *Richmondi* have restricted ranges and ambiguous morphological descriptions and are not particularly useful concepts.

1.1.4
Domestication and Diffusion of Cotton

We do not know the specific purpose for which each of the four species was originally used (e.g., candlewicks, fishing nets, wound dressings, hammocks), but we can be reasonably sure that it was the elongated seed coat trichomes, the "lint", that attracted the attention of our agricultural ancestors. The presence of terpenoid aldehydes in the large oil- and protein-rich seeds rendered them toxic to all but nonruminant animals, and it was not until modern methods of extracting the terpenoid aldehydes were developed that the seed oils and protein meals became economically important. And while there is some evidence that the terpenoid aldehydes were used medicinally, all four species, historically and currently, were and are seen primarily as a source of a high-quality fiber for textiles.

We know very little about the domestication of the Old World cottons. There are few known archeological sites. Cloth fragments and yarn dating to 4,300 bp have been reported from archeological sites in India and Pakistan (Gulati and Turner 1928), and they are probably attributable to *G. arboreum*. Inferences as to the geographic origins of the Old World cotton species are largely based on the distribution of the cultigens at various stages of agronomic development, the hypothesized predomestication range, and what is known about human diffusion during the early stages of domestication.

G. herbaceum occurs as a cultigen throughout northern Africa, Iran, Afghanistan, western and northern India, and western China. The only demonstrably wild *G. herbaceum* (ssp. *africanum*) is found in southern Africa, where it occurs in undisturbed native forests and grasslands. Assuming that domestication occurred within the range of the wild progenitors, Hutchinson (1954) nominated southern Africa as the site of original domestication for *G. herbaceum*. That the most primitive *G. herbaceum* cultivars, race *acerifolium*, are distributed along the coasts boarding the Indian Ocean trade routes, the most likely route of human-mediated dispersal, supports this conjecture. Following this initial dispersal, Hutchinson (1954) suggests further diffusion into western Africa and temperate regions to the north was accompanied by further agronomic development. The annualized *G. herbaceum* cultigens adapted for the temperate zones were later introduced into India where they displaced the resident perennial *G. arboreum* cultivars.

G. arboreum is the only one of the four cotton species for which a probable wild progenitor has not been identified (Hutchinson et al. 1947). The two most likely sites of domestication are thought be Madagascar, where a primitive arborescent form has been found in xerophytic woodlands, and the Indus valley, where the earliest documented evidence for Old World cultivated cottons has been found (Gulati and Turner 1928). Without a definitive point of origin, we can only review the more recent diffusions of *G. arboreum* germplasm (Hutchinson 1954). The primitive race *indicum* cultivars appear to have arisen in western India and then moved into peninsular India and along the east coast of Africa along the Indian Ocean trade routes, although if Madagascar was the original site of domestication the movement may have been in the opposite direction. The race *indicum* cottons are the most agronomically primitive forms of *G. arboreum* and probably represent the remnants of the first wave of diffusion. At a later date there was clear diffusion of material from the Indus Valley region westward across northern Africa and eastward into China and eastern Asia.

The archeological record of the domestication and diffusion of the New World tetraploid cottons is as fragmentary and complex as it is for the Old World species. Our understanding of the domestication and diffusion, however, has benefited from a number of comparative genetic studies that have provided some insight.

G. barbadense was most likely domesticated along the coasts (and inland steams and rivers) of northwest South America (Hutchinson et al. 1947). Seed, fiber, fruit, yarn, fishing nets, and textiles have been recovered from coastal archeological sites in Peru and Ecuador that date to 5,000–5,500 bp (Damp and Pearsall 1994). Molecular evidence also implicates this

region as the site of original domestication (Percy and Wendel 1990). Following domestication, the first wave of diffusion appears to have been a trans-Andean expansion into northern South American. Subsequently germplasm appear to have moved into Mesoamerica, the Caribbean, and the Pacific Islands, followed by a tertiary post-Columbian diffusion of germplasm into Argentina and Paraguay. Modern elite *G. barbadense* cultivars trace their origins to the Sea Island cottons developed on the coastal islands of Georgia and South Carolina that probably originated from west Andean Peruvian germplasm (Hutchinson and Manning 1945; Percy and Wendel 1990). The Sea Island industry of the USA eventually collapsed under boll weevil pressure by 1920 (Niles and Feaster 1984), but the Sea Island lineage contributed to the development of the Egyptian cottons which were later reintroduced into the USA as the basis of the Pima gene pool.

G. hirsutum L., the most widely cultivated cotton in the world, is known by various common names, including, among others, Acala or Upland cotton, short staple cotton, Mocó cotton, and Cambodia cotton (Johnson 1926). Figure 1 shows the modern domesticated representation of this species. The oldest archeological remains of *G. hirsutum* date to 4,000–5,000 bp from the Tehuacan Valley in Mexico (P. Frxyell and J. Vreeland, personal communication). *G. hirsutum* has two centers of genetic diversity, southern Mexico – Guatemala and the Caribbean (Wendel et al. 1992). The Mesoamerican (Mexico – Guatemala) center of diversity corresponds with the center of morphological diversity and encompasses the only known wild form of *G. hirsutum*, the coastal Yucatan race *yucatanense* populations (Brubaker and Wendel 1994; Ulloa et al. 2006). Race *punctatum*, containing cottons that are agronomically intermediate between the wild race *yucatanense* populations and the more advanced indigenous cultigens, is spread throughout the Yucatan peninsula, along the Mexican Gulf coast, and into some of the Caribbean islands. This is the most likely route of the first major diffusion. From these original *punctatum* cultivars several localized races were developed: *Richmondi* (south coast of the Isthmus of Tehuantepec), *morilli* (central Mexico highlands), *palmeri* (the Mexican states of Guerrero and Oaxaca), and *latifolium* (Guatemala and Honduras) (Hutchinson 1951). Race *marie-galante*, which is widespread throughout southern Central America, northern South America, and the Caribbean most likely arose from some form of Mesoamerican *G. hir-*

Fig. 1. Modern cotton, *Gossypium hirsutum* L. **a** Cotton plant, **b** cotton flower, **c** immature cotton boll, and **d** open cotton boll with seed and fiber

sutum, but the details have not been satisfactorily elucidated. The superior agronomic characteristics of the *latifolium* cultivars mark a later diffusion of germplasm into the Mexican highlands (Hutchinson 1951).

The development of the modern elite *G. hirsutum* cultivars, or Upland cottons, started with the introduction of a tremendous range of Caribbean and Mesoamerican cultivars, including *G. barbadense* accessions into the southern USA (Ware 1951; Ramey 1966; Niles and Feaster 1984; Meredith 1991). From this diverse gene pool, two categories of cultivars arose: "green-seed" and "black-seed." As the textile industry in the USA developed, however, these cultivars proved inadequate and from 1806 forward they were replaced by cultivars developed from newly introduced *latifolium* accessions collected from the Mexican highlands (Niles and Feaster 1984). RFLP evidence and allozyme evidence suggest that these Mexican highland stocks were refined *latifolium* cultivars that had been transported northward earlier

from southern Mexico and Guatemala (Wendel et al. 1992; Brubaker and Wendel 1994). Subsequent to their introduction into the cotton belt of the USA, further augmentation of the modern Upland gene pool involved a series of additional, deliberate introductions, beginning in the early 1900s, in response to the devastation brought on by the boll weevil. Eventual selection for locally adapted cultivars led to the development of regionally adapted cultivars whose modern derivatives account for the majority of Upland cotton grown worldwide (Niles and Feaster 1984; Meredith 1991).

That each of the four *Gossypium* species with this unique seed trichome characteristic was independently domesticated in four different locations in the New World and the Old World is extraordinary but is well established by genetic analyses and archeological records (reviewed by Brubaker et al. 1999). As a consequence of domestication, the four lint-bearing *Gossypium* species (as apposed to the "wild cottons") have spread well beyond their indigenous ranges.

G. hirsutum, for example, the species that currently dominates world cotton cultivation (providing over 90% of the annual cotton crop) has been spread from its original home in Mesoamerica to over 40 countries. It can be found as far north as 37°N in the USA, and as far south as the 32nd parallel in Australia and South America (Niles and Feaster 1984).

Although each species arose via the natural processes of evolution, under domestication their morphology, genetic composition, and indigenous ranges have been significantly altered. The identity of the wild progenitors of *G. barbadense*, *G. herbaceum*, and *G. hirsutum* are still uncertain and all are rare in nature. The wild antecedent of *G. arboreum* has probably disappeared. What mostly remain are feral or commensal populations, all of which have been affected by human manipulation, and their annualized modern derivatives.

This impressive human-mediated dispersal, however, is overshadowed by the tremendous genetic and phenotypic changes that resulted from selection under domestication (Stephens 1967). The wild progenitors of each domesticated species are sprawling perennial shrubs or small trees with small impermeable seeds bearing a sparse coat of coarse hairs. The domesticated cottons that we now know are compact annualized shrubs with large readily germinable seeds bearing a thick coat of long, strong white fibers, the cotton of commerce. This transformation began with the original indigenous domesticators and has only increased in intensity as the importance of cotton as a textile fiber has increased.

1.1.5
Traditional Breeding

Following the domestication of cotton, selection under cultivation in discrete areas of Mesoamerica led to the development of the landraces recognized by the cotton scientists of the last century. Today, only vestiges of the landraces survive as occasional garden plants that are maintained mainly for sentimental reasons (Ulloa et al. 2006). Cook (1906) observed that the Mayan Indians in Chiapas, Mexico, and Kekchi Indians in Guatemala would intercrop cotton with peppers. Once the early bearing cottons were harvested, the entire cotton crop would be culled to allow the peppers to mature properly. Unintentionally this practice would have quickly selected for the genotypes that bore large crops early regardless of environmental clues.

Traditional breeding has dramatically transformed the cotton plant over the last 70 years. Modern cotton cultivars are high-yielding, day length neutral flowering, and early-cropping plants with easily ginned and abundant fiber. These improved characteristics resulted from human selection from perennial ancestors with shorter, sparser fiber (Fryxell 1984). Many cotton (*Gossypium* spp.) research programs today require measurements of agronomic and fiber-quality traits such as lint percentage, boll weight, 2.5 and 50% fiber span length, fiber bundle strength, and fineness (micronaire reading, fiber maturity, fiber perimeter, etc.) in order to breed for the fiber properties the textile industry requires for a new generation of textile technologies (Ulloa 2006). The cotton research community has established fiber testing methods (breeder, spinning, areolometer, sticky, and high-volume instrument, HVI) for the above traits, which are run in-house, or through public or private institutions, e.g., the International Textile Center, Lubbock, TX, USA, and Starlab. Knoxville, TN, USA. A brief review of the most important fiber quality traits is presented here before we turn our attention to how the genetics of these traits is being dissected and selected (Sects 1.4, 1.5). Fiber span length at 50 and 2.5% can be measured with a digital fibrograph instrument. Span length is the distance spanned by the indicated percentage of fibers in the fiber sample as assayed by the fibrograph instrument. Fiber span length at 2.5% estimates the length of the longest 2.5% of fibers in a sample, and the distance is presented in millimeters. Fiber strength is the strength of a bundle of fibers measured by the stelometer. Elongation is an estimate of the elasticity of the bundle sample. Micronaire reading is a measure of fiber fineness and maturity. Arealometers measure fiber fineness and shape based on the resistance a given sample offers to the flow of air at two pressures. Fineness and shape provide estimates of maturity and perimeter. The HVI measures cotton quality parameters such as fiber strength, staple length, and length uniformity, and its measurements are used for marketing proposes (Anderson 1999). A relatively new fiber testing method has slowly been incorporated, which uses the advanced fiber information system (AFIS) for measuring neps, fiber length and diameter, and trash (Bragg and Shofner 1993; Hossein et al. 1994). Zellweger Uster. (Knoxville, TN, USA) manufactures the USTER^{TM}

AFIS. The AFIS provides an alternate method of collecting fiber information on cotton quality by separating a sample into single fibers and measuring the properties of each fiber.

The first source breeders turn to for genetic variability is the primary germplasm pool. Around the world there are large collections of *G. hirsutum* and *G. barbadense* landrace accessions, but they are as yet poorly evaluated (Percival 1987). The elite lines contain a number of traits that originated in the primary germplasm pool. Among these are blight resistance genes from *G. hirsutum* and *G. barbadense* (Endrizzi et al. 1985), the nectariless trait from *G. tomentosum* (Meyer and Meyer 1961), root-knot nematode resistance from a Mexican wild landrace (*G. hirsutum*) named Jack John landraces (Shepherd 1974, 1982), and resistance to Fusarium and Verticillium wilt from *G. hirsutum* landraces and from *G. darwinii* (Bell 1984).

Breeders have also used secondary germplasm pool species, i.e., the A- and D-genome diploid species effectively. Introgressed traits include bacterial blight resistance genes from *G. arboreum*, *G. herbaceum*, and *G. anomalum* (Endrizzi et al. 1985); the cytoplasmic male sterility (CMS) and restorer genes from G. *harknessii* (Meyer 1975) and *G. trilobum* (Stewart 1992); and fiber quality genes from *G. armourianum* (Meyer 1957; Ndungo et al. 1988). Introgression of high fiber strength genes was achieved via the triple hybrid (*G. hirsutum* × *G. arboreum* × *G. thurberi*) (Demol et al. 1972; Harrell and Culp 1979).

1.1.6
Current Priorities for Breeding in Cotton

The genetic diversity in the cotton crop has been used as an indicator of a possible plateau in breeding progress and the potential threat to sustaining high yields. Over the last 30 years the cotton germplasm base has narrowed considerably. Continued growth and competitiveness of the cotton industry around the world is dependent upon improving varieties with high yield, better fiber quality, and pest resistance. In terms of maintenance of elite genes, very high constraints are placed on today's cotton breeders. There is strong pressure against the use of exotic germplasm sources in breeding owing to the large blocks of genes that are also introgressed during recombination between two parental lines because of linkage drag. As cotton landraces disappear, the accessions that are

preserved ex situ will be our only access to the genetic diversity that once resided in situ in Mesoamerica, South America, Africa, Arabia, India, and Asia. Public cotton germplasm enhancement programs in the world are limited in number but they are critical if we want to preserve the genetic diversity of the cotton species.

Yield, quality, and resistance to pests and diseases continue to be primary motivations for plant breeders, particularly in response to changes in agricultural and textile processing and technologies. Water-use efficiency and drought tolerance are critical priorities as the cost of water increases and its availability decreases. There are considerable social pressures to make cotton production less environmentally damaging. Changes in yarn manufacturing technologies are driving how "quality" is defined. The introduction of air-jet spinning means that fiber length uniformity is as important as fiber strength in attracting price premiums at the gin. The challenge breeders face is that all of the high-priority breeding objectives are phenotypically complex traits controlled by many interacting genes. Progress will depend on genetic sources and the ability to dissect traits into genes for selection. The complex evolutionary history of cotton complicates this task, but we do have new tools to address these issues. As the number and diversity of molecular markers increase, our ability to use genetic linkage mapping to locate and identify the genes controlling the complex phenotypes that we want to manipulate also increases. This chapter will review what has been done, what is ongoing, and where genetic linkage mapping can take us in the future.

1.2
Construction of Genetic Linkage Maps

In just over a decade, our understanding of the structure of the cotton genome has progressed tremendously. Prior to 1994, genetic linkage mapping in cotton was limited to linkage detection among morphological markers and the placement of morphological markers on chromosomes using cytological stocks (reviewed by Endrizzi et al. 1985). In 1985, the diploid A-genome map comprised 18 morphological markers in seven linkage groups, and the tetraploid AD-genome map comprised approximately 80 markers distributed across 18 linkage groups (Endrizzi et al.

Table 1. Summary of published molecular genetic linkage maps for *Gossypium* species

Mapping populations	References	No. of loci	No. of linkage groups	Length (cM)	Average distance between markers (cM)	Comment
57 *G. hirsutum* × *G. barbadense* F_2 progeny	Reinisch et al. (1994)	683 RFLP loci	41	4,675	6.8	First molecular genetic linkage of tetraploid AD genome First tentative assignments of linkage groups to chromosomes Nomenclature for At and Dt subgenomes established
(*G. arboreum* × *G. trilobum*) × *G. hirsutum* F_2 progeny	Altaf et al. (1997)	216 loci 194 AFLPs 19 RAPDs 3 morphological markers	11	522	16.8	Pilose vestiture is linked to markers on linkage group T1
96 *G. hirsutum* (HS46 × MARCABUSCAG8US-1-88) $F_{2.3}$ progeny	Shappley et al. (1998)	120 RFLP loci	31	865	7.2	First intraspecific molecular genetic linkage map of the *G. hirsutum* genome
58 *G. herbaceum* [A_2 (=A_1)-97] × *G. arboreum* ($A_2$47) F_2 progeny	Brubaker et al. (1999)	161 loci 6 isozymes 155 RFLPs	18	856	5.3	First molecular genetic linkage map of the diploid A genome A-genome translocations confirmed Nomenclature for A-genome linkage groups established
62 *G. trilobum* (s.n.) × *G. raimondii* (s.n.) F_2 progeny	Brubaker et al. (1999)	306 RFLP loci	17	1,486	4.9	First molecular genetic linkage map of the diploid D-genome Nomenclature for D-genome linkage groups established
119 *G. hirsutum* (MD5678n ex Prema) $F_{2.3}$ progeny	Ulloa and Meredith (2000)	81 RFLP loci	17	701	8.7	First molecular genetic linkage map used for QTL detection: 26 QTLs for agronomic and fiber quality traits detected

RFLP restriction fragment length polymorphism, *AFLP* amplified fragment length polymorphism, *RAPD* randomly amplified polymorphic DNA, *QTL* quantitative trait locus, *SSR* simple sequence repeat, *STS* sequence-tagged site, *BC_1* backcross one, *RIL* recombinant inbred line

Table 1. (continued)

Mapping populations	References	No. of loci	No. of linkage groups	Length (cM)	Average distance between markers (cM)	Comment
118 G. hirsutum (NM24016 × TM1) $F_{2,3}$ progeny	Ulloa et al. (2000)	199 loci 125 RAPDs 68 SSRs	28	1,058	5.3	2 QTLs for stomatal conductance identified
171 G. hirsutum (TM-1) × G. barbadense (3-79) F_2 progeny	Kohel et al. (2001), Yu et al. (1998)	355 loci 216 RFLPs 139 RAPDs	50	4,766	13.4	First genetic linkage map of genetic and cytogenetic standards for G. barbadense and G. hirsutum (Kohel 1973)
58 G. hirsutum (TM-1) × G. barbadense (Hai 7124) haploid and doubled-haploid progeny	Zhang et al. (2002)	489 loci	43	3,315	6.78	First use of a doubled-haploid population for molecular genetic linkage mapping in cotton
Genetic joinmap	Ulloa (2002)	283 RFLP loci; 1 morphological marker	47	1,503	2.9	First composite genetic linkage map of the AD tetraploid genome
Population A: 96 G. hirsutum (HS46 × MARCABUSCAG8US-1-88) $F_{2,3}$ progeny (Shappley et al 1998)		120 RFLP loci	31	865	7.2	
Population B: 119 G. hirsutum (MD5678 × Prema) $F_{2,3}$ progeny (Ulloa and Meredith 2000)		81 RFLP loci	17	701	8.7	
Population C: 199 G. hirsutum (HQ95-6 × MD51ne) $F_{2,3}$ progeny (Ulloa 2002)		82 RFLP loci; 1 morphological marker	24	830	10.0	
Population D: 155 G. hirsutum (119-5 subokra × MD51ne) $F_{2,3}$ progeny (Ulloa 2002)		56 RFLP loci	16	520	9.3	

Table 1. (continued)

Mapping populations	References	No. of loci	No. of linkage groups	Length (cM)	Average distance between markers (cM)	Comment
96 G. hirsutum [TM-1 × Li1 (Ligon lintless)] F2 progeny	Karaca et al. (2002)	23 loci 22 SSRs 1 morphological marker	8	218	9.5	*Ligon lintless (Li1)* mapped to chromosome 22
94 G. nelsonii (Gos-5024) × G. australe (Gos-5005) F2 progeny	Brubaker and Brown (2003)	389 AFLP loci 213 G. australe 176 G. nelsonii	13 linkage group assemblages 17 G. australe 21 G. nelsonii	G. australe map = 931 G. nelsonii map = 773	G. australe map = 4.4 G. nelsonii map = 4.4	First molecular genetic linkage map of the diploid G genome
75 G. hirsutum (Guazuncho 2) × [G. hirsutum (Guazuncho 2) × G. barbadense (VH8-4602)] BC1 progeny	Lacape et al. (2003)	888 loci 465 AFLPs 229 SSRs 192 RFLPs 2 morphological markers	37	4,397	5.0	Molecular genetic map of tetraploid AD genome expanded and refined Final length of tetraploid genome estimated to be 5,500 cM
94 G. hirsutum (Acala 44) × G. barbadense (Pima S7) F2 progeny	Mei et al. (2004)	392 loci 333 AFLPs 12 RFLPs 47 SSRs	42	3,287	8.4	7 QTLs for fiber-related traits detected
82 G. hirsutum (TMS-22) × G. tomentosum (WT936) F2 progeny	Waghmare et al. (2004)	589 RFLP loci	52	4,259	8.45	First molecular genetic linkage map of the tetraploid cotton genome to use G. tomentosum as a parent
57 G. hirsutum (race palmeri) × G. barbadense (K101) F2 progeny	Rong et al. (2004)	2,584 STS loci	26	4,448	1.72	First fully resolved high-density molecular genetic map of the tetraploid AD genome Nomenclature for AD-genome linkage groups updated

Table 1. (continued)

Mapping populations	References	No. of loci	No. of linkage groups	Length (cM)	Average distance between markers (cM)	Comment
62 *G. trilobum* × *G. raimondii* F_2 progeny	Rong et al. (2004)	763 STS loci	13	1,493	1.96	First fully resolved high-density molecular genetic map of the diploid D genome Nomenclature for D-genome linkage groups updated
69 *G. hirsutum* (Handan 208) × *G. barbadense* (Pima 90) F_2 progeny	Lin et al. (2003, 2005)	566 loci 205 SSRs	41	5,141	9.08	13 QTLs for fiber quality detected
163 *G. hirsutum* (*G. anomalum* introgression line 7235 × TM-1) F_2 progeny	Shen et al. (2005), Zhang et al. (2003)	86 SSR loci	21	667	7.8	8 QTLs for fiber traits detected
169 *G. hirsutum* (HS427-10 × TM-1) F_2 progeny	Shen et al. (2005)	56 SSR loci	17	558	10.0	7 QTLs for fiber traits detected
142 *G. hirsutum* (PD6992 × SM3) F_2 progeny	Shen et al. (2005)	73 SSR loci	22	588	8.1	1 QTL for fiber traits detected
167 *G. arboreum* (SMA4; PI529740) × *G. herbaceum* (A_1-97; PI529670) F_2 progeny	Desai et al. (2006)	275 RFLP loci	12	1,147	4.2	First fully resolved molecular genetic map of the diploid A genome Nomenclature for A-genome linkage groups updated
183 *G. hirsutum* (TM-1) × *G. barbadense* (3-79) RIL progeny	Park et al. (2005)	224 SSR loci	40	1,277	6.2	First genetic linkage map used RIL
183 *G. hirsutum* (TM-1) × *G. barbadense* (3-79) RIL progeny	Frelichowski et al. (2006)	407 SSR loci	43	2,126	4.5	First genetic linkage map used RIL
140 [*G. hirsutum* (TM-1) × *G. barbadense* (Hai7124)] × TM-1 BC_1 population	Han et al. (2004, 2006), Song et al. (2005)	907 SSR loci	30	5,060	5.6	This map also included publicly developed SSRs from http://www.mainlab.clemson.edu/cmd/projects

1985). Now almost fully resolved molecular genetic maps for the diploid A genome and D genomes and the tetraploid AD genome are available (Rong et al. 2004; Desai et al. 2006; Table 1), and all 26 AD linkage groups (and their A and D homoeologs) have been linked to the cytological map (Rong et al. 2004; Frelichowski et al. 2006; Wang et al. 2006). In addition, polymerase chain reaction (PCR) based markers, such as microsatellites, have been placed on 23 of the 26 cotton chromosomes (http://www.mainlab.clemson.edu/cmd/). With molecular markers distributed every 2–5 cM (on average) for the diploid A and D genomes and the tetraploid AD genome, genetic linkage maps can be used to parse and manipulate the cotton genome in ways that were inconceivable 10 years ago. In particular, molecular genetic linkage maps allow us to dissect and characterize the complex agronomic traits (see QTL discussion to follow) that will be the focus of cotton breeding over the next decade. The purpose of this section is to briefly review the history of the genetic linkage mapping in cotton, before turning to the more critical topics of how genetic linkage maps are being used, will be used over the next decade, and how they will integrate with all the other new molecular tools that have emerged.

1.2.1
History of Genetic Linkage Mapping in Cotton

The first published molecular genetic linkage map for any of the *Gossypium* species was published in 1994 (Reinisch et al. 1994). Over the following 11 years, 24 major mapping efforts in *Gossypium* emerged (Table 1). Twenty of these have been maps of the tetraploid genome, while only five have been diploid maps. Of diploid maps, all but one have been of the diploid A and D genomes that were recombined in the AD tetraploids. There is only one map of a *Gossypium* tertiary genome species. The first genetic map was based on an interspecific *G. hirsutum* × *G. barbadense* F_2 and the markers used were RFLPs (Reinisch et al. 1994). It set the trend that many subsequent maps were to follow. Half of the maps listed in Table 1 incorporate RFLPs and all but three populations (two backcross one, BC_1, families and one doubled-haploid family) are derived from selfed F_1s. That most of the mapping families are interspecific families reflects the relative paucity of available molecular markers during the past decade and the lack of genetic variation within the cultivated germplasm. The selection of in-

terspecific mapping families maximizes the number of available markers that will be polymorphic, even when wild accessions are used. As the overall number of mapped sequence-tagged sites (STSs) continues to climb and as SNPs become technically easier to resolve, there is likely to be an increase in the number of intraspecific maps which offer more flexibility and wider applicability to cultivated cotton germplasm.

As was predicted from cytological observations (reviewed by Endrizzi et al. 1985), synteny and colinearity are highly conserved across the A, D, or AD genomes (Brubaker et al. 1999; Rong et al. 2004). Apart from the well characterized A-genome translocations no large scale rearrangements are evident. While there are certainly some small-scale events (inversions), the number of these events is not so large that it has complicated analyses (Brubaker et al. 1999; Rong et al. 2004). Most populations have some regions of gametic or zygotic transmission bias (evident as segregation distortion), but again this has not proven insurmountable. In fact it should be noted, that as more complementary DNA (cDNA) and expressed sequence tags (ESTs) are mapped, careful analyses of those that fall within these regions of distorted segregation could lead to a better understanding of the genetic interactions that result in the F_2 breakdown that is often observed in later generations (Jiang et al. 2000). The overall conservation of synteny and colinearity across the *Gossypium* genomes makes the interspecific cross as analytically amenable as the intraspecific cross and technically easier to map because more markers are available.

Mapping Populations

Breeding priorities will also continue to promote a bias for F_2 interspecific mapping families, most notably for the tetraploid cotton genome. The severe genetic bottlenecks that the cultivated cotton genomes experienced during domestication clearly limited the amount of genetic diversity available for selection. It is likely that many useful alleles are still to be found outside the current cultivated gene pools. Introgressing these traits requires interspecific crosses. Genetic linkage maps of these introgressant populations will be invaluable in documenting and tracking the introgression of alien chromatin into donor genomes.

The other types of populations used include BC_1 families, recombinant inbred lines (RILs), and doubled-haploid families. The first of these is attractive because they facilitate the use of anonymous dominant markers (AFLPs and RAPDs), but the amount of human intervention required to construct the populations makes them less appealing logistically. Furthermore the growing trend to use genetic linkage maps for identification of QTLs will reduce their application, unless there is a clear need to look at the genetic contribution of only one of the parents. As discussed later, the large number of codominant markers that are now becoming available will reduce the technical attractiveness of dominant markers, and hence the need for BC_1 families.

The use of doubled-haploid families would be of tremendous utility across any number of genetic studies, but most critically for complex traits with significant genotype × environment interactions. Unfortunately the methodologies for generating haploids and doubling the chromosomes in cotton are still tedious and inefficient, and this approach has only been used once (Zhang et al. 2002). Until the technical methodologies of generating haploids and doubling chromosomes are optimized, this approach will not be widely used.

Marker Type

While the mapping of morphological traits continues (Endrizzi and Nelson 1989; Percy 1999; Kohel and Bird 2002), the advent of molecular markers (e.g., RFLPs, RAPDs, AFLPs, SSRs, SNPs) has dramatically altered the utility and application of genetic linkage mapping in cotton. Although isozymes can rightly claim to be the first "molecular" marker, it was not until DNA sequence differences among organisms could be visualized directly that the true era of molecular markers began. The advantage of molecular markers is that, unlike morphological markers, they are only limited by the number of nucleotide differences among individuals. Even a single nucleotide substitution can be used as a molecular marker. Molecular markers are phenotype-neutral, whereas morphological markers may be difficult to maintain and have deleterious effects on other traits. Linkage analyses with morphological markers and correlations with chromosome behavior (cytogenetics) helped the development of the framework for mapping molecular markers. With a nearly unlimited pool of genetic markers, cotton geneticists could construct linkage maps of entire genomes that could be used to dissect complex traits (Ulloa and Meredith 2000; Ulloa et al. 2000, 2002, 2005; Lin et al. 2003, 2005; Mei et al. 2004;

Shen et al. 2005), locate genes (Guo et al. 1998; Lan et al. 1999; Karaca et al. 2002; Rungis et al. 2002; Liu et al. 2003; Zhang and Stewart 2004; Feng et al. 2005; Ulloa et al. 2005; Zhang et al. 2005), and compare the structure of related genomes (Brubaker et al. 1999; Rong et al. 2004; Desai et al. 2005).

Half of the maps listed in Table 1 used RFLPs, but in the future all commonly used mapping markers will be PCR-amplified codominant makers. Initially RFLPs represented the state of the art. Their unambiguous genetic interpretations made them ideal genetic markers for mapping. The Southern hybridizations required to detected RFLPs are labor-intensive, however, and RFLPs have been replaced by PCR-amplified markers that are amenable to multiplexing on modern fluorescent capillary electrophoresis machines. Many of the mapped RFLPs have now been converted to STSs and will reappear as PCR-amplified markers (Rong et al. 2004). There has also been an incredible increase in the number of publicly available SSRs that are transferable across a number of the *Gossypium* genomes. The first *Gossypium* genetic linkage map to incorporate SSRs was published in 2000, and there has been a notable rise in their application since then (Table 1).

Dominant markers have been used in some studies, but they are rarely the marker of choice. The analytical penalty of applying dominant markers to F_2 families is significant (Brubaker and Brown 2003) and in this regard codominant markers will always be preferred.

1.2.2
What Have We Learned from Genetic Linkage Mapping in Cotton?

Constructing genetic linkage maps is a tedious technical exercise that only becomes useful when applied to second-generation questions. How genetic linkage maps have been used to address these second-generation questions will be the focus of the remainder of this chapter, but before moving on it is worthwhile highlighting a few of the things that genetic linkage mapping, particularly comparative linkage mapping, has revealed about the cotton genome.

The conservation of synteny and colinearity across the cotton genomes is high, except for the well-characterized translocations that differentiate the two A-genome species (*G. arboreum* and *G. herbaceum*) from each other and from the tetraploid At

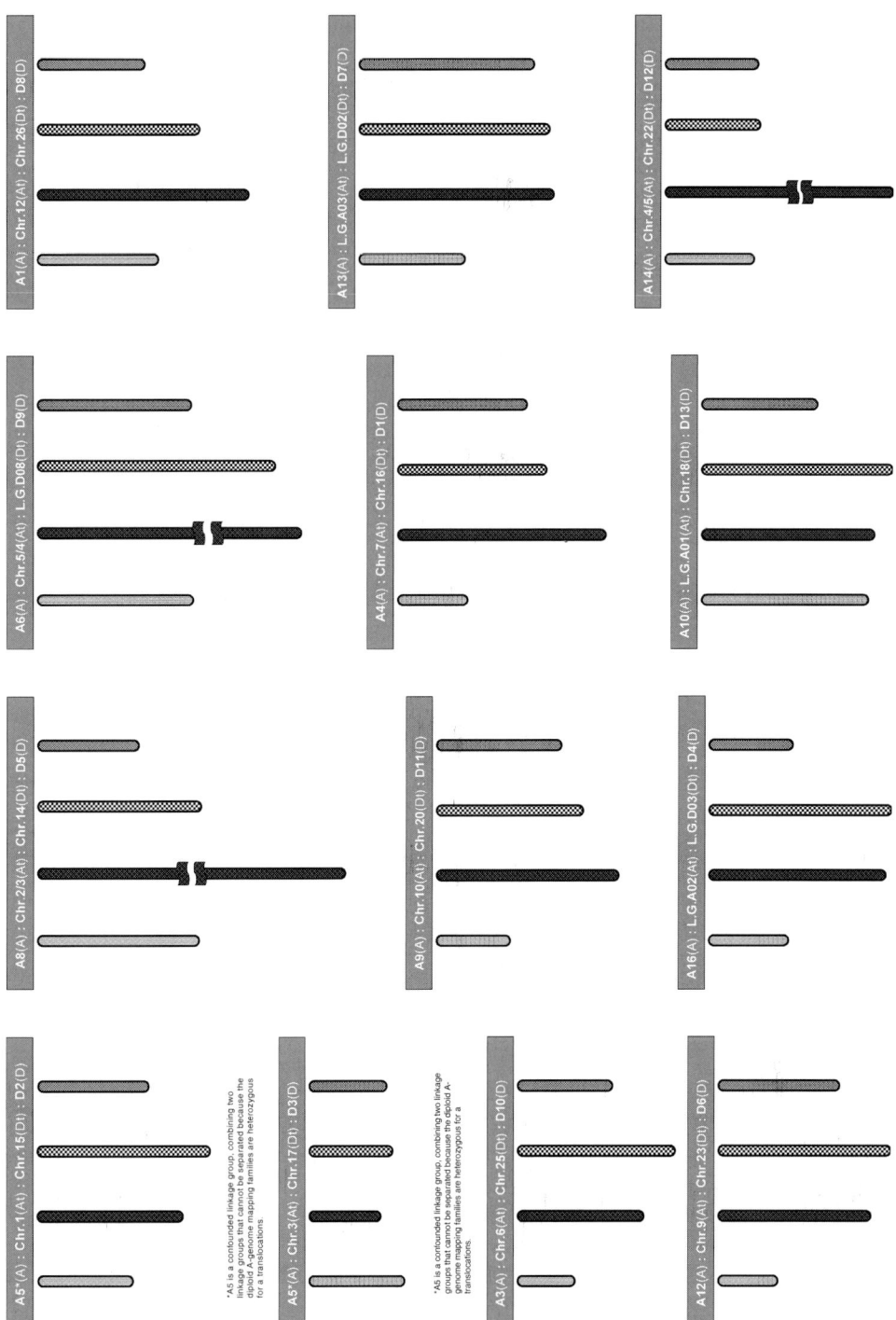

Fig. 2. The homoeologous relationships among the diploid A and D genomes and the tetraploid At and Dt subgenomes summarizing the linkage group nomenclature as of publication. The lengths of the linkage groups in the figure are proportional to the genetic lengths of the linkage groups as reported by Desai (2005; diploid A genome) and Rong et al. (2004; diploid D genome and tetraploid At and Dt genomes)

subgenome (Brubaker et al. 1999; Rong et al. 2004). Locus duplication patterns reveal 13 expected homeologous chromosome pairs in the tetraploid cotton (At subgenome from 1–13 and Dt subgenome from 14–26). Homeologous groups are arranged as a follow: chromosome 1 – chromosome 15, chromosome 2 – chromosome 14, chromosome 3 – chromosome 17, chromosome 4 – chromosome 22, chromosome 5 – LGD08/chromosome 9, chromosome 6 – chromosome 25, chromosome 7 – chromosome 16, LGA02/chromosome 8 – LDG03/chromosome 24, chromosome 9 – chromosome 23, chromosome 10 – chromosome 20, LGA03/chromosome 11 – LDG02/chromosome 21, chromosome 12 – chromosome 26, and LGA01/chromosome 1 – chromosome 18 (Rong et al. 2004; Frelichowski et al. 2006; Wang et al. 2006). This level of conservation corresponds with the evidence that the cotton genomes are quite static structurally, particularly in the transition from diploidy to allopolyploidy (Liu et al. 2001), and was predicted by early cytological observations (reviewed by Endrizzi et al. 1985). This contrasts with other plant genomes that have undergone more extensive restructuring (Quiros et al. 1994; Song et al. 1995; Lagercrantz and Lydiate 1996; Feldman et al. 1997; Liu et al. 1998a, b). This gross structural conservation will increase the transferability of information across genomes.

The conservation of synteny and colinearity across the cotton genomes is not absolute, however. There is clear evidence for small-scale inversions and intragenomic duplications (Reinisch et al. 1994; Brubaker et al. 1999; Ulloa et al. 2002; Rong et al. 2004). The presence of local structural differences was predicted by early cytological observations when it was noted the some homoeologous chromosome bivalents were characterized by reduced chiasmata frequencies (Menzel et al. 1978, 1982, 1984; Hasenkampf and Menzel 1980). While some of these are visible in map comparisons, many more may be hidden in regions of suppressed recombination, when the parents of the mapping family are heterozygous for the alternate localized marker orders.

While the presence of intragenomic duplications may represent activity of genomic processes (e.g., retrotransposon movement), the observation that many of these duplicated loci occur as discrete duplicated blocks within a single genome is tantalizing, in that it may represent the vestiges of a historical polyploid event. The presence of these multilocus intragenomic duplications is consistent with the hypothesis that the diploid cotton species are themselves paleopolyploids (Reinisch et al. 1994; Muravenko et al. 1998; Brubaker et al. 1999; Rong et al. 2004).

Finally one of the more intriguing aspects of the cotton genomes is that physical length is not a predictor of the genetic size. The D genome is approximately 980-Mbp long, the A genome contains about 1,860 Mbp, and the length of the AD genome is additive, about 2,835 Mbp (Grover et al. 2004). Genetically, however, the At and Dt subgenomes are about 50–93% longer than their extant diploid models (Reinisch et al. 1994; Brubaker et al. 1999; Rong et al. 2004). At the same time, the diploid A and D genomes, which differ in physical size by a factor of about 2, are recombinationally equivalent (Brubaker et al. 1999; Desai et al. 2006), as are the At and Dt tetraploid subgenomes (Brubaker et al. 1999; Rong et al. 2004). A diagrammatic representation of the homoeologous relationships among the diploid A and D genomes and the tetraploid At and Dt subgenomes summarizing the linkage group nomenclature is presented in Fig. 2. The lengths of the linkage groups in the figure are proportional to the genetic lengths of the linkage groups as reported by Desai et al. (2006) for the diploid A genome and Rong et al. (2004) for the diploid D genome and tetraploid At and Dt genomes. These observations suggest that this difference in recombinational lengths is attributable to polyploidy. As more diploid *Gossypium* genomes are be mapped with a comparative set of markers it will be of interest to see how their genetic lengths compare with those of the A and D genomes.

1.2.3
What Is the Future for Genetic Linkage Mapping in Cotton?

Genetic linkage mapping is primarily a tool for gene discovery. As is evident in Table 1, QTL mapping is being used to dissect complex genetic traits (Ulloa and Meredith 2000; Lin et al. 2005; Shen et al. 2005; Ulloa et al. 2005). At the same time the saturated fully resolved tetraploid AD genetic linkage map is also being used as the framework for BAC-based physical mapping of the tetraploid genome (Rong et al. 2004). The association of molecular markers with agronomic traits that are costly or laborious to score will provide cheap, easily scored surrogates for marker-assisted breeding (MAB) programs. While the current maps are being applied to these downstream issues, future maps will be further saturated and additional genomes will be mapped. With the abundance of transferable codominant markers available and new platforms for resolving genotypes, the ease and speed with which genomes can be mapped will only increase.

1.3
Gene Mapping

1.3.1
Simple Phenotypes in *Gossypium* spp.

This section focuses mainly on the genetic mapping of genes influencing qualitative traits, for which the phenotypic variation falls into discrete classes in the progeny and follow a simple Mendelian inheritance. Genetic mapping of quantitative traits will be discussed in the next section.

Approximately 200 simply inherited genes have been identified and characterized in the diploid (*G. arboreum* and *G. herbaceum*) or tetraploid (mostly in *G. hirsutum* and *G. barbadense*) species. A summary of the classical genes identified with their symbols and origins was presented in Endrizzi et al. (1984). Most of these genes are morphological mutants that have arisen through spontaneous mutation, although a few were recovered from irradiation or from natural variation between species and described by means of interspecific hybridization. Through the collective efforts of various researchers, many of the loci have been assembled into multiple genetic mutant stocks, each carrying numerous dominant or recessive genes. The classical morphological genes identified, classified, and annotated involve the whole cotton plant morphology, stems, leaves, flowers, bolls, and seeds (summarized by Endrizzi et al. 1984). An example of developed mutants that carry more than one marker gene is the line Texas 586, which carries nine dominant marker loci that fall into seven classical linkage groups, while another line, Texas 582, carries five recessive genes that each fall into different linkage groups; all but one marked by Texas 586. These lines have greatly facilitated the construction of the classical genetic linkage map, which now contains 61 loci covering 16 linkage groups in the tetraploid genome and 19 loci in seven linkage groups in the diploid genome. Numerous mutant loci, though not associated with a specific linkage group, have subgenome and chromosome affinities identified through the use of aneuploid stocks (Stelly 1993).

Because of the cost associated with molecular mapping and the fact that most morphological mutants have no agronomic utility, there have been few efforts to map them onto the molecular genetic map. Many of these classical traits have been selected for placement on the molecular map solely as a tool for aligning the various linkage groups to chromosomes assigned by the classical map. Since many of the genetic mapping populations have been derived from *G. hirsutum* and *G. barbadense*, morphological traits that have been used in this manner include those that can easily differentiate the two species. In particular, *G. barbadense* (Pima, Egyptian, or Sea Island cottons) is characterized by a petal spot, yellow petals, yellow pollen, and green/brown fuzz, while the cultivated forms of *G. hirsutum* (Upland) have no petal spot, cream petals, cream pollen, and white fuzz. Except for the fuzz color locus (Lacape et al. 2003), the chromosomal locations mapped by molecular markers for all these loci are largely congruent with those assigned using cytogenetic stocks (Wright 1998; Lacape et al. 2003; Desai 2005; Wang et al. 2006). For example, Petal spot (R_2) was mapped near the RFLP locus *pAR1182b* on chromosome 7 in three *G. barbadense* × *G. hirsutum* populations (Wright 1999; Lacape et al. 2003; Desai 2005) and one *G. arboreum* × *G. herbaceum* population (Desai 2005). Pollen color was associated with the RFLP locus A1535b (Wright 1999) and the AFLP locus E6M5 (Lacape et al. 2003) in two separate *G. barbadense* × *G. hirsutum* populations, both of which belong to the linkage group assigned to chromosome 5, which is also where these classical markers were located.

Petal color (*Y1*) was mapped to LGA01 in two *G. barbadense* × *G. hirsutum* F_2 populations (Wright 1999; Lacape et al. 2003), consistent with its assignment to the A subgenome (Endrizzi et al. 1984). However, in a *G. arboreum* × *G. herbaceum* F_2 population, the yellow petal locus was associated with the interval between *Gate4AC11* and *pAR10E04a* on a linkage group orthologous to the tetraploid LGA03 (Desai 2005). The incongruent chromosome assignment for the *Y1* locus in the tetraploids and the *Ya* locus in the diploids suggests that the two petal color loci are not homologous, and that petal color is governed by different genetic loci in diploid and tetraploid *Gossypium* species.

1.3.2
Simple Phenotypes Molecular Mapped in *Gossypium* spp.

The mutant traits that have been specifically targeted for genetic mapping include those that are related to agricultural productivity and lint quality. A listing of the mutants in which diagnostic DNA markers have been developed is provided in Table 2 and they are discussed in the following sections.

Disease resistance genes

Genetic studies on the plant–pathogen relationship of resistance to bacterial blight caused by the pathogen *Xanthomonas campestris* pv. *malvacearum* (Smith) Dye have identified at least 16 resistance genes in diploid and tetraploid cottons (Endrizzi et al. 1984). These resistance genes are inherited as single Mendelian factors and are given the symbol "*B*" or "*b*" with the appropriate subscript to indicate resistance to particular pathogenic isolates.

Wright et al. (1998) constructed a detailed RFLP map from crossing four different resistant *G. hirsutum* parents to a single *G. barbadense* 'Pima S7' parent to determine the chromosomal locations and subgenomic distributions of genes that confer resistance to the bacterial blight pathogen. By analyzing the disease rating as quantitative characters, they mapped three major loci with each explaining over 88% of the total phenotypic variation. They concluded that the blight resistance genes B_2 and B_3 are located on the linkage group LGD08 with the former mapped near the RFLP locus G1292 and the latter near the locus pGH510a. The third blight resistance gene, B_{12}, was mapped near the RFLP locus pAR043 on chromosome 14 (Table 2). Although several other QTLs were also identified in this study that possibly correspond to other bacterial blight resistance genes previously described using classical genetic stocks, the inheritance of these loci deviated from a simple segregation model and each locus accounted for only a small portion of the total phenotypic variance. Fine mapping of the B_2, B_3, and B_{12} loci is now in progress at the University of Texas in Lubbock, TX, USA. This work is expected to provide closely linked diagnostic DNA markers for use in cotton improvement.

In addition, Bezawada et al. (2003) reported the association of SSRs with root-knot nematode resistance in an F_2 population. One SSR marker, BNL 1421, explained 8% of the gall variation index. This low per-centage is due to distant linkage to the locus responsible for the trait. However, the overall molecular and phenotypic data confirmed that the cultivar Clevewilt 6-1 was the source of the recessive gene for resistance (Table 3).

More recently, Wang et al. (2005) screened and mapped progeny (F_1, F_2, BC_1F_1, and $F_{2:7}$ RILs from intraspecific crosses and an interspecific F_2) developed by *G. hirsutum* 'Acala NemX', crossed with *G. barbadense* 'Pima S7' using 284 SSR markers, which covered 23 of the 26 cotton chromosomes, in order to identify microsatellite markers (SSR) closely linked to root-knot nematode resistance gene *rkn1*. The codominant SSR marker, CIR316, was tightly linked (2.1–3.3 cM, depending on the population used) to *rkn1*, producing amplified DNA fragments of approximately 221 bp (CIR316a) and 210 bp (CIR316c) in Acala NemX and susceptible Acala SJ-2, respectively. Additional markers, including BNL1231 with loose linkage to *rkn1* (map distance 25.1–27.4 cM), BNL1066, and CIR003, allowed the *rkn1* gene to be mapped to cotton linkage group A03 and/or chromosome 11 (Table 2).

Fertility Restorer Genes

CMS and its fertility restoration (*Rf*) system have been studied intensively in cotton. As in most other crop species, CMS is produced in cotton by placing the nuclear genome of the target species in the cytoplasm of a genetically isolated species. This results in the production of nonfunctional pollen without affecting the female fertility. The two CMS systems developed in Upland cotton include those based on the diploid cytoplasms of *G. harknessii* (Meyers 1975) and *G. trilobum* (Stewart 1992). According to Zhang and Stewart (2001), the nuclear restorer gene Rf_1 can restore the fertility of the CMS line carrying either cytoplasm, while the Rf_2 can only restore fertility to lines carrying the *G. trilobum* cytoplasm. Rf_1 and Rf_2 are linked in repulsion with a distance of about 1 cM (Zhang and Stewart 2001).

The development of molecular markers linked to nuclear restorer genes in cotton was first conducted using RAPD markers (Guo et al. 1998; Lan et al. 1999). Testing of an F_2 population developed from crossing the near-isogenic lines HAF277 and DELCOT277, which carry the Rf_1 gene and CMS, respectively, allowed a single RAPD marker, $R6592$, to be determined 2.3 cM from the fertility restorer gene (Lan et al. 1999). Zhang and Stewart (2004) identified a RAPD marker

Table 2. Gene linked to DNA markers in diploid and tetraploid cottons

Gene symbol	Trait	Marker type	Linked markers	Chromosome location	References
B_2	Blight resistance	RFLP	G1219	LGD08	Wright et al. (1998)
B_3	Blight resistance	RFLP	pGH510a	LGD08	Wright et al. (1998)
B_{12}	Blight resistance	RFLP	pAR043	Chr. 14	Wright et al. (1998)
Fz	Tufted seed[a]	RFLP	A1691b	LGA03	Rong et al. (2005)
h_a	Glabrous-lintless[a]	RFLP	G1261b	LGA03	Rong et al. (2005)
Li_1	Ligon lintless	SSR	MP4030	Chr. 22	Karaca et al. (2002)
l_2^o	Okra leaf	RFLP	A1485	Chr. 15	Jiang et al. (2000)
P	Pollen color	RFLP	A1535b	Chr. 5	Wright (1999)
P	Pollen color	AFLP	E6M5	Chr. 5	Lacape et al. (2003)
R_2	Petal spot	RFLP	pAR1182b	Chr. 7	Lacape et al. (2003)
R_2	Petal spot	RFLP	pAR1182b	Chr. 7	Desai (2005)
R_2	Petal spot[a]	RFLP	pAR139	Chr. 7	Desai (2005)
R_2	Petal spot	RFLP	pAR1182b	Chr. 7	Wright (1999)
Rf_1	Fertility restorer	RAPD	R6592	–	Lan et al. (1999)
Rf_1	Fertility restorer	RAPD	UBC169(700)	–	Zhang and Stewart (2004)
Rf_2	Fertility restorer	RAPD	UBC188(500)	–	Zhang and Stewart (2004)
Rf_1	Fertility restorer	RAPD	NAU/RAPD/Rf$_1$3$_{1480}$	Chr. 4	Liu et al. (2003)
Rf_1	Fertility restorer	STS	UBC147$_{1400}$	–	Feng et al. (2005)
$rkn1$	Root-knot nematode resistance	SSR	CIR316	Chr.11	Wang et al. (2005)
$T1$	Leaf pubescence	RFLP	pGH667	Chr. 6	Wright et al. (1999)
$t2$	Leaf pubescence	RFLP	pGH303	Chr. 25	Wright et al. (1999)
Ya	Petal color	RFLP	Gate4AC11	LGA03	Desai (2005)
$Y1$	Petal color	AFLP	E4M7	LGA01	Lacape et al. (2003)
$Y1$	Petal color	RFLP	pAR338b	LGA01	Wright (1999)

[a] Mapped in diploid

(*UBC188*) that was about 3 cM from Rf_2. Both the *UBC188* and *R6592* markers, along with several other possibly more tightly linked loci, have been converted to STS markers to improve their reproducibility and utility in the development of restorer parental lines in cotton CMS breeding (Feng et al. 2005).

In addition to the RAPD and STS markers discussed above, three SSR markers closely linked to the Rf_1 gene were also identified (Liu et al. 2003). By testing one of the SSR markers on monosomic and telosomic genetic stocks, the authors concluded that the Rf_1 locus was located on the long arm of chromosome 4 (Liu et al. 2003); however, this chromosome assignment has recently been challenged because the Rf_1 gene was transferred from *G. harknessii*, a diploid species that has been assigned to the D genome (Feng et al. 2005). Accordingly, the Rf_1 locus was introgressed into *G. hirsutum* through homologous chromosome recombination with a D-subgenome chromosome (Liu et al. 2003). Therefore,

despite the extensive mapping efforts, the chromosome assignments of the *Rf* genes have yet to be determined conclusively.

Okra Leaf Locus

The degree of leaf lobing in *G. hirsutum* can range from "broad" such as those commonly found in most *G. hirsutum* cultivars to deeply cut edge known as "okra leaf" found in some Australian cultivars and *G. hirsutum* landrace *palmeri* (Endrizzi et al. 1984). The two forms of lobing are caused by allelic variation at the L_2 locus, located on chromosome 15, with the okra leaf (l_2^o) showing dominance over the broad leaf allele (l_2). Jiang et al. (2000) conducted a QTL mapping experiment using an interspecific *G. barbadense* × *G. hirsutum* F_2 population segregating for four leaf morphology traits that included okra leaf. The distal region on chromosome 15 was found to harbor large-effect QTLs for leaf-lobe length and width, both of

Table 3. Information of gene mapping and location of detected QTL on cotton mapping populations

Interspecific mapping population	No. of families	Map groups	Groups with QTL	QTL no.	Putative QTL and genomes Chromosome and/or group A	D	Trait	References
Clevewilt 6-1 × Stoneville	F_2 77		1	1	A01 BNL1421	26 BNL1669	Root-knot nematode	Bezawada et al. (2003) 15.4 cM
TM-1 × Li F_2	F_2 96	8	1	1		22 BNL4030 BNL673	Ligon lintless (Li_1) chromosome assignment Single dominant gene model	Karaca et al. (2002) 12.5 cM
Siv'on × F-177 F_2	F_2 214		13	12	6	25	Osmotic potential	Saranga et al. (2002)
NM24016 × TM-1 F_2	F_3 118	28	2	2	1 S3 OPA13	20 BNL 3800 BNL3257	Stomatal conductance	Ulloa et al. 2000 5.6 cM 6.1 cM
G. hirsutum mutant × Seberry F_2	F_2 180	26	11	21	1, 6, A01, A06	15, 17, 20, 25, D03, D07	Leaf morphology	Jiang et al. (2000) 3,664 cM
PS 7 × Empire B2b6 F_2 × Empire B2 × Empire B3 × S295	F_2 199–150		5	4	1, 6, A05	23, 25	Trichomes, pubescence leaves and stems	Wright et al. (1999) 921.6 cM 1746.8 cM 2,100.9 cM 1,266.6 cM
PS 7 × Empire B2b6 F_2 × Empire B2 × Empire B3 × S295	F_2 199–150	48 45 49 48	1 1 5 1	1 1 5 1	5	20, G1219 20, pGH510a 14, 20, D04 14	*Xanthomonas campestris* Races 2 and 4 Races 2 and 4 Races 2 and 4 Races 2, 4, 7, and 18	Wright et al. (1998) 921.6 cM 1,746.8 cM 2,100.9 cM 1,266.6 cM

which influence leaf shape. Since the okra leaf mutation was placed on chromosome 15 on the classical map (Endrizzi et al. 1994), the region of this chromosome near pGH624a A1485 was inferred to be the likely location for the l_2 locus (Jiang et al. 2000). Interestingly, they also observed that the parental allele effects at this locus vary for different lobes, with some exerting positive but others exerting negative effects, suggesting that the l_2 locus may be a complex genetic locus. Fine-scale mapping of this locus may shed light on whether the l_2 locus contains a single gene or a cluster of genes.

Fiber Development Mutants

In both diploid and tetraploid cottons, fiber development mutants have been discovered and some have been assigned to cytologically identifiable chromosomes using aneuploid stocks (Endrizzi et al. 1994; Percy and Kohel 1999; Karaca et al. 2002; Kohel et al. 2002). In tetraploid *G. hirsutum*, *ligon lintless-1* (*Li1*) and *ligon lintless-2* (*Li2*) were reported to be monogenic and dominant, causing extreme reductions in fiber length (to less than 10 mm) on mature seeds (Griffee and Ligon 1929; Kohel et al. 1992). Also, *N1* and *n2* were indicated to be allelic (Percy and Kohel 1999) and determined the presence or absence of fuzz, thus causing naked seed. *Fbl* is an incompletely dominant mutation exhibiting no lint or fuzz fibers (Kearney and Harrison 1927). In the diploid *G. arboreum*, the h_a allele is a recessive mutation that confers glabrous plants and lintless seed but an epistatic recessive mutation (*fz*) produces tufted seed in plants heterozygous for h_a (Beasley and Egli 1977).

Determining the map locations and identifying tightly linked DNA markers are the first two critical steps toward the isolation of these loci by map-based cloning or comparative approaches. Thus far in the tetraploids, only the *G. hirsutum* locus *Li1* gene has been mapped on chromosome 22, approximately 13 cM from the SSR marker MP4030 (Karaca et al. 2002). Recently, the chromosome locations of the diploid *G. arboreum* glabrous plant, lintless seed (h_a), and tufted seed (*fz*) loci have been determined using an F_2 interspecific *G. arboreum* × *G. herbaceum* population and a comprehensive RFLP map consisting of 275 loci. The *fz* locus was mapped near the terminus of a linkage group that is orthologous to chromosome 6 of the tetraploid genome, about 4 cM away from the marker G1261b, while the h_a locus mapped to the middle region of the same linkage group, flanked

by Gate1BB03 and A1691b at 2.3 and 0.7 cM, respectively. With several laboratories currently pursuing the genetic mapping of fiber mutant genes (Karaca et al. 2002; Han et al. 2004), it is likely that their chromosome locations will be revealed in the near future.

Because cotton fibers are single-celled trichomes anatomically that differentiate from the outermost testa layer (protoderm) of the ovule (Wilkins and Jernstedt 1999), the genes that govern trichome density on the leaves and stems of mature plants have also been genetically mapped. Wright et al. (1999) detected two major leaf trichome density QTLs on chromosomes 6 and 25, which they inferred to be the *t1* and *t2* loci, respectively, previously described by Lee (1985). The *t1* locus explained about 46% of the total phenotypic variation and was tightly linked to the RFLP locus pGH667, while the *t2* locus explained about 23% of the total phenotypic variation and was tightly linked to the RFLP locus pGH303. Interestingly, the map location of *t1* corresponds very closely to the h_a locus (Desai 2005), suggesting the possibility of a close genetic association between leaf-/stem-borne epidermal hairs (trichomes) and seedborne epidermal hairs (lint fiber).

Molecular Mapping of Genes Expressed in Fibers

A cotton fiber EST gene discovery project (Arpat et al. 2004) provided a valuable new resource for developing DNA markers for fiber genes. The genetic complexity of the cotton fiber transcriptome is very high, and accounts for as much as 50% of the cotton genome (Arpat et al. 2004; Wilkins and Arpat 2005; Wilkins et al. 2005). This complexity translates to an estimated 36,000 or so homoeologous genes in the A and D genomes of tetraploid species that are expressed in fibers. To date, approximately 1,135 fiber cDNA clones have been mapped as RFLPs to homoeologous chromosomes (Rong et al. 2004). In addition, Han et al. (2004) published a genetic map based on 99 SSR markers (NAU) developed from fiber ESTs. Thus, despite the economic importance of cotton fibers, only approximately 6% of the genes expressed in fibers have been mapped, and there is a significant opportunity for further mining of the fiber EST database for the development of PCR-based markers to facilitate genetic mapping of fiber genes.

Recently, Park et al. (2005) reported a genetic map, which located 108 genes expressed during a key developmental stage of fiber morphogenesis (Arpat et al.

2004) to 13 chromosomes of the A subgenome and to ten chromosomes of the D subgenome. On the basis of gene functions, mapped MUSS and MUCS markers revealed that the majority of the mapped markers were associated with genes belonging to two major GO categories (molecular function and/or biological process). Four assigned to the category cellular component (subcellular structures, locations, and macromolecular complexes) are mainly associated with nuclear and viral envelopes. Protein transport and folding, cell adhesion, transcriptional regulation, and steroid metabolism are represented by six markers categorized under biological process. The markers in the molecular function category primarily belong to metabolism-related genes, including the seven novel markers for kinase, hydrolase, catalytic enzyme, oxidoreductase, and endopeptidase inhibitor activities, in addition to metal ion binding (Park et al. 2005). Although the D-genome diploid species, and by inference the progenitor of the allotetraploid Dt-subgenome, do not produce spinnable fibers, many EST loci were mapped on the D genome (Han et al. 2004; Ulloa et al. 2006).

1.3.3
Multigenic Qualitative Traits

Although all morphological mutant loci are assumed to have discrete phenotypes, some degree of variation is often observed for many of these traits when segregating in F_2 populations, implicating the interaction of several genes. For such traits, the use of a quantitative genetics approach may be more appropriate in studying their inheritance.

An example of this phenomenon can be observed in the expression of petal spot, which is characterized by anthocyanin pigmentation at the base of the flower petals. In *G. hirsutum* and *G. barbadense* F_2 populations, some variation for size and intensity of petal spot is observed. Yet, when this trait was scored as a binary character (present or absent) in two independently derived populations, it mapped to chromosome 7 (linked to the DNA marker pAR1182b) as predicted by the analysis of cytological stocks (Wright 1999; Lacape et al. 2003). Reassuringly, when petal spot was treated as a quantitative trait in a different *G. hirsutum* × *G. barbadense* F_2 family, a QTL was identified on chromosome 7, with the highest-likelihood peak falling at the same pAR1182b locus (Desai 2005). However, a minor QTL explaining about

11% of the phenotypic variation was also detected on a Dt-subgenome linkage group that is not homologous to the QTL mapping to pAR1182b. Endrizzi et al. (1984) predicted that modifier genes may influence the size and intensity of petal spot, and it is plausible that the minor QTL reported by Desai (2005) may represent one such modifier.

In addition to detecting minor "modifier" genes that may otherwise escape detection, scoring highly heritable morphological traits as quantitative phenotypes rather than discrete genetic locus can also improve the reliability of genetic mapping data. Wright et al. (1998) reported that when the bacterial blight disease rating was scored as a discrete phenotype, many of the genetic loci were detected outside the likelihood interval mapped by QTL analysis and mapped to large gaps or linkage group termini. This genetic mapping outcome is indicative of possible phenotypic misclassification in the dataset because each false-positive would be interpreted as a double recombinant between the phenotype and the two flanking loci in the linkage analysis. Therefore, forcing discrete resistance scores into the QTL peak intervals results in an inflated map distance between flanking markers. These observations reinforce the need for quantitative phenotypes to obtain reliable map positions (Table 3).

1.4
Quantitative Trait Loci

1.4.1
History of QTLs

The identification and characterization of genes controlling agronomic traits of interest in plant breeding has long been a focus of scientists in the agricultural community. Traditionally, morphological characteristics of an individual parent have been used to predict the characteristics of its progeny. Quantitative genetic theory was developed on the premise that the variability in expression of a quantitative trait for the most part is the result of multiple segregating genetic loci interacting with the environment (Fisher 1918). Most biometric techniques are designed to estimate the number of underlying genetic factors responsible for the variability of continuously expressed traits. These techniques require population structure and sizes that are seldom used by plant breeders (Zeng 1992). Molecular markers (e.g, RFLP, AFLP, microsatellite) are important tools for generating genetic linkage maps and have provided a significant increase in genetic knowledge of many cultivated plant species (Tanksley and Hewitt 1988). These markers make it possible to investigate the numbers, magnitudes, and distributions of QTLs. Genetic maps consist of identifiable markers on the genome at known locations and can be used in the search for genes affecting traits of interest (O'Brien 1993). The association between markers and traits of interest can be revealed from studies based on measurements of the trait in mapped populations (Ulloa and Meredith 2000). Early studies in maize revealed the identification of QTLs with the use of isozyme markers. These studies were able to detect major QTLs in small segregating families (Edwards et al. 1992). Combining QTL mapping with other biotechnological techniques, such as physical mapping and whole genome sequencing, provides an opportunity to identify the genes responsible for the expression of traits of interest (Tanksley et al. 1995; Paterson et al. 2003; Ulloa et al. 2005). The immediate impact of identifying specific portions of the genome will be on the enhancement of breeding programs, and in the long term finding the location of genes affecting QTLs will enhance the characterization and manipulation of these genes. The purpose of this section is to briefly review the history of the QTL studies in cotton; how QTL information is being used, and how it will be used over the next decade to integrate with all the other new molecular tools that have emerged.

In cotton, there have been two basic QTL mapping strategies. One approach is the use and development of interspecific mapping families. This maximizes the number of polymorphic markers but small-scale differences in colinearity and gametic/zygotic selection will skew segregation ratios and hence genetic distances in some regions (Reinisch et al. 1994; Wright et al. 1998; Ulloa et al. 2000; Kohel et al. 2001; Paterson et al. 2003; Mei et al. 2004; Frelichowski et al. 2006; Lacape et al. 2005; Park et al. 2005). The second approach is to use intraspecific mapping populations (Shappley et al. 1998; Ulloa and Meredith 2000, 2002; Shen et al. 2005; Ulloa et al. 2005). Although this minimizes the number of cryptic cytostructural differences that can complicate the detection of linkage and the skew genetic distances, it is more difficult to find sufficient polymorphic markers to saturate the genome (Ulloa et al. 2005).

The estimated numbers and locations of reported QTLs for fiber quality from interspecfic and intraspe-

cific mapping populations are summarized in Tables 4 and 5. The number of progeny evaluated in these experiments ranged from 96–569, and the family types include F_2-derived lines, backcrosses, and recombinant inbreds. To compare and summarize QTL research in cotton, we concentrated on fiber quality traits. Fiber length, strength, and fineness are the critical components of fiber quality, and for this reason, we focused our attention to these QTLs. Tables 4 and 5 summarize the relevant genetic maps.

1.4.2
QTL Methodologies

There are two basic approaches for detecting the specific genomic regions that are associated with quantitative characters, i.e., QTL: single-marker methods which compute test statistics at each marker (e.g., t test and F test) and multiple-marker methods which order markers across genomes (requires genetic maps) and calculate test statistics at each position in the genome, for example, interval mapping (Lander and Botstein 1989) and composite interval mapping (CIM; Jansen 1994; Zeng 1994). The CIM procedure employs simple interval mapping (Tinker and Mather 1995) of a QTL and analyses the variance for other QTLs used as cofactors, using partial regression coefficients. Of the two methods, CIM is more powerful and can minimize the bias that would normally be associated with a QTL that is linked to the position being tested. The computer program QTL Cartographer uses simple linear regression, stepwise regression, interval mapping, and CIM that extends the regression equation to include more markers and uses the remaining markers as cofactors in order to remove the effects of multiple QTLs. The linkage groups are scanned to determine whether the likelihood ratio (LR) test statistic is increasing or decreasing. In brief if we assume L_1 is the likelihood that the QTL is located in the interval flanked by the markers and L_0 is the likelihood there is no QTL in the interval (i.e., the null hypothesis or H_0), the logarithm of odds (LOD) ratio can be defined as

$$LOD = -\log(L_0/L_1).$$

The LR test statistic calculated by QTL Cartographer according to Basten et al. (1997) is

Table 4. Information summary for QTLs and location of fiber quality QTL detected in interspecific mapping populations (G. hirsutum × G. barbadense)

Columns under "Putative QTL and genomes, chromosome, and/or group": Length (A, D), Strength (A, D), Finness (A, D)

Interspecific mapping population	No. of families	Map groups	Groups with QTL	QTL no.	Length A	Length D	Strength A	Strength D	Finness A	Finness D	References
TM-1 × 3-79 RIL	RIL 183	RIL 43	8	25	2, 9, 10	18	3	15, 18	3, 5, 12	17, 18	Frelichowski et al. (2005) 2,126 cM
TM-1 × 3-79 RIL	RIL 183	RIL 40	5	9	2				3	15, 18	Park et al. (2005) 1,277 cM
(Guazuncho 2 × VH8-4602) × Guazuncho 2 BC$_1$	BC$_2$S$_1$ 200	26	23	80	3, 5, 6, A01	14, 23, 25, 26, D02	3, 5, A01	16, 18, 23, 25, D02	3, 4, 5, 6, 9, 10, A02, A03	16, 18, 20, 22, 25, D03, D08	Lacape at al. (2005) For QTL and Nguyen et al. (2004) 5,519 cM
Acala 44 × PS 7 F$_2$	F$_2$ 94	42	4	7	9		9, G05		G35		Mei et al. (2004) 3,287 cM
Siv'on × F-177 F$_2$	F$_3$ 214	41	29	79	9, A01, A02, A03, A05	18, 20	1, 4, A01, A02, A03, A05	14, 17, 20, 22, 23, 25, D02, D03, D04	2, 4, 5, 6, 9, A01, A05, A06	14, 15, 20, 23, 25, D01, D02, D03, D07	Paterson et al. (2003) for QTL and Reinisch et al. (1994) 4,675 cM
7235 × TM-1 F$_{2:3}$	F$_{2:3}$ 186	1	1	2			10				Zhang et al. (2003) 156 cM
Siv'on × F-177 F$_2$	F$_3$ 214	41	13	161							Saranga et al. (2002), Paterson et al. (2003)
TM-1 × 3-79 F$_2$	Aneuploid. 1	1	1	9		16				16	Ren et al. (2002)
TM-1 × 3-79 F$_2$	F$_2$ 171	50	13	13	4	18, 22	3	14,15, 15, 25	1, 2, 3, 12	16, D01	Kohel et al. (2001) 4,766 cM
Tamcot SP37 × PS 7 F$_2$	F$_2$ 98	10	10	10	3	25	10	20, 26			Brooks et al. (2001)
CAMD-E × Seaberry F$_2$	F$_2$ 271	27	11	14		D03			10	20, 22, D02	Jiang et al. (1998) 3,767 cM
TM-1 × 3-79 F$_2$	F$_2$ 171	40		11							Yu et al. (1998) 3,855 cM

Table 5. Information summary for QTLs and location of fiber quality QTL detected in intraspecific mapping populations (*G. hirsutum × G. hirsutum* L.)

Interspecific mapping population	No. of families	Map groups	Groups with QTL	QTL no.	Putative QTL and genomes, chromosome, and/or group Length A	Length D	Strength A	Strength D	Finness A	Finness D	References
7235 × TM-1 RIL	258	22	10	16	5	15, 25		D03		25, D08	Saen et al. (2005) 656.7 cM
7235 × TM-1 F_2	F_2 163	21	6	21	7	23, 25, D02, D03		23, D03		23, 25, D03	Saen et al. (2005) 656.7 cM
HS 427-10 × TM-1 F_2	F_2 169	17	8	23	3, 10	23	3, 10, A02	16, 23	10	16, 26	Saen et al. (2005) 557.8 cM
CPD 6992 × TM-1 F_2	F_2 142	22	6	11	7	16, 18	A02, A03	16	A03	14	Saen et al. (2005) 538.0 cM
G. hirsutum JOINMAP $F_{2:3}$	$F_{2:3}$ 569	47	14	92	3, 7, 9, 10, 12	14, 20, 26	3, 7, 9, 10, 12	14, 20, 26	3, 7, 9, 10, 12	14, 20, 26	Ulloa et al. (2005) 1.502.6 cM
119-5 sub-okra × MD51ne $F_{2:3}$	$F_{2:3}$ 155	16	6	19		RFLP markers located on (Ulloa et al. 2005)					Ulloa and Meredith (2002) 520.4 cM
HQ95-6 × MD51ne $F_{2:3}$	$F_{2:3}$ 199	24	10	18		RFLP markers located on (Ulloa et al. 2005)					Ulloa and Meredith (2002) 830.1 cM
MD5678 × Acala Prema $F_{2:3}$	$F_{2:3}$ 119	17	9	26	See Ulloa et al. (2005)						Ulloa and Meredith (2000) 730.7 cM
HS 46 × MARCABUCA-1-88 $F_{2:3}$	F_5 96	31	24	100	See Ulloa et al. (2005)						Snappley et al. (1998) 855.0 cM

$$LR = -2\ln(L_0/L_1) = -2ln10^{-LOD}$$
$$= 2(ln10)LOD = 4.605LOD,$$

and thus

$$LOD = -\log\{exp[-(LR/2)]\}$$
$$= (1/2)(loge)LR = 0.217LR.$$

If a threshold LR value of 9.21 is set for detection of a QTL, this corresponds to a LOD score of 2.0 under the Mapmaker/QTL statistical test. In addition, a QTL can be detected when the tail probability of the F statistic (from the least-squares regression) is significant at the 0.01 probability level, assuming one and n-1 degrees of freedom in the numerator and denominator, respectively.

1.4.3
QTLs Detected Using Interspecific Mapping Families

Yu et al. (1998) reported putative QTLs for fiber quality traits on several linkage groups using an interspecific mapping population (TM-1 × 3-79) and Mapmaker/QTL (Lander et al. 1987) and SAS. The framework map was based on 171 F_2 progeny and consisted of 219 RFLP, RAPD, and SSR markers. These QTLs collectively explained about 35–50% of the total genetic variance for the fiber quality traits considered. In the same year, Jiang et al. (1998) suggested that chromosomal locations and subgenomic distributions of most QTLs influencing the quality and quantity of fiber were located on the "D" subgenome, derived from an ancestor that does not produce spinnable fibers. One QTL explained 14.7% of fiber span length variation and was located on linkage group D03. The *G. barbadense* allele increased fiber length. Three fiber strength QTLs were detected, two that reduced fiber strength located on chromosomes 20 and 22, and a third QTL on linkage group D02 (Table 4).

On the basis of the single-marker analysis of microsatellite markers previously located on chromosomes 3, 20, and 25, Brooks et al. (2001) reported QTLs for fiber span length and fiber bundle strength (Table 4). Five microsatellite loci collectively accounted for 11.7–16% of the fiber length variation, while seven QTLs explained 11.7–20.6% of the fiber strength variation. In addition, Kohel et al. (2001) located putative QTLs on different chromosomes and/or linkage groups developed from an interspecific mapping population derived from a cross between TM-1

(*G. hirsutum*) and 3-79 (*G. barbadense*), which collectively explained 30–60% of the total phenotypic variance for fiber span length and fiber strength, using Mapmaker/QTL with interval mapping analysis (Table 4).

In 2003, a more extensive study was undertaken by Paterson et al. (2003) using a subset of RFLP markers previously mapped by Reinish et al. (1994). QTL analyses of genotype × environment interactions affecting cotton fiber quality were performed. Interactions of QTLs with environment were evaluated on the basis of two criteria. The first criterion was based on a single-marker analysis or a single-point analysis of variance using SAS (Joyner 1985), which is a common method to evaluate statistical interactions when data from multiple environments are used. The second criterion was based on detected QTLs (significantly meeting LOD scores of 3.0 and above) using interval mapping as implemented in Mapmaker/QTL (Lander and Botstein 1989), In this case, the criterion for identifying QTLs influenced by the environment included detection in some but not all environments and a LOD difference of more than 2 (100-fold) between environments. Paterson et al. (2003) reported 79 QTLs under well-watered and water-limited conditions for fiber quality traits such as fiber length, length uniformity, elongation, strength, fineness, and color. However, only 17 QTLs were detected in the water-limited treatment and two were only specific to the well-watered treatment, indicating that the genetic control of cotton fiber quality was markedly affected by growing season (years) and water-treatment regimes. One genomic region appeared to contain a QTL involved with fiber architecture, affecting both fiber strength and fineness. Two QTLs, one near pAR418a on LGD03 and the other one near A1658b also on LGD03, were associated with main-effect QTLs, being associated across various treatments. Alleles from two loci for increased fiber length were from the long-fibered *G. barbadense* parent; alleles from three loci were from the short-fibered *G. hirsutum* parent, and one with heterotic effect for reduced fiber length. Alleles from 16 loci for increased fiber strength were from the long-fibered *G. barbadense* parent. The heterozygote condition showed lower fiber strength at two loci. Of all the alleles for increased fiber fineness (lower micronaire reading value), 14 loci were from the long-fibered *G. barbadense* parent and four loci were from the *G. hirsutum* parent. In addition, the observations above suggest that improvement of fiber quality under water stress will be more complicated than under well-watered conditions. *G. barbadense* has markedly superior fiber length, strength, and fineness, and QTLs for these traits were indeed derived from this parent. QTLs genome coverage was reported for 15 chromosomes (six from the A genome and nine from the D genome) and nine linkage groups (Table 4).

Mei et al. (2004), using the QTL cartographer version 1.15 CIM module (Bastein et al. 2001) and the statistical software SAS version 8.1, detected seven QTLs. Six QTLs were fiber-related and five of these mapped to A-subgenome chromosomes (Table 4). Three QTLs for seed number, seed weight, and fiber strength clustered in the same region of chromosome 9. However, when Mapmaker/QTL was used to run QTL analysis on the same data, 14 QTLs were detected (LOD = 3.0) in the D genome and four in the A genome. With the use of a permutation test (Churchill and Doerge 1994) and CIM (Jansen and Stam 1994; Zeng 1994), which is a more powerful statistical test, improvement of the efficiency and reliability of the QTLs detected in the study was accomplished. QTLs located in the A-subgenome chromosomes confirm the significant contributions of the A subgenome to fiber development and production during cotton domestication. The combination of A and D genomes stimulates production of fibers superior to those produced by the A genome progenitor. This end-product result may be from the expression and interaction between homoeologous A and D subgenomes (Mei et al. 2004). More recently, Lacape et al. (2005) reported QTL analyses of 11 fiber properties measured on three backcross generations of an interspecific mapping population, using the computer software QTL Cartographer version 1.13 (Basten et al. 1999). Collectively, 80 QTLs surpassed the permutation-based likelihood (LOD) thresholds (3.2–5.7). Nine QTLs which met the permutation-based thresholds plus six additional QTLs were considered for fiber length. For fiber strength, six permutation-based threshold QTLs and six others with lower LOD values were considered. For fineness, 14 of the 21 reported QTLs met the permutation-based threshold. In addition, QTLs were detected in clusters for several traits within a close vicinity in a genome region and/or chromosome sections as previously reported in other studies (Shappley et al. 1998; Ulloa and Meredith 2000). Slight overrepresentation (58%) of mapped QTLs was observed for the QTL distribution between chromosomes or linkage groups for the D genome compared with the A genome. As expected, the positive allelic contribu-

tion for fiber quality traits came from *G. barbadense*, targeting 19 regions on 15 different chromosomes for possible marker-target regions and/or a marker-assisted introgression breeding strategy (Lacape et al. 2005; Ulloa et al. 2005; Table 4).

In the first preliminary QTL analysis, on a genetic map developed from RILs for fiber-related traits, Park et al. (2005) reported one QTL on chromosome 2 for fiber length (50% span length) at a significant LOD level (2.9), which was higher than the empirical threshold (2.8) determined by 1,000 permutation tests at $P < 0.05$, using MapQTL 4.0 (Van Ooijen and Maliepaard 1996) and WinQTLCart 2.0 (Wang et al. 2004) computer programs (Table 4). The MUCS620 marker linked to this 50% SL fiber QTL is homologous to an *endo-β-1,4-glucanses* ($E = 0.0$), an enzyme that plays an important role in cell extension during rapid polar elongation of developing fibers (Arpat et al. 2004). For other fiber traits, permutation tests revealed LOD = 2.8 (fiber elongation and fiber span length at 2.5%) and LOD = 2.9 (fiber bundle strength and fiber fineness) as the significant thresholds, although no QTL was detected above those LOD scores. At LOD $\geq$ 2.0, however, a total of eight potential QTLs for fiber elongation, fiber bundle strength, fiber span length at 2.5%, and fiber fineness were detected on chromosomes 2, 3, 15, and 18, and LG.A01 (Table 4). From single-marker analysis using MapQTL 4.0 and WinQTLCart 2.0, all of these loci were associated with the fiber traits at the $P < 0.01$ significance level. QTLs detected for the same chromosomes as above were reported in previous studies for fiber elongation, fiber bundle strength, and fiber fineness (Kohel et al. 2001; Paterson et al. 2003; Ulloa at al. 2005). The location of fiber QTLs to the same chromosomes in different mapping populations supports QTL assignments in this study, despite the low statistical support from permutation tests (Park et al. 2005). Frelichowski et al (2005) using the same interspecific RIL population from the above study (Park et al. 2005) mapped 407 marker loci and 43 linkage groups, which included BNL, CIR, JESPR, NAU, MUSB, MUCS, and MUSS microsatellite markers (http://www.mainlab.clemson.edu/cmd/). The genetic map covered 2,126.3 cM, approximately 45% of the cotton genome, with an average distance between two loci of 4.9 cM. Twenty-three of 26 chromosomes were assigned to these linkage groups using hypoaneuploid deficiency analysis and previously mapped SSR markers. Some marker loci originally placed in genome specific linkage groups A01 and A03/D02 were reassigned for the first time to chromosomes 18 and 8, respectively. Limited QTL analyses detected 25 putative QTLs, suggesting that loci on chromosomes 2, 3, 12, 15, and 18 may affect variation in fiber quality traits and eventually these regions can be targeted for gene discovery. Assignment of our MUSB (BAC-derived) markers to chromosomes will facilitate integration of linkage and physical maps by the structural alignment of BAC clones into the cotton genome.

1.4.4
QTLs Detected Using Intraspecific Mapping Families

The first report of QTLs for fiber quality traits using an intraspecific mapping population was by Shappley et al. (1998). They used a mixed-model approach to search for QTLs along the linkage groups in 2.0-cM steps. In the mixed-model approach, QTLs are considered fixed effects and molecular markers are treated as random effects. A LR value threshold of 6.63 or above was chosen, which provided significance with a probability of 0.001, with one degree of freedom for detecting QTLs (Table 5). One hundred QTLs to 60 maximum-likelihood positions in 24 linkage groups were reported. The additive and dominant genetic estimates for each trait show the relative importance of the various QTLs for any given trait. For most traits, alleles at different QTLs from either parent could contribute to increased performance for the trait. In addition, several QTLs influenced more than one trait. Although several QTLs were detected for fiber length none showed a major effect. Four QTLs for fiber strength were found to have an additive genetic effect of more than 9.8 kN m/kg. Three reported QTLs for fiber fineness on three different linkage groups were found to have an additive genetic effect of more than one micronaire unit (Shappley et al. 1998). QTLs were later reported from three additional mapping populations (Ulloa and Meredith 2000, 2002), using three different computer programs, Mapmaker/QTL (Lander et al. 1987), MapQTL (van Ooijen and Maliepaard 1996), and QTL Cartographer (Basten et al. 1999). Mapmaker/QTL uses interval mapping to detect and locate a QTL. MapQTL conducts QTL analysis by interval mapping and CIM. It also calculates the estimated mean of the distribution of the quantitative trait associated with each parental source as well as the heterozygote. The threshold value of 9.21 for the LR test

statistic was set for declaring a QTL. This corresponds to a LOD score of 2.0 (Table 5).

Ulloa and Meredith (2000) reported 26 QTLs on nine linkage groups, explaining 3.4–44.6% of the variation for the considered traits. Three QTLs were detected for fiber length, three for fiber strength, and four for fiber fineness. The majority of the RFLP alleles associated with long fiber span length, fineness, and strong fiber were contributed from Acala Prema parent, while the majority of the alleles associated with high yield and high lint percentage were contributed from the MD5678ne (Upland) parent. The results reveal that gene introgression at the DNA level within *G. hirsutum* was successfully accomplished for the agronomic and fiber quality traits in the $F_{2.3}$ families. However, these small additive and dominant effects did not preclude other types of gene action for the expression of these agronomic and fiber quality traits. Two linkage groups 1 and 3 locate and share strongly the above RFLP allelic expression, suggesting that genes for fiber traits may recombine as blocks during meiosis (Ulloa and Meredith 2000). Multiple traits can be correlated owing to linkage and pleiotropy, or the correlated traits may be components of a more complex variable. Two components of bundle fiber strength are fiber length and perimeter (Meredith 1992; Ulloa and Meredith 2000). Fine fiber (e.g., small perimeter) results in more fibers per bundle, which can confer greater fiber strength. Additionally longer fibers promote fiber-to-fiber contact, tending to increase fiber strength. Micronaire reading and fiber length are components of lint percentage. Linkage groups containing QTLs all of which are components of one or more other traits may represent the above scenario. Unexpected sources of variation may also be attributed to epistasis. Favorable (or unfavorable) alleles may be present in a parental line but not expressed in its genetic background. When crossed to another individual, however, epistatic interaction may affect the expression of the alleles (Ulloa and Meredith 2000). Further research is needed on recombination rates to monitor these events.

Ulloa et al. (2005) reported QTLs from the above four cotton populations (Shappley et al. 1998; Ulloa and Meredith 2000, 2002) on the linkage groups generated by JoinMap (Stam and Van Ooijen 1995; Table 5). On the basis of 111 shared RFLP loci, the percentage of common heterozygous loci between populations varied from 9.0–41.0%. One hundred and forty-five confirmed QTLs from the four populations were located on 21 linkage groups of the composite map. Eight linkage groups carried QTLs at the same locus or within the vicinity of the linkage group from at least two of the four different populations (Table 5). Detected QTLs, from a single population, were identified on the composite map. For example, Fig. 3 shows the QTL positions from four different populations on chromosome three near the RFLP locus (centimorgan distance) with the higher LOD score. Linkage group 4 (Ulloa et al. 2002), which is a part of chromosome 5, carried QTLs from two different populations within the vicinity of locus C115A1V. Locus C115A1V was polymorphic in two populations, and the heterozygote condition – family means with both alleles (one from each parent) – had the highest value, for example, for lint yield 1,074 kg/ha and for fiber strength 234.0 kN m/kg. Overall results revealed the presence of 63 QTLs on five different chromosomes of the A subgenome, chromosomes 3, 7, 9, 10, and 12, and 29 QTLs on the three different D subgenomes, chromosomes 14 long arm, 20, and the long arm of 26. Linkage group 1 (chromosome 3) harbored 26 QTLs, covering 117 cM with 54 RFLP loci. Linkage group 2 (long arm of chromosome 26) harbored 19 QTLs, covering 77.6 cM with 27 RFLP loci (Ulloa et al. 2005).

Shen et al. (2005) with the use of SSR makers reported QTLs for fiber qualities in three diverse cotton upland lines, which were used to develop three intraspecific mapping populations (Table 5). Windows QTL Cartographer with CIM was used to identify QTLs (LOD $\geq$ 3). LODs between 2 and 3 were used to nominate possible QTLs. A total of 39 (17 significant and 22 suggestive) QTLs were detected from the three populations, including suggestive QTLs, 11 QTLs for fiber length, ten for fiber strength, and nine for micronaire. At least five QTLs could be detected in two successive generations (F_2 – $F_{2:3}$) and three QTLs could be identified in two populations. In the first population, a QTL located on chromosome 25 in both F_2 and $F_{2:3}$ generations explained 14.1 and 13.1% of the fiber length variation, respectively. For fiber fineness (micronaire reading), four QTLs, three located on chromosome 25 and one on chromosome 15, were observed to have the same direction of additive effect originating from one of the parents (TM-1), explaining 50.7–63.9% of the total trait variation. In the second population, a QTL influencing fiber length on chromosome 10 was detected in both F_2 and $F_{2:3}$ generations with positive additive effects originating from the parent TM-1. A QTL affecting fiber strength was also detected on chromosome 10 in both gener-

ations in which TM-1 allele increased fiber strength by 0.72–1.03 cN/tex. A total of three QTLs were detected for fiber fineness. In the third population, one significant QTL for fiber length, three QTLs for fiber strength, and two QTLs for fiber fineness were identified (Table 5).

1.4.5
QTLs Detected on Mapping Population for Specific Gene(s)

There are several studies in cotton (Wright et al. 1998, 1999; Jiang et al. 2000; Ulloa et al. 2000; Saranga et al. 2002; Karaca et al. 2002; Bezawada et al. 2003) where the QTL approach has been used to identify and map a gene or genes which confer(s) resistance to a particular pest or trait (Table 3). However, the lack and/or limited number of suitable molecular markers in cotton have delayed this approach and its application to breeding. Although the a studies did not provide strong markers for breeding, they provided information to further understand the cotton genome and the power of QTL analysis. This was previously discussed (Sect. 1.3), and herein only information related to QTLs will be provided. Wright et al. (1998, 1999) used an established RFLP map (Reinisch et al. 1994) to identify and map genes which confer resistance to several races of the pathogen *Xanthomonas campestris* pv. *malvacearum* (Smith) (*Xcm*), and mapped genes affecting pubescence of cotton (Table 3).

QTL analyses of leaf morphology were investigated on an interspecific population (Jiang et al. 2000). QTLs were detected on several chromosomes affecting leaf size and shape, suggesting there were genes modifying the expression of the classical Mendelian okra-leaf shape locus which is found on chromosome 15. Jiang et al. (2000) reported 21 QTLs for this morphological trait with LOD > 3.0 and $P < 0.001$. The parental allelic contribution contained some alleles with positive as well as negative effects for leaf size from the same donor. Sixty-three percent of the above QTLs mapped to the D-subgenome chromosomes. Stomatal conductance (Ulloa et al. 2000) and osmotic potential (Saranga et al. 2002) were also addressed by a QTL approach. Cotton is routinely grown in hot areas, and in the USA hot and irrigated areas of the Southwest. Extended periods of extremely high temperatures are common in these areas during the critical stage of peak flowering, and as a consequence cotton yield is affected. Cotton plants use transpiration to cool them-

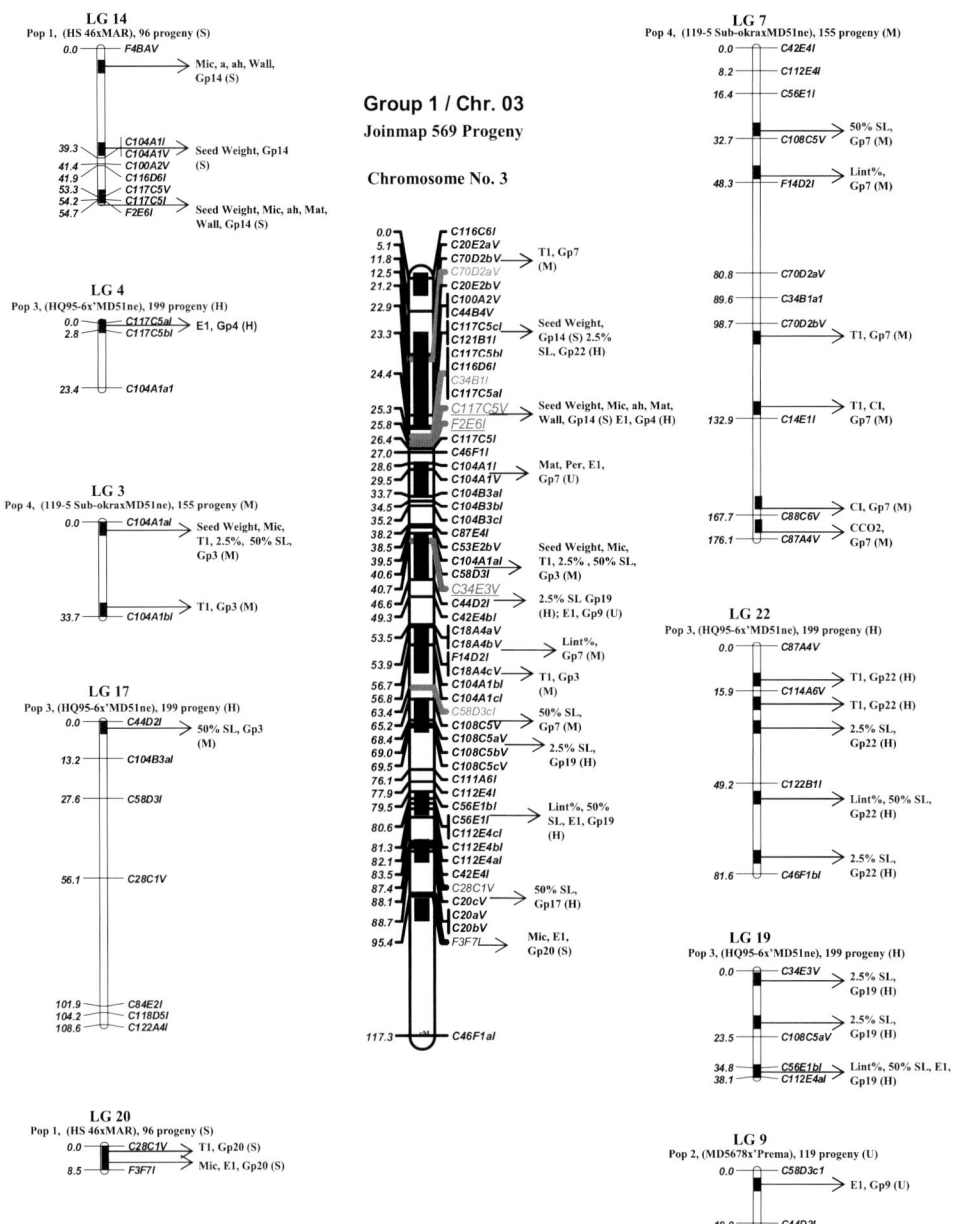

Fig. 3. Amalgamation of joinmap linkage group no. 1 and/or chromosome 3 compared with groups from the four different bulk-sampled plots of an $F_{2:3}$ *G. hirsutum* population. Map distances between adjacent markers are in centimorgans. The map was constructed by using the JOINMAP (Stam and Van Ooijen 1995) computer program, with Kosambi function and a logarithm of odds (*LOD*) of 3.0. The location of quantitative trait loci (*QTLs*) on a single population for agronomic and fiber quality traits was detected with LOD > 2.0 by at least two different QTL programs (Mapmaker/QTL, QTL Cartographer, and MapQTL). *Pop 1* mapping population no. 1 (Shappley et al. 1998b), *Pop 2* mapping population no. 2 (Ulloa and Meredith 2000), *Pop 3* mapping population no. 3 (Ulloa et al. 2002a, b), *Pop 4* mapping population no. 4 (Ulloa et al. 2002a, b). Restriction fragment length polymorphism loci identified to specific cotton chromosome (Saha et al. 2000). *Chr* chromosome, *G* linkage group, *Mic* micronaire reading (fineness), *T1* fiber strength, *E1* fiber elongation, *2.5% SL* and *50% SL* fiber span lengths at 2.5 and 50%, *Lint%* fiber lint percentage.

selves but the efficacy of transpiration cooling varies and is genetically determined. QTL analysis of replicated F_3 families permitted the identification of two QTLs on two linkage groups influencing stomatal conductance (Ulloa et al. 2000; Table 3). The QTL analysis of stomatal conductance obtained from the 118 $F_{2.3}$ progeny means was performed by two computer programs, QTL Cartographer version 1.12 (Basten et al. 1997) and QGENE version 2.26 (Nelson 1996). In addition, QTL mapping was used to test the association between productivity and quality under water deficit, and only reduced plant osmotic potential was clearly implicated in improved cotton productivity under arid conditions (Saranga et al. 2002). Selection for low stomatal conductance significantly reduced cotton lint yield (Ulloa et al. 2000). Chromosomal regions were identified with DNA markers that controlled this trait for possible usage in breeding programs to assist in selection for this difficult-to-measure physiological trait (Table 3).

1.4.6
Comparison of QTLs

The results of comparison of QTL consistency from cotton populations were similar to those of studies in other crops (Beavis 1994; Austin and Lee 1996; Groh et al. 1998; Melchinger et al. 1998; Monforte and Tanksley 2000; Subudhi et al. 2001). Consistency of QTLs in different environments and/or genetic backgrounds has been found in some instances to be as low as 20% between two experiments. The inconsistency of QTL positions across populations can be attributed to several factors. One factor is that genetic variability between populations for a given trait is not the same for all populations. Two other factors are the precision with which the trait is measured, and population size. A QTL explaining 8% of the trait variation with $h^2 = 0.4$ can be overestimated up to 388% for $N = 100$ and up to 44% for $N = 300$ (Utz and Melchinger 1994). Beavis (1994) suspected that increased power through modifications in data analysis techniques will have a minor impact on analyses of polygenic traits because the most important factors are primarily a function of the experimental design, i.e., the numbers and types of progeny as well as the field plot designs that are used to evaluate the traits. The paradox is that there is little power to identify a QTL in a trait that exhibits low heritability. To realize the full potential of molecular markers in breeding for polygenic traits, it may

be necessary to develop new MAB methods, rather than augmenting existing methods with MAS (Beavis 1994). The decision of which QTL regions to transfer with MAS and/or to consider in a selection index should be based on QTL effects verified in an independent validation sample (Melchinger et al. 1998).

Even though consistencies of QTLs were found in some instances to be as low as 20% between two experiments, QTL examination for agronomic and fiber quality traits revealed highly recombined and gene abundant regions on cotton chromosomes (Ulloa et al. 2005; Tables 4, 5). The maps presented herein were developed from interspecific and intraspecific cotton populations from parents that arose from artificial selection and breeding at different locations, with differing germplasm pools, including Acala, Midsouth Upland, Texas MAR germplasm, Pima, and Sea Island. The compilation of genetic linkage maps and the joinmap from different breeding gene pools enabled partial dissection of the A and D subgenomes in *G. hirsutum* with genomic and cDNA-probe RFLP markers, EST microsatellites (the results describe locations of expressed genes), and unknown genomic-sequences for molecular markers. Relative to other types of markers, the cDNA-derived RFLPs and EST microsatellites may be more likely than some other marker types to be physically and recombinationally close to the comprehensive array of economically important genes. Two major linkage groups of the joinmap (Ulloa et al. 2005), chromosome 3 and chromosome 26 long arm, in two different chromosomes belonging to the A and D subgenome, respectively, indicate that cotton chromosomes may have regions of high (hot spot) and low (cold spot) meiotic recombination like many other eukaryotic organisms (Gill et al. 1996; Petes 2001; Tables 4, 5, Fig. 3). The presence of unevenly distributed high marker-rich recombinationally active regions in two different A- and D-subgenome-specific chromosomes suggests that no correlation was present between the length of the chromosome and the number of loci because overall the A genome has larger chromosomes relative to the size of the D-genome chromosomes. However, the variation in marker density along linkage groups was primarily due to the presence of dense marker regions (Paterson et al. 2003; Mei et al. 2004; Rong et al. 2004; Lacape et al. 2005; Ulloa et al. 2005).

On the basis of genome-specific chromosomes identified in *G. hirsutum* (A and D), the A subgenome contains 43% (Paterson et al. 2003) to 68% (Ulloa et al. 2005) of QTLs on different chromosomes. In contrast, the D subgenome contains 32% (Ulloa et al. 2005) to 57% (Paterson et al. 2003) of QTLs on different chromosomes. Although the D-genome diploid species do not have spinnable fiber, the D subgenome does contain QTLs positively affecting fiber and morphological traits (Reinisch et al. 1994; Wright et al. 1998; Jiang et al. 2000). These loci may have come under selection only after polyploidy formation, and therefore harbor greater allelic diversity among tetraploid forms. Conversely the homoeologous associations between the two subgenomes in the tetraploid species support the assumption that the A-genome diploids, because they have spinnable fibers, may already have contained favorable alleles at some major loci affecting fiber traits before the polyploids evolved as a result of prior natural selection (Paterson et al. 2003). Notably, the union of A and D genomes results in the production of fibers superior to those produced by the A-genome progenitor. This end-product result may be from the expression and interaction between homoeologous A and D subgenomes (Mei et al. 2004).

The uneven distribution of molecular markers (hot and cold spots) in the tetraploid cotton species can be used as a basis for genotypic selection among individuals, accelerating introgression of desired chromosomal segments into a genetic background. In addition, on the basis of these studies, the QTLs for various cotton traits tend to reside at, or near, the same locus for certain linkage groups from the diverse populations (Ulloa and Meredith 2000). The detection of common QTLs suggests the possibility of existing elite fiber genes for quality and possibly of the same origin (Shen et al. 2005; Ulloa et al. 2005). Most genes of elite fiber properties originated from a parent with high fiber quality (triple hybrid) and/or from introgression of a DNA segment from *G. barbadense*. Across a variety of traits, QTLs are more likely to be clustered into specific genomic regions than distributed haphazardly across the genome. Recently, a *G. hirsutum* and *G. barbadense* backcross self-population was used to dissect the molecular basis of genetic variation governing 15 parameters reflecting fiber span length with the use of RFLP markers. A total of 22 genomic regions were identified in which both homoeologous loci were segregating, and could be tested for QTL analysis. Some of the findings indicated that fiber span length may be governed by many minor genes with predominantly additive effects (Chee et al. 2005). The future availability of PCR-based DNA makers linked to QTL-rich regions common to many

Gossypium species will assist breeders in transferring and maintaining valuable traits from exotic sources during cultivar development. In addition, QTLs for fiber quality traits on these linkage groups with chromosomal positioning on certain regions located between two molecular markers (an average of less than 1 cM = 0.4–0.6 Mb) could be the target for map-based cloning. New populations are being developed to validate existing QTLs (from different gene pools and RILs) and for the identification of loci lineages for *G. hirsutum* from its diploid progenitors (the A and D genomes). Increased marker density in the joinmap or on a consensus map would eventually facilitate both genetic and physical mapping of the cotton genome.

1.5
Marker-Assisted Breeding

The development of genetic maps is the foundation for MAB of quantitative traits. Both *G. hirsutum* L. intraspecific (Shappley et al. 1998; Ulloa et al. 2002, 2005) and interspecific *G. hirsutum* L. × *G. barbadense* L. populations (Reinisch et al. 1994; Jiang et al. 1998, 2000; Nguyen et al. 2002; Lacape et al. 2003; Paterson et al. 2003; Mei et al. 2004) have been used to produce genetic maps. This has been succeeded by QTL mapping studies, which include traits for fiber quality (Jiang et al. 1998; Shappley et al. 1998; Ulloa and Meredith 2000; Kohel et al. 2001; Paterson et al. 2003; Mei et al. 2004; Lacape et al. 2005; Ulloa et al. 2005), agronomic traits (Shappley et al. 1998; Ulloa and Meredith 2000; Ulloa et al. 2005), fertility restoration (Liu et al. 2003), leaf morphology (Jiang et al. 2000), disease resistance (Wright et al. 1998; Rungis et al. 2002), stomatal conductance (Ulloa et al. 2000), gossypol content (Vroh Bi et al. 1999), and nematode resistance (Bezawada et al. 2003).

MAB requires a certain level of information and technical capability, DNA markers tightly linked to the QTLs and genes of interest, and a high-throughput marker screening system. The high-throughput marker systems are still being developed and improved, but the evolution of these marker systems is both positive and negative. Breeding programs need markers that can be easily and quickly applied, but the RFLP markers and the maps based on them will only remain useful as the RFLPs are converted into PCR-based markers.

1.5.1
Challenges Facing MAB in Cotton

There are a number of obstacles to using MAB in cotton and most involve the utility of the markers themselves. Within the Upland cotton germplasm, there is little accessible genetic diversity on the basis of what has been seen with isozymes (Wendel et al. 1992) and RFLP markers (Brubaker and Wendel 1994; Chee et al. 2004). More recent markers, such as AFLP and SSR markers have produced a similar result (Ulloa et al. 1999; Abdalla et al. 2001; Iqbal et al. 2001; Gutiérrez et al. 2002; Ulloa et al. 1999). This lack of accessible genetic diversity in such a large genome (approximately 4,400–4,500 cM; Lacape et al. 2003; Rong et al. 2004) makes it difficult to identify enough polymorphic markers regardless of the marker system to adequately and uniformly cover the entire genome.

Given the number of markers required to address these needs, it is clear that the available pool of markers may not be sufficient for infraspecific genetic families. The RFLP marker system has the largest number of publicly available markers (Rong et al. 2004), but it is a marker system that is noted neither for high throughput nor for ease of use in a breeding program, and the ongoing marker development effort within the cotton community needs to continue. A recent step in this direction is the establishment of an ad hoc bioinformatics resource, the Cotton Microsatellite Database at http://www.mainlab.clemson.edu/cmd/ (sponsored by Cotton Incorporated). Currently considerable attention is focused on the development of SNP marker systems, as they allow breeders to access sequence differences directly. Markers based on electrophoretic differences, including AFLPs and SSRs, can mask underlying genetic variation. However, the complex polyploid cotton genome will continue to complicate marker discovery and application.

Genotype × environment interactions are key obstacles in developing cotton cultivars in a conventional breeding program and this will not be overcome by MAB. The number of QTLs affecting fiber quality is greatly increased by water stress (Paterson et al. 2003) and will complicate the development of drought-resistant cultivars; however, MAB will facilitate the development of drought-resistant cultivars. Once QTLs have been identified and validated, accurate selection (using marker surrogates) will no longer be dependent on environmental conditions or require large populations. Guo et al. (2003) have begun to use MAB to develop high-strength cotton cultivars using a population derived from G. *hirsutum* '7235' (an introgression line of G. *anomalum* Wawra ex Wawra & Peyritch backcrossed to G. *hirsutum* 'Acala 3080') and G. *hirsutum* 'TM-1'. They have identified markers linked to a major gene and a few modifers for fiber strength and are using these markers to develop new cultivars (Zhang et al. 2003). Although, crossovers between the markers and the QTLs have been observed, the surrogate markers for their major fiber strength QTL (T1) have remained associated across several segregating generations (Guo et al. 2003).

1.5.2
The Future of MAB in Cotton

Bacterial blight (*Xanthomonas campestris* pv. *malvacearum* (Smith 1901) Dye 1978b) is a potentially devastating disease in cotton. Although some of the resistance genes/QTLs have been mapped (Wright et al. 1998; Rungis et al. 2002), fine mapping of the resistance genes is under way to locate markers with tighter linkages. In rice (*Oryza sativa* L.), cultivars with four resistance genes for bacterial blight (*Xanthomonas oryzae* pv. *oryzae* (Ishiyama) Dye) have been assembled and were found to have a wider spectrum and higher level of resistance than lines with single resistance genes (Huang et al. 1997). Joseph et al. (2004), while stacking four resistance genes, found that two of the genes were as good as having three or four pyramided, which illustrates the complications of epistasis in gene pyramiding.

Cotton leaf curl virus (CLCuV) is a disease of epidemic proportion in Pakistan and India. RFLP (Aslam et al. 1999) as well as RAPD and SSR markers [National Institute for Biotechnology & Genetic Engineering (NIBGE), Faisalabad, Pakistan, http://www.nibge.org/plantres.htm] have been found linked to CLCuV resistance. As the DNA markers were developed, they were utilized concurrently with the conventional breeding program in the development of an elite strain (NIBGE-1) with resistance to CLCuV. The NIBGE-1 cultivars are currently in national field trials. The RFLP and RAPD markers are being sequenced to develop STS and SCAR markers, respectively, to facilitate further breeding efforts.

A multiplex MAB program using germplasm lines designed with the same goal in mind as the multiple adversity resistance program started by Byrd and continued by El-Zik at Texas A&M University would be of particular value to the cotton industry. These diseases could be approached as phenotype sets to directly develop resistance and then concurrently develop an understanding of the individual components of the disease. Fusarium wilt-root-knot nematode is a disease complex caused by *Fusarium oxysporum* Schlechtend.:Fr. f. sp. *vasinfectum* (Atk.) W.C. Snyder & H.N. Hans. and *Meloidogyne incognita* (Kofoid & White) Chitwood. The seed and seedling diseases are caused by a complex that includes *Colletotrichum gossypii* Southworth, *Fusarium* spp., *Pythium* spp., *Rhizoctonia solani* Kühn, and *Thielaviopsis basicola* (Berk. & Broome) Ferraris. Root rot can be a Verticillium wilt (caused by *Verticillium dahliae* Kleb.) or Phymatotrichum root rot (caused by *Phymatotrichopsis omnivora* (Duggar) Hennebert). Leaf spot can be caused by *Alternaria macrospore* A. Zimmerm., *A. alternata* (FR.:FR.) Keissl., *Cercospora gossypina* Cooke, *Cochliobolus spicifer* R.R. Nelson, *Myrothecium roridum* Tode:Fr, *Rhizoctonia solani* Kühn, and *Stemphylium solani* G. F. Weber.

Although it is clear MAB can accelerate the development of improved cultivars, the lack of genetic diversity among elite germplasm lines remains an obstacle. The frequency of polymorphic markers in 114 obsolete US Southeast and Midsouth cultivars/germplasm lines (1890–1900) ranged from 0–13%. Of the over 6,300 pairwise comparisons, only 0.08% had a 1:8 chance or better of the RFLP marker being polymorphic, while the average cultivar/germplasm lines had about a 1:20 chance or better. This may also be somewhat inflated since some of the included cultivars/germplasm lines were from the Triple Hybrid crosses and the two USDA-ARS Pee Dee introgression programs in Florence, SC, USA, which were obviously more polymorphic as a group (Chee et al. 2004).

MAB has been cited as a potential tool for stacking herbicide resistance genes in cotton in a US patent, no. 6,376,754. The markers useful for pyramiding the Bt and herbicide-resistance genes in cotton would be from the actual transgenes for a direct selection of the desired traits instead of an indirect selection with linked markers that is more commonly associated with MAB. Although the tools are available for pyramiding (stacking) of linked markers, we are not aware of any completed efforts outside of the transgenic cultivars.

Transgenic cotton breeding has been a huge success; 79.5% of the acreage of Upland cotton in the US was planted as transgenic cotton in 2004.

Mississippi had the highest acreage at 99.5%, while California had the lowest with 52.2% (cotton varieties planted, USA, 2004 crop, http://www.ams.usda.gov/cottonrpts/MNPDF/mp_cn833.PDF). This success has also paralleled a societal discourse of concern (and some fear) over the use of any transgenic crops and led to the touting of MAB as the "responsible" way to use biotechnology by a number of leading environmental organizations following the policy paper (http://www.biotech-info.net/marker_assisted_breeding.html) of the Soil Association, a registered charity (no. 206862) in the UK.

1.5.3
Marker-Assisted Introgression

While the genetic homogeneity of elite *G. hirsutum* cultivars limits the utility of MAB, MAB could prove powerful in efforts to improve Upland cotton cultivars through interspecific introgression. In addition to the fact that a much higher percentage of markers will be polymorphic, *G. hirsutum* and *G. barbadense* have distinct fiber traits. *Gossypium hirsutum* primarily is known for its yield, while *G. barbadense* is known for its fiber quality: length, strength, and fineness. Two approaches are obvious: improving the yield of high-quality cotton or improving the quality of high-yielding cotton. *G. hirsutum* and *G. barbadense* cultivars produce over 90% of the cotton in the world (Kohel 1999), so the obvious start is with these species. Each of these species is considered a germplasm resource for the other since they are interfertile. As expected from QTL research of interspecific crosses, the majority of the fiber quality improvements come from *G. barbadense* although some do come from the *G. hirsutum* parent (Jiang et al. 1998; Chee et al. 2005; Lacape et al. 2005). Also, some of the extra-long staple character within the *G. hirsutum* germplasm is historically considered to have come from inadvertent interspecific crosses but definitive corroboration is unavailable (Ware 1950).

However, there are difficulties in combining the traits from these two species. Ongoing development of a near-isogenic introgression line series of *G. barbadense* into *G. hirsutum* via MAB illustrates a difficulty in ensuring that the complete *G. barbadense* genome is incorporated into the *G. hirsutum* background. As expected on the basis of the level of

F_2 breakdown observed in *G. hirsutum* × *G. barbadense* families (Stephen 1946), portions of the *G. barbadense* genome are very difficult to stabilize in the *G. hirsutum* background (Jiang et al. 2000). Sealand cultivars, a product of the introgression of Sea Island *G. barbadense* into Upland cultivars developed at the USDA-ARS Pee Dee program, also show little introgressed DNA on the basis of RFLP marker evidence (Chee et al. 2003). Both of these occurrences follow a phenomenon noted as genetic breakdown or developmental incompatibility that occurs in subsequent generations after the original cross and subsequent backcrosses (Stephens 1946). This clearly implies that there are genetic interactions in these families that significantly reduce the fitness of individuals, and that it will be impossible to introgress some *G. barbadense* gene regions into *G. hirsutum*. Preparation of additional interspecific maps of *G. hirsutum* with the three other undomesticated tetraploid *Gossypium* species (*G. mustelinum* Miers ex Watt, *G. tomentosum* Nuttall ex Seemann, and *G. darwinii* Watt) and concurrent MAB programs are also in progress (A.H. Paterson, personal communication). Using MAB to monitor introgression is a key component in mining the related tetraploid species for desired traits to improve cotton production.

Fiber quality and yield, as mentioned before, are the two most important traits of cotton production. We are going to focus on fiber quality to illustrate the value MAB has in developing cultivars with improved value. There are a number of fiber properties that are used to measure fiber quality; fineness, length, strength, and elongation are four of the most important. There are also a number of ways to estimate these fiber properties, which can confound the development of QTLs of particular fiber properties. As an example, micronaire, the most common estimator of fiber fineness, can be confounded by the relationship of fiber fineness and the maturity of the fibers (May 1999). Some researchers therefore have used other tests to elucidate maturity and fineness, such as the arealometer, while others have not. Since combining different estimators together is not reasonable, comparisons across all of the QTL studies is not appropriate in some cases. Another reason that a comparison across the QTL research is often inappropriate is that the parents of each population could have different marker and trait polymorphisms. Each population is a unique sample of the germplasm and the QTLs will only be of direct value in that particular population.

Fiber fineness is a complex property within the complexity of overall fiber quality. Shappley et al. (1998) identified ten QTLs for micronaire out of 15 that affected other fineness or maturity measures. Of these six other fineness or maturity measures, only two QTLs of 36 were unique to their given property. Jiang et al. (1998) reported only one QTL for fineness and it was associated with all the measures. Kohel et al. (2001) found six QTLs for fiber fineness. Ulloa and Meredith (2000) found ten QTLs – four for micronaire, two for fiber maturity ratio, and four for fiber perimeter. There was one QTL colocating for micronaire and fiber perimeter and one QTL for fiber maturity ratio and fiber perimeter. Paterson et al. (2003) found 25 QTLs for micronaire, Mei et al. (2004) identified one fineness QTL, and Lacape et al. (2005) found 14 QTLs for measures of maturity and fineness using a maturometer (FMT3, Shirley Dev., UK) – six QTLs for micronaire, four for fiber maturity ratio, seven for linear fineness, and four for standard fineness. Of the QTLs of Lacape et al. (2005), one QTL colocated for micronaire, fiber maturity ratio, and linear fineness, two QTLs colocated for micronaire and fiber maturity ratio, one QTL colocated for linear fineness and standard fineness, and one QTL was associated with micronaire, linear fineness, and standard fineness. Inconsistencies for fiber properties measurements may be relative to specific instrument and analytical procedures of each laboratory.

The number of QTLs affecting fiber length, strength, and elongation percentage are not as complex as the fiber fineness/maturity complex. From the same populations described earlier, fiber length was measured by 2.5% of the span length, 50% of the span length, or upper half mean. Those studies that used one of the possible length measures identified one to six QTLs and in studies where two of the length measures were used, the number of putative QTLs ranged from three to nine, with only a single colocating QTL. Fiber length can be confounded by the distribution of the individual fiber lengths (estimated by fiber uniformity). In the previous population (Kohel et al. 2001), fiber strength was measured by a stelometer or by HVI; no study used more than one strength measure. The number of identified QTLs ranged from two to six, except for in the study of Chee et al. (2005), who identified 21 QTLs. The number of putative QTLs for fiber elongation percentage for these populations ranged from one to 18 (excluding since they did not measure fiber elongation percentage).

These numbers of QTLs for a given property are controlled by the genetic distance between the parents of the population of interest and so these component fiber properties could have a complex, polygenic inheritance or a simple inheritance depending on the parental genotypes, i.e., the number of loci segregating in the family. Also, QTLs are not always positive effectors, so improved fiber quality would depend on selecting for some of the markers and selecting against others. It is common to find useful alleles in both parents, which could result in transgressive fiber property phenotypes. Unexpected epistatic interactions with genes donated by the second parent can also modify the phenotypic expression effect, even to the point that changing backgrounds could change the effect of a QTL from positive to negative. QTLs can also be described by how much they contribute to the overall measure, which quantifies the level of importance of the particular QTL in MAB. As mentioned before, one factor is that genetic variability between populations for a given trait is not the same for all populations. Another factor is the precision with which the trait is measured, and another factor is the population size. In cotton, as with other crops, the selection of a QTL or genome segment to transfer is predicated on validation of the QTL effects in independent experiments. Eventually with enough molecular markers to saturate the entire genome and QTLs validated in different genetic families across multiple environments and years, the true utility of MAB will be realized, and cotton geneticists/breeders will be able to sample all important regions of the cotton genome, identify the location of the genes controlling critical agronomic traits (e.g., yield, fiber quality and-water use efficiency, and disease and pest resistance), and manipulate the expression of these traits to achieve their breeding aims more efficiently and cost-effectively.

1.6
Map-Based Cloning

Although the tetraploid cotton genome, which is estimated to range in size from 2,835 Mb (Grover et al. 2004) to 3,000 Mb of DNA (Michaelson et al. 1991; Gomez et al. 1993), is considerably larger than that of all major crop species such as tomato (950 Mb), rice (450 Mb), soybean (1,810 Mb), and maize (2,500 Mb) (Goldberg 1978; Arumuganathan and Earle 1991), the average DNA content of 1 cM is estimated to be only around 400 kb (Paterson and Smith 1999). This is because the tetraploid cotton genome is recombinantly more active than the genome of most other crop species (Rong et al. 2003). The recombinational to physical distance ratio is less than that of tomato, and only slightly larger than that of the model plant species *Arabidopsis*, species where genes have been successfully cloned by chromosome walking. Therefore, the large genome size of cotton should not be a deterrent to the success of map-based cloning in cotton.

The application of map-based cloning to target a specific phenotype in cotton is lagging behind that of other crop species. This will change in the future as the requisite molecular tools continue to accumulate. The two basic tools required for chromosome walking are high-density genetic maps and large-insert DNA libraries, both of which have been developed in cotton. The current molecular maps of cotton contain about 4,000 low-copy DNA landmarks, which correspond to more than 2,000 individual DNA probes mapped as RFLPs and SSRs (Rong et al. 2003). A total of 295 genes have been plotted on the cotton map as QTLs. These QTLs condition 26 traits related to plant growth, development, and morphology (Wright et al 1999; Jiang et al. 2000), reproductive biology (Lan et al. 1999), fiber yield and quality (Jiang et al. 1998; Ulloa and Meredith 2000; Paterson et al. 2003; Park et al. 2005; Ulloa et al. 2005; Frelichowski et al. 2006), and response to biotic (Wright et al. 1998) and abiotic (Ulloa et al. 2000; Saranga et al. 2001) stresses. With a total recombination length of 5,200 cM, the intervals between markers are estimated to be about 2 cM, or an average physical distance of about 800 kb. While this physical distance between markers may be sufficiently close to permit chromosome walking, additional markers are needed to meet the criteria for chromosome landing (Tanksley et al. 1995).

Large-insert libraries in the form of BACs have been developed from at least four genotypes for tetraploid cotton (*G. hirsutum*) and from one diploid progenitor *G. raimondii* (Table 6). Three of the cotton libraries have more than 4 times coverage of the haploid genome. The average insert size of these libraries varies from about 100 to 150 kb, with a mean of around 140 kb across all libraries. The construction of additional BAC libraries for the diploid species *G. herbaceum*, *G. longicalyx*, and *Gossyp-*

Table 6. Summary of bacterial artificial chromosome libraries developed from *Gossypium* species

Genotype	Insert size	Number of clones	Genome coverage[a]	References
G. hirsutum 'Tamcot HQ95'	93	51,354	2.3X	Yu et al. (2000)
G. hirsutum 'Auburn 623'	140	44,160	2.7X	Yu et al. (2000)
G. hirsutum 'ESP'	158	38,784	2.3X	Yu et al. (2000)
G. hirsutum 'ESP'	142	38,400	2.4X	Yu et al. (2000)
G. hirsutum 'ESP'	148	38,400	2.5X	Yu et al. (2000)
G. hirsutum 'TM-1'	130	61,440	3.6X	Yu et al. (2000)
G. hirsutum 'TM-1'	150	55,296	3.7X	Yu et al. (2000)
G. hirsutum 'TM-1'	152	46,080	3.1X	Yu et al. (2000)
G. raimondii (D5)	140	23,040	4.6X	Yu et al. (2000)
G. hirsutum 'Maxxa'	137	129,024	8.3X	Tomkins et al. (2001)

[a] Haploid genome equivalent of the AD tetraploid

ioidies kirkii are in progress (http://www.nsf.gov/bio/pubs/awards/bachome.htm).

The integration of the physical and the genetic map for Upland cotton is being pursued using three BAC libraries constructed from TM-1 and one from Acala Maxxa, the genetic standard of Upland cotton (Kohel et al. 2003). The TM-1 and Maxxa BAC clones are anchored on the genetic map developed from the RILs (RIL-F7) of a population from a cross between TM-1 and the genetic standard line for *G. barbadense* 3-79. More than 1,000 markers from the existing cotton genetic map have been physically located to TM-1 BAC contigs and about 1,000 new markers, isolated from the contigs, are being genetically mapped by use of the RILs. Fingerprints from more than 42,000 clones of TM-1 BAC libraries have been produced and the fingerprints were assembled into contigs. The initial results showed that the cotton physical contigs representing respective subgenomic origins (A vs. D) can be readily assembled, sorted, and anchored to the cotton genetic map, thus making it feasible to develop a robust genomewide map that integrates physical and genetic mapping information for the cultivated AD cotton. In addition, Frelichowski et al. (2005, 2006) developed microsatellites, alternatively called SSRs, from end sequences of a BAC library constructed from *G. hirsutum* 'Maxxa' (Tomkins et al. 2001). From 2,603 BAC-end sequences, 1,316 PCR primer pairs (MUSB markers) were designed to flank SSR motif sequences and 1,164 (88%) MUSB markers successfully amplified DNA from three species with an average of three alleles per marker, and 365 markers (21%) were polymorphic between *G. hirsutum* 'TM-1' and *G. barbadense* 3-79. Selected SSR markers, polymorphic between *G. hirsutum* and *G. barbadense*, were assayed on 183 RILs and a set of monosomic lines developed from the cross of *G. hirsutum* 'TM-1' × *G. barbadense* 3-79 for assignment of the markers to cotton chromosomes and linkage groups (Table 4). The appeal of the BAC-end derived SSR markers is that the molecular marker cotton map may enable alignment of the original full-length cotton sequences and help develop the cotton genome framework more fully. The average size of the full BAC sequences is 140 kb and one outcome of this research is that actual genetic distances (bases) can be compared with genetic distances. Also QTL associations with markers from known BAC sequences may facilitate the identification and characterization of genes associated with important traits in cotton.

The development of EST libraries has provided a new avenue by which gene isolation and characterization can be accomplished without the need for map-based cloning. Currently, the sequences of more than 14,000 unique genes isolated from developing fibers have been assembled from the cultivated diploid species *G. arboreum* (Arpat et al. 2004). Armed with such a large resource of known gene sequences, cloning by PCR has been increasingly common in cotton (Wang et al. 2004; Wu et al. 2004; Zhang and Liu 2005) and can be easily performed when the target gene is conserved across species. Functional analysis is simplified as gene function can be predicted on the basis of sequence comparison to orthologous genes from well-studied species such as *Arabidopsis* and rice, or can be deduced by differences in expression profiles between different tissue types. Such a comparative functional genomic approaches hold great potential in uncovering genes that are transcribed during fiber development.

1.7
The Future

The next phase of cotton genetics and molecular breeding will be characterized by multidisciplinary and multiinstitutional collaborative initiatives that combine expertise in classical and molecular genetics, molecular biology, functional and structural genomics, biotechnology, and bioinformatics to develop publicly available genomics tools and resources. The research priorities of these collaborative initiatives will be directed by the input and feedback of the broader cotton research community and will ultimately facilitate and advance the development of integrated, molecular consensus maps for MAB, and the development of bioinformatic tools with user-friendly interfaces for the deposition and retrieval of data from public Web sites. Fiber, biotic and abiotic stress EST single-gene sets using cDNA microarrays in standard and mutant (gene knockouts) germplasm will be used to develop genomewide gene expression profiles of genes. The expression profiles will facilitate the classification and categorization of genes with similar expression patterns within a developmental framework, regardless of whether the gene function is known.

These efforts are likely to be augmented by more extensive and innovative use of mutants. Mutagenized populations have been instrumental in defining the function of a gene through genetic analysis and phenotypic screenings. There are about 145 mutant phenotypic markers that have been described in allotetraploid cotton. Most of these mutants have originated as spontaneous mutants from genetic nurseries or have been selected from exotic germplasm sources (Percy and Kohel 1999). The application and utilization of these mutants is still very limited in cotton. In addition, the use of chemical and radiation mutagenesis to induce genetic variation in cotton has been of interest to some breeders and researchers, but its full utilization has also been limited.

Interest has been renewed recently with the introduction of targeted-induced local lesions in genomes

(TILLING). TILLING chemical (ethyl methanesulfonate) mutagenesis induces point mutations (mostly C/G and T/A transitions) in the genome. EMS induces a high density of point mutations in the genome, providing a potential knockout and an allelic series for a candidate gene. TILLING relies on the detection of base-pair mismatches in the formation of heteroduplexes from heteroallelic DNA (MacCallum et al. 2000a, b; Colbert et al. 2001). Early efforts at chemical mutagenesis of cotton were no doubt impeded to a large degree by the large polyploid genome and oil content of the seed. More recently, however, Auld et al. (1998) reported the first exhaustive and successful attempt at chemical (ethyl methanesulfonate) mutagenesis in tetraploid cotton.

The most versatile methods of choice for genomewide mutagenesis are chemical, radiation, and transposon mutagenesis (Meissner et al. 2000). There are other approaches that need to be explored in cotton in order to determine gene function via creation and characterization of mutant alleles and phenotypes. Mutant families in cotton species can exhibit a range of mutant phenotypes, providing a valuable resource for identifying mutant alleles and determining the role such genes play in determining fiber yield and fiber quality. Mutant alleles, identified on the basis of SNPs, also provide an immediate molecular marker that can be applied to mapping populations to map loci that may not be identified with more traditional DNA markers.

In cotton like in other crops, SNPs are being developed as the ultimate DNA marker. SNPs represent single base-pair changes. Their importance has increased recently with the availability of DNA sequence databases and comparative analyses of sequence diversity (http://snp.cshl.org). Compared with other markers, SNPs are more abundant, for example, in humans a SNP is found every 1,000–2,000 bp (Sachidanandam et al. 2001) and in *Arabidopsis* every 3,300 bp (Drenkard et al. 2000). These types of makers are easily subjected to high-throughput automation. Because of their high density in the genome, SNPs can provide a marker that is tightly linked to a gene or even located at the gene of interest. In addition, this feature has renewed interest in so-called association studies based on linkage disequilibrium (LD) to identify either marker tags for genes of medical or agronomic importance or causal mutations responsible for a specific phenotype in other organisms, for example, human (Templeton et al. 2000) and maize (Thornsberry et al. 2001).

LD is an alternative paradigm for mapping of markers and genes. Recent studies on the extent of LD have been performed for the human (Reich et al. 2001), *Drosophila* (Langley et al. 2000), and maize genomes (Remington et al. 2001). Cotton researchers are now poised to make significant advances in functional analysis of the cotton genome to aid in developing a functionally anchored genetic map and association analysis to link genes to phenotypes and QTLs via marker development and implementation.

Improved resolution and integration of cotton genome maps are desirable in order to reduce or close gaps in genome regions with low or very high rates of recombination. Several methods have been suggested to increase map resolution. One strategy is to unit different genetic maps using common makers (Ulloa et al. 2002). Physical mapping is another valuable alternative of mapping in genome regions with very high or very low rates of recombination, because such regions are difficult to resolve by segregation analysis. In regions of high rates of recombination, a large number of polymorphic markers are required to close gaps in the linkage map. Radiation hybrid mapping (RHM) uses radiation treatment to increase chromosome rearrangement and is used not only to increase recombination events, but it may also offer the advantages of very high rates of polymorphism between donor and recipient cell lines. RHM has been used extensively to map the genomes of human and certain animal species (Cox et al. 1990; Walter et al. 1994). Only recently has RHM been used in cotton. Gao (2003) developed a novel approach for RHM based on widecross in vivo hybridization, demonstrating that a radiation hybrid wide-cross whole genome (WWRH) can be used to map the cotton genome as a complement to traditional linkage mapping. The major limitation of RHM or WWRH is that it maps only one chromosome at a time. Another approach for high-resolution mapping is with the development and use of chromosome-specific RILs which are being developed by chromosome substitution lines from *G. barbardense* 3-79 into *G. hirsutum* 'TM-1' for agronomic and fiber trait studies (Saha et al. 2004). These new genomic resource will provide additional approaches for improvement of the cotton crop with again a major limitation to map only one chromosome at a time.

The cotton community has indicated that 5,000 DNA markers will be necessary to provide the requisite tools for genome analysis, evaluation of germplasm collections, and MAS, and these will soon be available, if they are not already so by the time this chapter is published. At present, the number of DNA markers available in the public domain numbers in the thousands, and they continue to accumulate. There are approximately 3,610 informative microsatellite markers (BNL, CIR, CM, JESPR NAU, MGHES, MUSB, MUCS, MUSS, and TMB) in cotton (http://www.mainlab.clemson.edu/cmd/). However, some of these DNA markers are derived from anonymous DNA sequences, and there is still a very real need to develop markers with biologically and agronomically relevant information, most notably fiber yield and quality, to create a functionally-anchored molecular map. A NSF-funded (USA) Cotton Genome Project (http://cfgc.ucdavis.edu; http://genome.clemson.edu) has developed a fiber EST sequence database of more than 45,000 fiber genes and that are available via GenBank. This valuable resource was the first attempt to engage in functional binning of fiber ESTs to study the expression of fiber genes, and their function in relation to the growth and development of cotton fibers, and hence, fiber yield and fiber quality. By the time this chapter is published, it is likely that over 250,000 ESTs will be publicly available. The challenge for the cotton community in the next decade is to develop a fully integrated genetic and physical map of the cotton genome that incorporates all expressed genes. This is the research tool required by the next generation of cotton breeders, geneticists, and molecular biologists.

While it is often said that cotton genomics is lagging relative to other agronomic crops, this past decade has seen substantial growth and development, conceptually and technically. Undoubtedly, this tremendous progress will continue. As noted already, this progress will require substantial collaboration across the cotton genomics community. A case in point is the cotton genomics Coordinated Agricultural Project (also known as CAP), which seeks to develop translation genomics for cotton, the process of translating basic data and information into useful tools and strategies for cotton improvement. This initiative has strong support from researchers in Australia, Brazil, China, India, Israel, and Uzbekistan. The CAP Executive Summary Workgroup has focused on four issues: (1) traits and breeding, which includes fiber uniformity, nematode resistance, and water stress resistance; (2) markers and map integration and consolidation, including a census of marker and mapping populations; (3) functional genomics, including a census of ESTs and microarrays; and (4) Web-based data and bioinformatics tools for flexible data exchange. One specific priority of the CAP group was to sequence the cotton genome. Currently there is considerable discussion as to which genome to sequence and how, but it is clear that this is a common and high-priority goal for the cotton genomics community, and we hope that the next decade is characterized by the publication of the cotton genome sequence and its use in cotton genomics research and molecular breeding.

References

Abdalla AM, Reddy OUK, El-Zik KM, Pepper AE (2001) Genetic diversity and relationships of diploid and tetraploid cottons revealed using AFLP. Theor Appl Genet 102:222–229

Altaf MK, Stewart JM, Wajahatullah MK, Zhang J, Cantrell RG (1997) Molecular and morphological genetics of a trispecies F_2 population of cotton. Proc Beltwide Cotton Conf; Jan 6–10, 1997, New Orleans, LA, USA, pp 448–452

Alvarez I, Cronn R, Wendel JF (2005) Phylogeny of the New World diploid (*Gossypium* L., Malvaceae) based on sequences of three low-copy nuclear genes. Plant Syst Evol 245:199–214

Anderson CG (1999) Cotton marketing. In: Smith CW, Cothren JT (eds) Cotton: Origin, History, Technology, and Production. John Wiley & Sons, Inc, NY, USA, pp 659–679

Anonymous (2005) Cotton: World Markets and Trade. United States Department of Agriculture, Foreign Agricultural Service, Circular Series FC-04-05, April 2005

Arumuganathan K, Earle ED (1991) Nuclear DNA content of some important plant species. Plant Mol Biol Rep 9:208–218

Aslam M, Jiang C, Wright R, Paterson AH (1999) Identification of molecular markers linked to Leaf Curl Virus disease resistance in cotton. Pakistan J Biol Sci 2:124–126

Arpat AB, Waugh M, Sullivan JP, Gonzales M, Frisch D, Main D, Wood T, Leslie A, Wing RA, Wilkins TA (2004) Functional genomics of cell elongation in developing cotton fibers. Plant Mol Biol 54:911–929

Auld, DL, Ethridge MD, Dever JK, Dotray PD (1998) Chemical mutagenesis as a tool in cotton improvement. Proc Beltwide Cotton Conf, Jan 5-9, 1998, San Diego, CA, USA, pp 550–552

Austin DF and Lee M (1996) Comparative mapping in $F_{2.3}$ and $F_{6.7}$ generations of quantitative trait loci for grain yield and yield components in maize. Theor Appl Genet 92:817–826

Basten, CJ, Weir BS, and Zeng ZB (1997) QTL cartographer: A reference manual and tutorial for QTL mapping. Dep of Statistics, NC State Univ, Raleigh, USA

Basten C, Wier B, Zeng ZB (1999) QTL cartographer, Version 1.13 Dep of Statistics, North Carolina St Univ, Releigh, NC, USA

Beasley JO (1940) The origin of American tetraploid *Gossypium* species. Am Nat 74:285–286

Beasley JO (1942) Meiotic chromosome behavior in species, species hybrids, haploids and induced polyploids of *Gossypium*. Genetics 27:25–54

Beasley CA, Egli E (1977) Fiber production in vitro from a conditional fiberless mutant of cotton. Dev Biol 57:234–237

Beavis WD (1994) The power and deceit of QTL experiments: lessons from comparative QTL studies. In:Wilkinson DB (ed) Proc 49th Annu Corn Sorghum Res Conf, 7–8 Dec 1994, Chicago, IL, Am Seed Trade Assoc, Washington, DC, pp 250–266

Bell AA (1984) Cotton protection practices in the USA and world. Section B: Diseases. In: Kohel RJ, Lewis CF (eds) Cotton, Agronomy Monograph 24. ASSA, Madison, WI, USA, pp 288–309

Bennett MD, Smith JB, Heslop-Harrison JS (1982) Nuclear DNA amounts in angiosperms. Proc R Soc Lond B 216:79–199

Bezawada C, Saha S, Jenkins JN, Creech RG, McCarty JC (2003) SSR Marker(s) associated with root knot nematode resistance gene(s) in cotton. J Cotton Sci 7:179–184

Bragg CK, Shofner FM (1993) A rapid direct measurement of short fiber content. Textile Res J 63:171–176

Brown MS, Menzel MY (1950) New trispecies hybrids in cotton. J Hered 41:291–295

Brubaker CL, Wendel JF (1993) On the specific status of *Gossypium lanceolatum* Todaro. Genet Resour Crop Evol 40:165–170

Brubaker CL, Wendel JF (1994) Reevaluating the origin of domesticated cotton (*Gossypium hirsutum*; Malvaceae) using nuclear restriction fragment length polymorphisms (RFLPs). Am J Bot 81:1309–1326

Brubaker CL, Paterson AH, Wendel JF (1999) Comparative genetic mapping of allotetraploid cotton and its diploid progenitors. Genome 42:184–203

Brubaker CL, Bourland F, Wendel J (1999) The origin and domestication of cotton. In: Smith C, Cothren J (eds) Cotton: Origin, History, Technology, and Production. Wiley, New York, USA, pp 3–31

Brubaker CL, Brown AHD (2003) The use of multiple alien chromosome addition aneuploids facilitates genetic linkage mapping of the *Gossypium* G genome. Genome 46:774–791

Chee PW, Lubbers EL, El-Zik KM, Gannaway JR, May OL, Wright RJ, Paterson AH (2003) Secondary Gene Pool Contributions in Domesticated Cotton. In: 2003 Agronomy Abstracts. ASA, Madison, WI, USA

Chee PW, Lubbers EL, Gannaway JR, Paterson AH (2004) Changes in genetic diversity of U.S. Upland cotton. In: Proc Beltwide Cotton Conf, 5–9 Jan 2004, San Antonio, TX Natl Cotton Council, Memphis, TN, USA

Chee PW, Draye X, Jiang C-X, Decanini L, Delmonte TA, Bredhauer R, Smith CW, Paterson AH (2005) Molecular dissection of phenotypic variation between *Gossypium hirsutum* and *G. barbadense* (cotton) by a backcross-self approach: III. Fiber length. Theor Appl Genet 111:772–781

Colbert T, Till BJ, Tompa R, Reynolds S, Steine MN, Yeung AT, McCallum CM, Comai L, Henikoff S (2201) High-throughput screening for induced point mutations. Plant Physiol 126:480–484

Comstock RE (1955) Theory of quantitative genetics: Synthesis. Cold Spring Harbor Symp on Quantitative Biol 20:93–102

Cook OF (1906) Weevil-resisting adaptations of the cotton plant United States Department of Agriculture Bureau of Plant Industry Bulletin 88. Government Printing Office, Washington, DC, USA

Cox DR, Burmeister M, Price ER, Kim S, Myers RM (1993) Radiation hybrid mapping: a somatic cell genetic method for constructing high-resolution maps of mammalian chromosomes. Science 250:245–250

Cronn RC, Zhao X, Paterson AH, Wendel JF (1996) Polymorphism and concerted evolution in a tandemly repeated gene family: 5S ribosomal DNA in diploid and allopolyploid cottons. J Mol Evol 42:685–705

Cronn R, Small RL, Haselkorn T, Wendel JF (2003) Cryptic repeated genomic recombination during speciation in *Gossypium gossypioides*. Evolution 57:2475–2489

Damp JE, Pearsall DM (1994) Early cotton from coastal Ecuador. Econ Bot 48:163–165

Demol J, Louant BP, Moreau JM (1972) Sur l'utilizsation de l'hybride trispécifique hirsutumarboreumthurberi (HAT) en amélioration cotonnière. III. Succès rencontrés et perspectives d'avenir. Bull Rech Agron Gembloux 7:41–58

Desai A (2005) Comparison of marker order synteny and floral trait evolution among diploid and tetraploid genomes of *Gossypium*. MSc Thesis, Univ of Georgia, USA

Desai A, Chee PW, Rong J, May OL, Paterson, AH (2006) Chromosome structural changes in diploid and tetraploid A genomes of *Gossypium*. Genome 49:336–345

Deussen H (1992) Improved cotton fiber properties: The textile industry's key to success in global competition. In: Benedict CR, Jividen GM (eds) Cotton Fiber Cellulose: Structure, Function, Utilization Conference. 1992 Natl Cotton Council, Memphis, TN, USA, pp 43–63

Drenkard E, Richter BG, Rozen S, Stutius LM, Angell NA, Mindrinos M, Cho RJ, Oefner PJ, Davis RW, Ausubel FM (2000) A simple procedure for the analysis of single nucleotide polymorphisms facilitates map-based cloning in Arabidopsis. Plant Physiol 124:1483–1492

Edwards GA, Mirza MA (1979) Genomes of the Australian wild species of cotton II. The designation of a new G genome for *Gossypium bickii*. Can J Genet Cytol 21:367–372

Edwards GA, Endrizzi JE, Stein R (1974) Genome DNA content and chromosome organization in *Gossypium*. Chromosoma 47:309–326

Edwards M, Helentjaris D, Wright S, Stuber CW (1992) Molecular-marker facilitated investigations of quantitative trait loci in maize. Theor Appl Genet 83:765–774

Endrizzi JE, Turcotte EL, Kohel JR (1984) Genetics, cytogenetics, and evolution of *Gossypium*. In: Kohel JR, Lewis CF (eds) Cotton, Agron Monogr 24. ASA, Madison, WI, USA, pp 81–129

Endrizzi JE, Turcotte EL, Kohel RJ (1985) Genetics, cytology, and evolution of *Gossypium*. Adv Genet 23:271–375

Endrizzi JE, Nelson R (1989) Linkage analysis and arm location of the open bud (*Ob1*) and yellow petal (*Y2*) loci in chromosome 18 of cotton. Genome 32:1041–1043

Feldman M, Liu B, Segal G, Abbo S, Levy AA, Vega JM (1997) Rapid elimination of low-copy DNA sequences in polyploid wheat: a possible mechanism for differentiation of homoeologous chromosomes. Genetics 147:1381–1387

Feng CD, Stewart MD, Zhang JF (2005) STS markers linked to the *Rf1* fertility restorer gene of cotton. Theor Appl Genet 110:237–243

Fisher RA (1918) The correlation between relatives on the supposition of Mendelian inheritance. Trans R Soc Edinburgh 52:399–433

Frelichowski JE, Ulloa M, Palmer M, Main D, Tomkins JP, Stelly DM, Cantrell RG, Kohel RJ, Yu J (2005) Genetic, physical and QTL mapping assessments of BAC-end derived microsatellite markers developed from cotton (*Gossypium. hirsutum* L.) Acala 'Maxxa'. In: Plant and Animal Genome XIII Conf, San Diego, CA, USA

Frelichowski JE, Palmer M, Main D, Tomkins JP, Cantrell RG, Stelly DM, Yu J, Kohel RJ, Ulloa M (2006) Genetic mapping of microsatellites derived from BAC-end sequences of *Gossypium hirsutum* Acala 'Maxxa'. Mol Genet Genom 275(5):479–491

Fryxell PA (1979) The Natural History of the Cotton Tribe. Texas A&M Univ Press, College Station, TX, USA

Fryxell PA (1992) A revised taxonomic interpretation of *Gossypium* L (Malvaceae). Rheedea 2:108–165

Galau GA, Wilkins TA (1989) Alloplasmic male sterility in AD allotetraploid *Gossypium hirsutum* upon replacement of its resident A cytoplasm with that of D species *G. harknessii*. Theor Appl Genet 78:23–30

Gao W (2003) Wide-cross whole-genome radiation hybrid (WWRH) mapping and identification of cold-responsive genes using oligo-gene microarray analysis in cotton, PhD Dissertation, Texas A&M Univ, USA

Gerstel DU (1953) Chromosomal translocations in interspecific hybrids of the genus *Gossypium* Evolution 7:234–244

Gerstel DU, Sarvella PA (1956) Additional observations on chromosomal translocations in cotton hybrids. Evolution 10:408–414

Goldberg RB (1978) DNA sequence organization in the soybean plant. Biochem Genet 16:45–68

Gomez M, Johnston JS, Ellison JR, Price HJ (1993) Nuclear 2c DNA content of *Gossypium hirsutum* L. accessions determined by flow cytometry. Biol Zent Bl 112:351–357

Griffee F, Ligon LL (1929) Occurrence of lintless cotton plants and the inheritance of the character 'lintless.' J Am Soc Agron 21:711–717

Groh S, Gonzalez-de-leon D, Khairallah MM, Jiang C, Bergvinson D, Bohn M, Hoisington DA, Melchinger AE (1998) QTL mapping in tropical maize: III. Genomic regions for resistance to *Diatraea* spp. and associated traits in two RIL populations. Crop Sci 38:1062–1072

Grover CE, Kim HR, Wing RA, Paterson AH, Wendel JF (2004) Incongruent patterns of local and global genome size evolution in cotton. Genome Res 14:1474–1482

Gulati AN, Turner AJ (1928) A note on the early history of cotton. Ind Cent Cotton Committee, Tech Lab Bull No 17

Guo W, Zhang T, Pan J, Kohel RJ (1998) Identification of RAPD marker linked with fertility-restoring gene of cytoplasmic male sterile lines in upland cotton. China Sci Bull 43:52–54

Guo W, Zhang T, Xinlian S, Yu JZ, Kohel RJ (2003) Development of a SCAR marker linked to a major QTL for high fiber strength and its usage in molecular-marker assisted selection in Upland cotton. Crop Sci 43:2252–2256

Gutiérrez OA, Basu S, Saha S, Jenkins JN, Shoemaker DB, Cheatham CL, McCarty JC Jr (2002) Genetic distance among selected cotton genotypes and its relationship with F_2 performance. Crop Sci 42:1841–1847

Han ZG, Guo WZ, Song XL, Zhang TZ (2004) Genetic mapping of EST-derived microsatellites from the diploid *Gossypium arboreum* in allotetraploid cotton. Mol Gen Genet 272:308–327

Han Z, Wang C, Song X, Guo WZ, Gou J, Li C, Chen X, Zhang TZ (2006) Characteristics, development and mapping of *Gossypium hirsutun* derived EST-SSR in allotetraploid cotton. Theor Appl Genet 112:430–439

Harrell SC, Culp TW (1979) Registration of Pee Dee 0259 and Pee Dee 2165 germplasm lines of cotton (Reg No GP39 and GP40). Crop Sci 19:418

Hasenkampf CA, Menzel MY (1980) Incipient genome differentiation in *Gossypium*. II. Comparison of 12 chromosomes in *G. hirsutum*, *G. mustelinum* and *G. tomentosum* using heterozygous translocations. Genetics 95:971–983

Hossein GM, Baldwin JC, Khan MA (1994) AFIS advancement in neps and length measurements. In: Heber DJ, Richter DA (eds) Proc Belt Cotton Res Conf, San Diego, CA, 5–8 Jan 1994, Nat. Cotton Council, Memphis, TN, USA, pp 1433–1436

Huang N, Angeles ER, Domingo J, Magpantay G, Singh S, Zhang G, Kumaravadivel N, Bennet J, Khush GS (1997) Pyramiding of bacterial blight resistance in rice: marker-assisted selection using RFLP and PCR. Theor Appl Genet 95:313–320

Hutchinson JB (1951) Intra-specific differentiation in *Gossypium hirsutum*. Heredity 5:161–193

Hutchinson JB (1954) New evidence on the origin of the Old World cottons. Heredity 8:225–241

Hutchinson JB (1959) The application of genetics to cotton improvement. Cambridge Univ Press, UK

Hutchinson JB, Manning HL (1945) The Sea Island cottons. Emp J Exp Agric 13:80–92

Hutchinson JB, Silow RA, Stephens SG (1947) The evolution of *Gossypium*. Oxford Univ Press, London, UK

Hutchinson JG, Ghose RLM (1937) The classification of the cottons of Asia and Africa. Ind J Agric Sci 7:233–257

Iqbal MJ, Reddy OUK, El-Zik KM, Pepper AE (2001) A genetic bottleneck in the 'evolution under domestication' of upland cotton *Gossypium hirsutum* L. examined using DNA fingerprinting. Theor Appl Genet 103:547–554

Jansen RC (1994) Controlling the type I and type II errors in mapping quantitative trait loci. Genetics 138:871–881

Jiang C, Writght RJ, EL-Zik KM, Paterson AH (1998) Polyploid formation created unique avenues for response to selection in Gossypium (cotton). Proc Natl Acad Sci USA 95:4419–4424

Jiang C, Wright RJ, Woo SS, DelMonte TA, Paterson AH (2000) QTL analysis of leaf morphology in tetraploid *Gossypium* (cotton). Theor Appl Genet 100:409–418

Johnson WH (1926) Cotton and its Production. McMillan, London, UK, 536 p

Joseph M, Gopalakrishnan S, Sharma RK, Singh VP, Singh AK, Singh NK, Mohapatra T (2004) Combining bacterial blight resistance and Basmati quality characteristics by phenotypic and molecular marker-assisted selection in rice. Mol Breed 13:377–387

Joyner S (1985) SAS/STAT Guide for personal computer. SAS Institute, Cary, North Carolina, USA

Kadir ZBA (1976) DNA evolution in the genus *Gossypium*. Chromosoma 56:85–94

Karaca M, Saha S, Jenkins JN, Zipf A, Kohel RJ, Stelly DM (2002) Simple sequence repeat (SSR) markers linked to the Ligon Lintless (Li1) mutant in cotton. J Hered 93(3):221–224

Kearney TH, Harrison GJ (1927) Inheritance of smooth seeds in cotton. J Agric Res 35:193–217

Kohel RJ (1999) Cotton germplasm resources and the potential for improved fiber productivity and quality. In: Basra AS (ed) Cotton Fibers: Developmental Biology, Quality Improvement, and Textile Processing. Food Products Press, Binghamton, NY, USA, pp 167–182

Kohel RJ, Yu J, Park Y-H, Lazo GR (2001) Molecular mapping and characterization of traits controlling fiber quality in cotton. Euphytica 121:163–172

Kohel RJ, Stelly DM, Yu J (2002) Tests of six cotton (*Gossypium hirsutum* L.) mutants for association with aneuploids. J Hered 93:130 132

Kohel RJ, Bird LS (2002) Inheritance and linkage analysis of the yellow pulvinus mutant of cotton. J Cotton Sci 6:115–118

Kohel RJ, Yu J, Dong J, Steele NL, Zhang H, Xu Z (2003) Integrated genome mapping for cotton improvement. Natl Cotton Council Beltwide Cotton Conf. Natl Cotton Council Am, Memphis, TN, USA

Lacape JM, Nguyen TB, Thibivilliers S, Bojinov B, Courtois B, Cantrell RG, Burr B, Hau B (2003) A combined RFLP-SSR-AFLP map of tetraploid cotton based on a *Gossypium hirsutum* x *Gossypium barbadense* backcross population. Genome 46:612–626

Lacape J-M, Nguyen T-B, Courtis B, Belot J-L, Giband M, Gourlot J-P, Gawryziak G, Roques S, Hau B (2005) QTL Analysis of cotton fiber quality using multiple *Gossypium hirsutum* x *Gossypium barbadense* backcross generations. Crop Sci 45:123–140

Lagercrantz U, Lydiate DJ (1996) Comparative genome mapping in *Brassica*. Genetics 144:1903–1910

Lan T, Cook CG, Paterson AH (1999) Identification of a RAPD marker linked to a male fertility restoration gene in cotton (*Gossypium hirsutum* L.). J Agric Genom 4:1–5

Lander ES, Green P, Abrahamson J, Barlow A, Daly MJ, Lincoln SE, Newburg L (1987) MAPMAKER: An interactive computer package for constructing primary genetic linkage maps of experimental and natural population. Genomics 1:174–181

Lander ES, Botstein D (1989) Mapping Mendelian factors underlying quantitative traits using RFLP linkage maps. Genetics 121:185–199

Langley CH, Lazzaro BP, Phillips W, Heikkinen E, Braverman JM (2000) Linkage disequilibria and the site frequency spectra in the su(s) and su(w(a)) regions of the *Drosophila melanogaster* X chromosome. Genetics 156:1837–1852

Lee JA (1985) Revision of the genetics of the hairiness-smoothness system of *Gossypium*. J Hered 76:123–126

Liu S, Cantrell RG, McCarty JC, Stewart McD (2000) Simple sequence repeat-based assessment of genetic diversity in cotton race stock accessions. Crop Sci 40:1459–1469

Liu Q, Brubaker CL, Green AG, Marshall DR, Sharp PJ, Singh SP (2001) Evolution of the FAD2-1 fatty acid desaturase 5' UTR intron and the molecular systematics of *Gossypium* (Malvaceae). Am J Bot 88:92–102

Liu L, Guo W, Zhu X, Zhang T (2003) Inheritance and fine mapping of fertility restoration for cytoplasmic male sterility in *Gossypium hirsutum* L. Theor Appl Genet 106:461–469

Lin Z, He D, Zhang X, Nie Y, Guo X, Feng C, Stewart JM (2005) Linkage map construction and mapping QTL for cotton fibre quality using SRAP, SSR and RAPD. Plant Breed 124:180–187

Lin ZX, Zhang XL, Nie YC, He DH, Wu MQ (2003) Construction of a genetic linkage map for cotton based on SRAP. Chinese Sci Bull 48:2063–2067

Liu B, Brubaker CL, Mergeai G, Cronn RC, Wendel JF (2001) Polyploid formation in cotton is not accompanied by rapid genomic changes. Genome 44:321–330

Liu B, Vega JM, Feldman M (1998a) Rapid Genomic changes in newly synthesized amphiploids of *Triticum* and *Aegilops* – II – changes in low-copy coding DNA sequences. Genome 41:535–542

Liu B, Vega JM, Segal G, Abbo S, Rodova H, Feldman M (1998b) rapid genomic changes in newly synthesized amphiploids of *Triticum* and *Aegilops*. I. changes in low-copy noncoding DNA sequences. Genome 41:272–277

McCallum CM, Comai L, Greene EA, Henikoff S (2000b) Targeted screening for induced mutations. Nat Biotechnol 18:455–457

May OL (1999) Genetic variation in fiber quality. In: Basra AS (ed) Cotton Fibers: Developmental Biology, Quality Improvement, and Textile Processing. Food Products Press, Binghamton NY, USA, pp 183–230

Mei M, Syed NH, Gao W, Thaxton PM, Smith CW, Stelly DM, Chen ZJ (2004) Genetic mapping and QTL analysis of fiber-related traits in cotton (*Gossypium*). Theor Appl Genet 108:280–291

Meissner R, Chague V, Zhu Q, Emmanuel E, Elkind Y, Levy A (2000) A high-throughput system for transposon-tagging and promoter trapping in tomato. Plant J 22:265–274

Melchinger AE, Utz HF, Schon CC (1998) Quantitative trait locus (QTL) mapping using different testers and independent population samples in maize reveals low power of QTL detection and large bias in estimates of QTL effects. Genetics 149:383–403

Menzel MY, Brown MS (1954) The significance of multivalent formation in three-species *Gossypium* hybrids. Genetics 39:546–557

Menzel MY, Brown MS, Naqi S (1978) Incipient genome differentiation in *Gossypium*. I. Chromosomes 14,15,16,19 and 20 assessed in *G. hirsutum*, *G. raimondii* and *G. lobatum* by means of seven A-D translocations. Genetics 90:133–149

Menzel MY, Hasenkampf CA, Stewart JM (1982) Incipient genome differentiation in *Gossypium*. III. Comparison of chromosomes of *G. hirsutum* and Asiatic diploids using heterozygous translocations. Genetics 100:89–103

Menzel MY, Naqi S, Brown MS (1984) Incipient genome differentiation in *Gossypium*. IV. The genome of *G. laxum*. J Hered 75:389–391

Meredith WR (1991) Contributions of introductions to cotton improvement. In: Shands HL, Wiesner LE (eds) Use of Plant Introductions in Cultivar Development, part 1. Crop Science Society of America, Madison, WI, USA, pp 127–146

Meredith WR Jr, Culp TW, Robert KQ, Ruppenicker GF, Anthony WS, Williford JR (1991) Determining future cotton variety fiber quality objectives. Text Res J 61:720–723

Meyer JR (1957) Origin and inheritance of D2 smoothness in upland cotton. Crop Sci 48:249–250

Meyer JR, Meyer VG (1961) Origin and inheritance of nectariless cotton. Crop Sci 1:167–169

Meyers VG (1975) Male sterility from *Gossypium harknessii*. J Hered 66:23–27

Michaelson MJ, Price HJ, Ellison JR, Johnston JS (1991) Comparison of plant DNA contents determined by Feulgen microspectrophotometry and laser flow cytometry. Am J Bot 78:183–188

Monforte AJ, Tanksley SD (2000) Fine mapping of a quantitative trait locus (QTL) from *Lycopersicon hirsutum* chromosome 1 affecting fruit characteristics and agronomic traits: breaking linkage among QTLs affecting different traits and dissection of heterosis for yield. Theor Appl Genet 100:471–479

Muravenko OV, Fedotov AR, Punina EO, Fedorova LI, Grif VG, Zelenin AV (1998) Comparison of chromosome Brdu-Hoechst-Giemsa banding patterns of the A(1) and (AD)(2) genomes of cotton. Genome 41:616–625

Ndungo V, Demol J, Marechal R (1988) L'amélioration du cotonnier *Gossypium hirsutum* L. par hybridation interspécifique. Publications agricoles No. 23. Faculté des Sciences Agronomiques de l'Etat, Gembloux, Belgium

Nelson C (1996) QGENE Macintosh software for DNA-marker-based genetic analysis. Version 2.26. Dept of Plant Breeding and Biometry, Cornell Univ, Ithaca, NY, USA

Nguyen T-B, Giband M, Brottier P, Risterucci A-M, Lacape J-M (2002) Wide coverage of the tetraploid cotton genome using newly developed microsatellite markers. Theor Appl Genet 109:167–175

Niles GA, Feaster CV (1984) Breeding. In: Kohel RJ, Lewis CF (eds) Cotton. Am Soc Agron, Madison, WI, USA, pp 201–231

O'Brien SJ (1993) Genetic Maps. O'Brian SJ (ed) Cold Spring Harbor Laboratory Press, Cold Spring Harbor, NY, USA

Park Y-H, Alabady MS, Sickler B, Wilkins TA, Yu J, Stelly DM, Kohel RJ, El-Shihy OM, Cantrell RG, Ulloa M (2005) Genetic mapping of new cotton fiber loci using EST-derived microsatellites in an interspecific recombinant inbred line (RIL) cotton population. Mol Genet Genom 274:428–441

Paterson AH, Smith RH (1999) Future horizons: Biotechnology for cotton improvement. In: Smith WC (ed) Cotton: Origin, History, Technology, and Production. John Wiley and Sons, Inc, NY, USA, pp 415–432

Paterson AH, Saranga Y, Menz M, Jiang C, Wright RJ (2003) QTL analysis of genotype x environmental interactions affecting cotton fiber quality. Theor Appl Genet 106:384–396

Percival AE (1987) The national collection of *Gossypium* germplasm. USDA Southern Crops Ser Bull 321. Dept Agric Comm, Texas A&M Univ, College Station, TX, USA

Percival AE, Wendel JF, Stewart JM (1999) Taxonomy and Germplasm Resources. In: Smith CW, Cothren J (eds) Cotton: Origin, History, Technology, and Production. John Wiley & Sons, Inc, NY, USA, pp 33–63

Percy RG, Wendel JF (1990) Allozyme evidence for the origin and diversification of *Gossypium barbadense* L. Theor Appl Genet 79:529–542

Percy RG, Kohel RJ (1999) Qualitative Genetics. In: Smith CW, Cothren J (eds) Cotton: Origin, History, Technology, and Production John Wiley & Sons, Inc, NY, USA, pp 319–360

Percy RG (1999) Inheritance of cytoplasmic-virescent *cyt-v* and dense-glanding *dg* mutants in American Pima cotton. Crop Sci 39:372–374

Phillips LL (1961) The cytogenetics of speciation in Asiatic cotton. Genetics 46:77–83

Phillips LL, Strickland MA (1966) The cytology of a hybrid between *Gossypium hirsutum* and *G. longicalyx*. Can J Genet Cytol 8:91–95

Quiros CF, Hu J, Truco MJ (1994) DNA-based marker maps of Brassica. In: Phillips RL, Vasil IK (eds) DNA-based Markers

in Plants. Kluwer Academic Publ, Dordrecht, The Netherlands, pp 199–222

Ramey HH (1966) Historical review of cotton variety development. Proc of 18th Cotton Improv Conf, Memphis, TN, USA

Reich DE, Cargill M, Bolk S, Ireland J, Sabeti PC, Richter DJ, Lavery T, Kouyoumjian R, Farhadian SF, Ward R, Lander ES (2001) Linkage disequilibrium in the human genome. Nature 411:199–204

Reinisch AJ, Dong JM, Brubaker CL, Wendel JF, Paterson AH (1994) A detailed RFLP map of cotton, *Gossypium hirsutum x Gossypium barbadense*: chromosome organization and evolution in a disomic polyploid genome. Genetics 138:829–847

Remington DL, Thornsberry JM, Matsuoka Y, Wilson LM, Whitt SR, Doeblay J, Kresovich S, Goodman MM, Buckler ES (2001) Structure of linkage disequilibrium and phenotypic associations in the maize genome. Proc Nat Acad Sci USA 98:11479–11484

Ren L, Gou W, Zhang T (2002) Identification of quantitative trait loci (QTLs) affecting yield and fiber properties in chromosome 16 in cotton substitution line. Acta Bot Sin 44:815–820

Rong J-K, Abbey C, Bowers JE, Brubaker CL, Chang C, Chee PW, Delmonte TA, Ding XL, Garza JJ, Marler BS, Park C-H, Pierce GJ, Rainey KM, Rastogi VK, Schulze SR, Trolinder NL, Wendel JF, Wilkins TA, Williams-Coplin TD, Wing RA, Wright RJ, Zhao X, Zhu L, Paterson AH (2004) A 3347-locus genetic recombination map of sequence-tagged sites reveals features of genome organization, transmission and evolution of cotton (*Gossypium*). Genetics 166:389–417

Rong J, Pierce GJ, Waghmare VN, Rogers C, Desai A, Chee PW, May OL, Gannaway JR, Wendel JF, Wilkins TA, Paterson AH (2005) Genetic mapping and comparative analysis of seven mutants related to seed fiber development in cotton. Theor Appl Genet 111:1137–1146

Rungis D, Llewellyn D, Dennis ES, Lyon BR (2002) Investigation of the chromosomal location of the bacterial blight resistance gene present in an Australian cotton (*Gossypium hirsutum* L.) cultivar. Aust J Agric Res 53:551–560

Saha S, Wu J, Jenkins JN, McCarty Jr JC, Gutierrez OA, Stelly DM, Percy RG, Raska DA (2004) Effect of Chromosome substitutions from Gossypium barbadense L. 3-79 into *G. hirsutum* L. TM-1 on agronomic and fiber traits. J Cotton Sci 8:162–169

Sachidanandam R, Weissman D, Schmidt SC, Kakol JM, Stein LD, Mullikin JC, Mortimore BJ, Willey DL, Hunt SE, Cole CG, Coggill PC, Rice CM, Ning ZM, Rogers J, Bentley DR, Kwok PY, Mardis ER, Yeh RT, Schultz B, Cook L, Davenport R, Dante M, Fulton L, Hillier L, Waterston RH, McPherson JD, Gilman B, Schaffner S, Van Etten WJ, Reich D, Higgins J, Daly MJ, Blumenstiel B, Baldwin J, Stange-Thomann NS, Zody MC, Linton L, Lander ES, Altschuler D (2001) A map of human genome sequence variation containing 1.42 million single nucleotide polymorphisms. Nature 409:928–933

Saranga Y, Menz M, Jiang C, Wright RJ, Yakir D, Paterson AH (2001) Genomic Dissection of genotype x environment interactions conferring adaptation of cotton to arid conditions. Genome Res 11:1988–1995

Saunders JH (1961) The Wild Species of *Gossypium* and their Evolutionary History. Oxford Univ Press, London, UK

Seelanan T, Schnabel A, Wendel JF (1997) Congruence and consensus in the cotton tribe. Syst Bot 22:259–290

Senchina DS, Alvarez I, Cronn RC, Liu B, Rong JK, Noyes RD, Paterson AH, Wing RA, Wilkins TA, Wendel JF (2003) Rate variation among nuclear genes and the age of polyploidy in *Gossypium*. Mol Biol Evol 20:633–643

Shappley ZW, Jenkins JN, Meredith WR, McCarty, JC Jr (1998) An RFLP linkage map of Upland cotton, *Gossypium hirsutum* L. Theor Appl Genet 97:756–761

Shappley ZW, Jenkins JN, Zhu J, McCarty JC (1998). Quantitative trait loci associated with agronomic and fiber traits of Upland Cotton. J Cotton Sci 4:153–163

Shen X, Zhang T, Guo W, Zhu X, Yuan Y, Zhang X (2005) Mapping fiber and yield QTLs with main, epistatic, and QTL x environment interaction effects in recombinant inbred lines of Upland cotton. Crop Sci 46:61–66

Shen X, Guo W, Zhu X, Yuan Y, Yu JZ, Kohel RJ, Zhang T (2005) Molecular mapping of QTLs for qualities in three diverse lines in Upland cotton using SSR markers. Mol Breed 15:169–181

Silow RA (1944) The genetics of species development in the Old World cottons. J Genet 46:62–77

Small RL, Ryburn JA, Cronn RC, Seelanan T, Wendel JF (1998) The tortoise and the hare: choosing between noncoding plastome and nuclear Adh sequences for phylogeny reconstruction in a recently diverged plant group. Am J Bot 85:1301–1315

Song K, Lu P, Tang K, Osborn TC (1995) Rapid genome change in synthetic polyploids of *Brassica* and its implications for polyploid evolution. Proc Nat Acad Sci USA 92:7719–7723

Sprague GF (1955) Problems in the estimation and utilization of genetic variability. Cold Spring Harbor Symp on Quantitative Biol 20:87–92

Stanton MA, Stewart JM, Percival AE, Wendel JF (1994) Morphological diversity and relationships in the A-genome cottons, *Gossypium arboreum* and *G. herbaceum*. Crop Sci 34:519–527

Stelly DM (1993) Interfacing cytogenetics with the cotton genome mapping effort. In: Proc Beltwide Cotton Conf, New Orleans, LA, 10–14 Jan 1993, Natl Cotton Council, Memphis, TN, USA, pp 1545–1550

Stephens SG (1946) The genetics of 'Corky'. The New World alleles and their possible role as an interspecific isolating mechanism. J Genet 47:150–161

Stephens SG (1950) The internal mechanism of speciation in *Gossypium*. Bot Rev 16:115–149

Stephens SG (1967) Evolution under domestication of the New World cottons (*Gossypium* spp). Ciência e Cultura 19:118–134

Shepherd RL (1974) Breeding rootknotresistant *Gossypium hirsutum* L. using a resistant wild *G. barbadense* L. Crop Sci 14:687–691

Shepherd RL (1982) Registration of Auburn 634. Crop Sci 22:642

Stewart MJ (1992) A new cytoplasmic male sterile and restorer. Proc Beltwide Cotton Conf, USA, p 610

Stewart JM (1995) Potential for crop improvement with exotic germplasm and genetic engineering. In: Constable GA, Forrester (eds) Challenging the Future. Proc World Cotton Res Conf 1 CSIRO, Melbourne, Australia, pp 313–327

Tanksley SD, Hewitt J (1988) Use of molecular markers in breeding for soluble solids content in tomato: A re-examination. Theor Appl Genet 75:811–823

Tanksley SD, Ganal MW, Martin GB (1995) Chromosome landing: a paradigm from map-based gene cloning in plants with large genomes. Trends Genet 11:63–68

Templeton AR, Weiss KM, Nickerson DA, Boerwinkle E, Sing CF (2000) Cladistic structure within the human Lipoprotein lipase gene and its implications for phenotypic association studies. Genetics 156:1259–1275

Thinker NA, Mather DE (1995) MQTL: software for simplified composite interval mapping of QTL in multiple environments. J Quant Loci ftp://gnome.agrenv.mcgill.ca/software/MQTL

Thornsberry JM, Goodman MM, Doebley J, Kresovich S, Nielsen D, Buckler ES (2001) Dwarf8 polymorphisms associate with variation in flowering time. Nat Genet 28:286–289

Tomkins JP, Peterson DG, Yang TJ, Main D, Wilkins TA, Paterson AH, Wing RA (2001) Development of genomic resources for cotton (*Gossypium hirsutum* L.): BAC library construction, preliminary STC analysis, and identification of clones associated with fiber development. Mol Breed 8(3):255–261

Turcotte EL, Percy RG (1990) Genetics of kidney seed in *Gossypium barbadense* L. Crop Sci 30:384–386

Ulloa M, Meredith Jr WR, Percy R, Moser H (1999) Genetic variability within improved germplasm of *Gossypium hirsutum* and *G. barbadense* cottons. Crop Sci abstr, Annu Mtg ASA, CSSA, SSSA, Madison, WI, USA, p 73

Ulloa M, Cantrell RG, Percy R, Lu Z, Zeiger E (2000) QTL Analysis of stomatal conductance and relationship to lint yield in an interspecific cotton. J Cotton Sci 4:10–18

Ulloa M, Meredith WR Jr (2000) Genetic linkage map and QTL analysis of agronomic and fiber quality traits in an intraspecific population. J Cotton Sci 4:161–170 Ulloa M, Percy R (2002) Comparison of genetic diversity within improved germplasm of cottons with RFLP, SSR, and RAPD markers. Crop Sci abstrs, Annul Mtg ASA, CSSA, SSSA, Madison, WI, USA, p 189

Ulloa M, Meredith WR Jr, Shappley ZW, Kahler AL (2002) RFLP genetic linkage maps from four F2.3 populations and a joinmap of *Gossypium hirsutum* L. Theor Appl Genet 104:200–208

Ulloa M, Meredith WR Jr (2002) QTL Examination on a RFLP joinmap from four intraspecific cotton populations.

In: Plant, Animal & Microbe Genome X Conf, San Diego, CA, USA, p 247

Ulloa M, Saha S, Jenkins JN, Meredith WR Jr, McCarty JC, Stelly MD (2005) Chromosomal assignment of RFLP linkage groups harboring important QTLs on an intraspecific cotton (*Gossypium hirsutum* L.) joinmap. J Hered 96:132–144

Ulloa M, Mc Stewart J, Garcia CE, Godoy AS, GaytanMA, Acosta NS (2006) Cotton genetic resources in the western states of Mexico: *In situ* conservation status and germplasm collection for *ex situ* preservation. Genet Resour Crop Evol 53(4):653–668

Ulloa M (2006) Heritability and correlation assessments for agronomic and fiber traits in an okra upland cotton population. Crop Sci 46:1508–1514

USTER AFIS^{TM} (1997) Advanced Fiber Information System (AFIS), Instruction Manual and software V4.1. Zellweger Uster, Inc, Knoxville, TN 37919, USA

Utz HF, Melchinger AE (1994) Comparison of different approaches to interval mapping of quantitative trait loci. In: Van Ooijen JW, Jansen J (eds) Proc 9th Meeting of the EUCARPIA, Section Biometrics in Plant Breeding: Applications of Molecular Markers. CPRO-DLO, Wageningen, The Netherlands, pp 195–204

Valderrama C (2004) Fifth year of record consumption. Cotton: review of the world situation (ICAC) 58:10–12

Van Ooijen JW, Maliepaard C (1996) MapQTL^{TM} Version 3.0, Sotfware for the calculation of QTL positions on genetic maps. Plant Research International, Wageningen, The Netherlands

Vroh Bi I, Maquet A, Baudoin J-P, du Jardin P, Jacquemin JM, Mergeai G (1999) Breeding for "low-gossypol seed and high-gossypol plants" in Upland cotton. Analysis of trispecies hybrids and backcross progenies using AFLPs and mapped RFLPs. Theor Appl Genet 99:1233–1244

Walter MA, Spillet DJ, Thomas P, Weissienbach J, Goodfellow PN (1994) A method for constructing radiation hybrid maps of whole genomes. Nat Genet 7:22–28

Wang C, Ulloa M, Roberts PA (2005) Identification and mapping of microsatellite markers linked to the root-knot nematode resistance gene *rkn1* in Acala NemX (*Gossypium hirsutum* L.). Theor Appl Genet 112(4):770–777

Wang K, Song X, Han Z, Guo W, Yu JZ, Sun J, Pan J, Kohel R, Zhang T (2006) Complete assignment of the chromosomes of *Gossypium hirsutum* L. by translocation and fluorescence in situ hybridization mapping. Theor Appl Genet 113:73–80

Wang S, Basten CJ, Zeng ZB (2001–2004) Windows QTL Cartographer 2.0. Dep of Statistics, North Carolina State Univ, Raleigh, NC, USA

Wang S, Wang JW, Yu N, Li CH, Lou B, Gou JY, Wang LJ, Chen XY (2004) Control of plant trichome development by a cotton fiber MYB gene. Plant Cell 16:2323–2334

Ware JO (1950) Origin, rise, and development of American Upland cotton varieties and their status at present. Mimeo Publication, University of Arkansas, College of Agriculture, Agric Exp Sta, Fayetteville, AR, USA

Ware JO (1951) Origin, rise and development of American Upland cotton varieties and their status at present. University of Arkansas, Fayetteville, Fayetteville, AR, USA

Wendel JF (1989) New World tetraploid cottons contain Old World cytoplasm. Proc Natl Acad Sci USA 86:4132–4136

Wendel JF, Percy RG (1990) Allozyme diversity and introgression in the Galapagos Islands endemic *Gossypium darwinii* and its relationship to continental *G. barbadense*. Biochem Syst Ecol 18:517–528

Wendel JF, Albert VA (1992) Phylogenetics of the cotton genus (*Gossypium*): Character-state weighted parsimony analysis of chloroplast-DNA restriction site data and its systematic and biogeographic implications. Syst Bot 17:115–143

Wendel JF, Schnabel A, Seelanan T (1995) An unusual ribosomal DNA sequence from *Gossypium gossypioides* reveals ancient, cryptic, intergenomic introgression. Mol Phylogenet Evol 4:298–313

Wendel JF, Cronn RC (2003) Polyploidy and the evolutionary history of cotton. Adv Agron 78:139–186

Wendel JF, Brubaker CL, Percival AE (1992) Genetic diversity in *Gossypium hirsutum* and the origin of Upland cotton. Am J Bot 79:291–1310

Wendel JF, Olson PD, Stewart JM (1989) Genetic diversity, introgression, and independent domestication of Old World cultivated cottons. Am J Bot 76:795–1806

Wilkins TA, Jernstedt JA (1999) Molecular genetics of developing cotton fibers. In: Basra AS (ed) Cotton Fibers. Haworth Press, New York, USA, pp 231–267

Wright R, Thaxton P, El-Zik K, Peterson A (1998) D-subgenome bias of *Xcm* resistance genes in tetraploid *Gossypium* (cotton) suggest that polyploidy formation has created novel avenues for evolution. Genetics 149:1978–1996

Wright R, Thaxton P, Peterson AH, El-Zik K (1999) Molecular mapping of genes affecting pubescence of cotton. J Hered 90:215–219

Wu CA, Yang GD, Meng QW, Zhang CC (2004) The cotton *GhNHX1* gene encoding a novel putative tonoplast Na+/H+ antiporter plays an important role in salt stress. Plant Cell Physiol 45:600–607

Yu J, Yong-Ha P, Lazo GR, Kohel JR (1998) Molecular mapping of the cotton genome: QTL analysis of fiber quality properties. Proc Beltwide Cotton Conf, San Diego, CA, USA, 5–9 Jan 1998. Natl Cotton Council Am, Memphis, TN, USA, p 485

Yu J, Kohel RJ, Zhang H-B, Dong J-M, Decanini LI (2000) Construction of a cotton BAC library and its applications to gene isolation. In: Plant and Animal Genome VIII Conf, San Diego, CA, USA

Zeng Z-B (1992) Correcting the bias of Wright's estimates of the number of genes affecting a quantitative character: A further improved method. Genetics 131:987–1001

Zeng Z-B (1994) Precision mapping of quantitative trait loci. Genetics 136:1457–1468

Zhang T, Yuan Y, Yu J, Guo W, Kohel RJ (2003) Molecular tagging of a major QTL for fiber strength in Upland cotton and its marker-assisted selection. Theor Appl Genet 106:262–268

Zhang J, Stewart McD (2001) Inheritance and genetic relationships of the D_8 and D_{2-2} restorer genes for cotton cytoplasmic male sterility. Crop Sci 44:1209–1214

Zhang J, Stewart McD (2004) Identification of molecular markers linked to the fertility restorer genes for CMS-D_8 in cotton (*Gossypium hirsutum* L.) Crop Sci 44:1209–1217

Zhang HM, Liu JY (2005) Molecular cloning and characterization of a beta-galactosidase gene expressed preferentially in cotton fibers. J Integr Plant Biol 47(2):223–232

Zhang JF, Stewart JM, Wang TH (2005) Linkage analysis between gametophytic restorer *Rf(2)* gene and genetic markers in cotton. Crop Sci 45:147–156

2 Forage Crops

Maiko Inoue[1], Masahiro Fujimori[2], and Hongwei Cai[1]

[1] Forage Crop Research Institute (FCRI), Japan Grassland Agricultural an Forage Seed Association, 388-5 Higashiakada, Nasushiobara, Tochigi 329-2742, Japan
e-mail: hcai@jfsass.or.jp
[2] Yamanashi Prefectural Dairy Experiment Station, 621-2 Nagasaka-Kamijo, Nagasaka, Yamanashi 408-0021, Japan

2.1
Introduction

2.1.1
Characteristics of Cross-Pollinated Forage Crops as Distinct from Those of Inbred Species

Forage crops are generally classified as either pasturage or silage crops. Silage crops include forage maize, sorghum, oats, and other grain crops from which the seeds are usually harvested and the remainder is used as animal feed. Almost all crops that are used for pasturage are either grasses or legumes, and the whole plant, including seeds, is usually used as animal fodder. In this chapter, we focus only on forage grasses and legumes. According to where they are grown, forage grasses and legumes can be classified as cool-season or warm-season crops, and according to the reproductive pattern, as autogamous, allogamous, or apomictic (Table 1).

Forage crops represent a large group of species, mostly cross-pollinated species with complex breeding systems. Cross-pollinated species differ from inbred species in the following ways:

- Plants of cross-pollinated forage crops are highly heterozygous. Each individual in an open-pollinated population may represent a different genotype.
- Inbreeding leads to depression of vigor and loss of fertility, although species and genotypes within a species vary considerably in these respects.
- Plants or genotypes of most species of forage grasses and legumes may be propagated vegetatively as clones.
- Self-incompatibility or apomixis may be a hindrance to the use of conventional breeding procedures in cross-pollinated species. However, these systems provide opportunities to use breeding techniques not possible with normal reproductive systems.

Polyploidy

It has been estimated that about 70% of the species in the grass family and 23% of the species in the legume family are polyploids (Muntzing 1956), including many forage grass and legume species. As a consequence of their larger cell size, polyploids often produce larger plants than their related diploids. For example, alfalfa (*Medicago sativa*, 4x), orchard grass (*Dactylis glomerata*, 4x), and timothy (*Phleum pratense*, 6x) are all autopolyploids. Usually, the fertility of an autopolyploid is reduced, and autopolyploids do not produce seeds as freely as diploids. Also, autopolyploids often show more complex genetic behavior than allopolyploids. Most of the natural polyploids are considered to be allopolyploids, including some forage crops such as tall fescue (*Festuca arundinacea*, 6x) and white clover (*Trifolium repens*, 4x). The genetic behaviors of allopolyploids are usually same as those of diploids.

A method to estimate linkage group relationships in autopolyploids by using a single-dose restriction fragment (SDRF) procedure has been developed in wild sugarcane (*Saccharum spontaneum* L.) (Wu et al. 1992). In this method, alleles that are present in single copies in the parental lines are scored individually on the basis of their presence or absence in the progeny, and linkage relationships are established for each individual chromosome. Homologous linkage groups can be then combined into consensus linkage groups.

Incompatibility

Incompatibility is a form of infertility caused by the failure of plants with normal pollen and ovules to set seed owing to some physiological hindrance that pre-

Table 1. Mode of pollination or seed set, chromosome number, genome size, and growth habit of some important cultivated species of forage grasses and legumes. (Modified from Poehlman and Sleper 1995)

Crop	Species	Chromosome number		Growth habit	Zone of cultivation	Genome size[a]
		x	2n			
Normally cross-pollinated forage grasses						
Bermudagrass	*Cynodon dactylon*	9	18, 36, 54	Perennial	W	1,436
Bromegrass	*Bromus inermis*	7	56	Perennial	C	11,564
Meadow fescue	*Festuca pratensis*	7	14	Perennial	C	2,181
Tall fescue	*Festuca arundinacea*	7	42	Perennial	C	5,929
Millet, pearl	*Pennisetum americanum*	7	14	Annual	W	2,352
Orchardgrass (cocksfoot)	*Dactylis glomerata*	7	28	Perennial	C	
Pangolagrass	*Digitaria decumbens*	15	30	Perennial	W	
Redtop	*Agrostis alba*	7	28, 42	Perennial	C	
Reed canarygrass	*Phalaris arundinacea*	7	14, 28	Perennial	C	
Rhodesgrass	*Chloris gayana*	10	20, 40	Perennial	W	343
Ryegrass , Italian	*Lolium multiflorum*	7	14	Winter annual	C	2,000[b]
Ryegrass, perennial	*Lolium perenne*	7	14	Perennial	C	2,034
Stargrass	*Cynodon nlemfuensis*	9	18, 36	Perennial	W	
Timothy	*Phleum pratense*	7	42	Perennial	C	4,067
Zoysiagrass	*Zoysia japonica*	10	40	Perennial	W	421
Zoysiagrass	*Zoysia matrella*	10	40	Perennial	W	
Normally cross-pollinated forage legumes						
Alfalfa , purple blossom	*Medicago sativa*	8	32	Perennial	C	1,715
Birdfoot trefoil	*Lotus corniculatus*	6	12, 24	Perennial	C	446, 1,029
Clover , alsike	*Trifolium hybridum*	8	16	Biennial	C	784
Clover, red	*Trifolium pratense*	7	14	Biennial	C	637
Clover, white	*Trifolium repens*	8	32	Perennial	C	956

C cool season, *W* warm season

[a] Royal Botanic Gardens, Kew, Plant DNA C-values database http://www.rbgkew.org.uk/cval/homepage.html

[b] Hutchinson et al. (1979)

Table 1. (continued)

Crop	Species	Chromosome number		Growth habit	Zone of cultivation	Genome size[a]
		x	2n			
Largely apomictic grasses						
Bahiagrass	*Paspalum notatum*	10	20, 40	Perennial	W	706
Bluegrass	*Poa pratensis*	7	28, 56, 70, 84	Perennial	C	5,263
Dallisgrass	*Paspalum dilatatum*	10	40, 50	Perennial	W	588
Guineagrass	*Panicum maximum*	8	16, 32, 40, 48	Perennial	W	
Weeping lovegrass	*Eragrostis curvula*	10	40, 50	Perennial	W	
Dioecious grasses						
Buffalograss	*Buchloë dactyloides*		40, 56, 60	Perennial		779

vents fertilization. Incompatibility is usually present in the grasses and legumes, and production of F_1 hybrid seeds by crossing pairs of self-incompatible genotypes has been used in breeding forage grasses and legumes. In the family Poaceae, Lundqvist et al. (1954) demonstrated in rye the first detailed two-locus system related to self-incompatibility. The two loci are designated S and Z. The two-locus system has been confirmed in perennial ryegrass, meadow fescue, and wild species of *Hordeum* and *Phalaris* (Baumann et al. 2000). A one-locus system was found in legumes such as some red clover and white clover species, although the two-locus system produces a higher percentage of compatible mating than the one-locus system.

Apomixis

Apomixis is common among forage grasses, citrus, mango, blackberry, guayule, and some ornamental shrubs. One effect of apomictic reproduction is maintenance of unusual chromosome numbers (Table 1). Table 1 displays some grass species that reproduce regularly by apomixis.

2.1.2
Botanical Descriptions

This section discusses the origin, use, genome size , and drought tolerance of some major forage crops (Table 1).

Among temperate forage grasses, tall fescue is cultivated throughout temperate regions for use as hay, pasturage, and silage, and in erosion control. Tall fescue originated in Europe and has a genome size of 5,929 Mb. Its drought tolerance ranges from moderate (endophyte-uninfected) to good (endophyte-infected). Italian ryegrass (*Lolium multiflorum*) is cultivated throughout temperate regions and is used for hay, pasturage, silage, and wildlife habitat. Italian ryegrass originated in Europe and has a genome size of 2,000 Mb. Its drought tolerance is moderate. Orchardgrass, also called cocksfoot, is cultivated throughout temperate regions and is used for hay and pasturage. Orchardgrass originated in Europe. Its drought tolerance is moderate.

Among tropical forage grasses, Bermudagrass (*Cynodon dactylon*) is cultivated throughout tropical regions for use as hay and pasturage. Cultivated forms of *C. dactylon* originated in southeastern Africa and have a genome size of 1,436 Mb. Its drought tolerance is excellent. Bahiagrass (*Paspalum notatum*) is cultivated throughout tropical regions for use as hay and pasturage. Bahiagrass is an apomictic grass that originated in Argentina, Brazil, Uruguay, and Paraguay. It has a genome size of 706 Mb. Its drought tolerance is excellent. Most tropical forage grasses have smaller genome sizes than temperate forage grasses.

Among temperate forage legumes, alfalfa, a highly productive forage species, is cultivated throughout temperate regions for premium-quality hay, silage, and pasturage. Alfalfa originated near the Caspian Sea in northern Iran and northeastern Turkey and has a genome of 1,715 Mb; its cultivation spread throughout the Mediterranean region and into Germany, France, and China at the time of the Roman Empire (Bolton 1962). Today, alfalfa is raised on all continents and is currently cultivated on more than 32 million hectares worldwide (Michaud et al. 1988). Its drought tolerance is excellent. Red clover (*T. pratense*) is cultivated throughout temperate regions and is used as hay and pasturage. Red clover originated in Europe and has a genome size of 637 Mb. Its drought tolerance is moderate to good. White clover is cultivated throughout temperate regions and is used for pasturage and wildlife habitat. White clover originated in the Mediterranean region and has a genome size of 956 Mb. Its drought tolerance is moderate.

2.1.3
Economic Importance

Forage crops, especially forage grasses and legumes, are widely used as animal forage and turf grasses worldwide. In 2004, the cultivated area of forage crops in the USA was 15,299,000 ha; of that, 5,364,000 ha was alfalfa (http://www.nass.usda.gov:81/ipedb/main.htm), whereas the area of forage crops in Japan was 801,200 ha in 2002.

2.1.4
Breeding Objectives

The objectives to be met by forage-crop breeding depend on the species and intended use. The basic goal is to produce a cultivar superior to existing cultivars for a specific trait or to produce a new cultivar invested with a new trait. Breeding objectives usually include:

– Forage and seed yield
– Winter hardiness
– Resistance to lodging and shattering
– Resistance to plant pathogens

– Resistance to insects and pests
– Tolerance to heat and drought
– Tolerance to soil water stress

2.1.5
Classical Mapping Efforts

Genetic maps of higher plants were initially based almost entirely on morphological and biochemical traits.

In the *Lolium* family, several isozyme markers were identified and used in the characterization of germplasm and linkage group detection (Hayward et al. 1988, 1994, 1995, 1998), and trisomic and doubled trisomic lines were constructed in *Lolium perenne* (Ahloowalia 1972; Meijer and Ahloowalia 1981).

Few isozyme markers have been reported in the *Medicago* group (Quiros and Morgan 1981; Quiros 1982). The first trisomics at the diploid level ($n = 17$) were reported by Kasha and McLennan (1967). Because cultivated alfalfas are mostly tetraploid, trisomic tetraploids ($2n = 31$) were also reported (Stanford 1959). Both diploid and tetraploid trisomics are useful for chromosome mapping of genes.

2.1.6
Classical Breeding Achievements

As described already, most forage crops are outcrossing species that possess some characteristics such as heterozygosity and self-incompatibility that differ from those of cereal crops. Therefore, the methods for breeding forage crops are also different from those used for other crops. The methods of mass selection and synthetic cultivars are the most-used breeding methods. So far, more than 3,000 forage-grass cultivars have been included in the Organization for Economic Cooperation and Development's list of varieties eligible for seed certification (OECD 2005).

2.1.7
Limitations of Classical Endeavors and Utility of Molecular Mapping

Although classical breeding methods are usually used in forage-crop breeding, most agriculturally important traits, including yield, quality, lodging resistance, and cold or drought tolerance, are quantitatively inherited and controlled by quantitative trait loci (QTLs). The use of molecular markers to assist the breeding process will allow breeding goals to be reached more efficiently and may eliminate the need for field assays by inoculation. For example, breeding for resistance to crown rust is an essential component of ryegrass-improvement programs, and development of a new cultivar takes more than 10 years. If the new cultivar is then overcome by a new race of pathogen, this effort and time are wasted. Thus, development of durable cultivars with superior resistance to crown rust is highly desirable. Pyramiding several resistance genes through the use of marker-assisted selection (MAS) in a cultivar is an excellent strategy to prevent breakdown of disease resistance.

2.2
Construction of Genetic Maps

The genomic study of most forage crops has lagged behind that of other major crops. This has been particularly true with respect to their genomic sequencing, compilation of expressed sequence tags (ESTs), and development of markers such as restriction fragment length polymorphism (RFLP) and simple sequence repeat (SSR) markers. The chief reasons for this lie in the outcrossing nature, relatively large genome, and relatively low economic value of forage crops compared with other crops.

2.2.1
Kinds of Mapping Population and Molecular Markers

Construction of a genetic linkage map first requires a population showing genetic segregation. As described already, almost all forage crops are outcrossing species with self-incompatibility traits that usually prevent development of an inbred line. So, unlike with most inbred species such as wheat, rice, and other major crops, the most popular segregating populations, including F_2, backcross (BC_1), doubled haploid (DH), and recombinant inbred lines are difficult to produce in the case of most forage crops. Therefore, several unique population types are usually used in constructing linkage maps of forage crops.

Pseudo-testcross F_1 Population
The F_2, DH, and BC_1 generations that researchers use for inbreeding plants are generally not avail-

able in highly heterozygous species. This includes most forage crops because of self-incompatibility, and forest trees because of their long life cycle. To construct linkage maps for those outcrossing species, F_1 progeny generated from the cross between unrelated multiple heterozygous individuals can be used (one-way pseudo-testcross by Ritter et al. 1990; two-way pseudo-testcross by Grattapaglia and Sederoff 1994). Working with the F_1 generation of a cross between two heterozygous individuals has allowed the direct analysis of segregation and the construction of genetic maps (Viruel et al. 1995; Maliepaard et al. 1998). But in this method, segregation tests and calculation of the recombination values may be complex and difficult because different segregating types (BC_1 and F_2 types) of marker are mixed. Furthermore, a locus may have multiple alleles that segregate in three or more individuals. Therefore, construction of a linkage map usually uses special software such as JoinMap (Stam 1993).

Map construction proceeds as follows:

- Among alleles whose genetic segregations are examined, only the markers of the maternal heteroallelic × paternal homoallelic types are selected. They are then analyzed using the same method as for the self-pollinated BC_1 algorithm to construct the maternal linkage map.
- Similarly, other markers that segregate as the maternal homoallelic × paternal heteroallelic types are selected, and the paternal linkage map is constructed.
- Finally, the two parental linkage maps are unified as a consensus map by positioning the codominant maternal heteroallelic × paternal heteroallelic markers (such as RFLPs and SSRs) from the two linkage maps.

Full-Sib Family
First, a one-pair (or few-pair) cross is made between two parental individuals; second, a random-pair cross is made between two individuals of progeny derived from the first cross. The next generation is used to construct the linkage map.

Half-Sib Family
First, a two-pair cross is made between four individuals of four unrelated parents; second, a random-pair cross is made between two individuals of progeny de-

rived from the two first crosses. The next generation is used to construct the linkage map.

Typical Markers Used in Molecular Linkage Map Construction

Several types of markers are usually used in molecular linkage map construction:

- Random amplified polymorphic DNA (RAPD)
- Amplified fragment length polymorphism (AFLP)
- RFLP
- SSR
- Single nucleotide polymorphism (SNP)
- Others such as sequence-tagged site (STS) and cleaved amplified polymorphism sequence (CAPS) markers

RAPD is a DNA polymorphism assay based on the amplification of random DNA segments with single primers of arbitrary nucleotide sequences (Williams et al. 1990). The RAPD assay detects single-base changes in genomic DNA. RAPD analysis requires an element of genomic DNA and is fast and economical. A universal set of primers that do not need sequence information of a target organism can be used for genomic analysis in a wide variety of species.

Nearly all RAPD markers are dominant. The most prevalent difficulty with the RAPD technique, however, is a lack of reproducibility and lack of locus specificity, particularly in polyploid species. It is not possible to distinguish whether a DNA segment is amplified from a locus that is heterozygous (one copy) or homozygous (two copies) with a dominant RAPD marker that segregates as 3:1. The information content of an individual RAPD marker is very low when it is used to guess the recombination value.

In the initial phase of molecular marker research, genomic libraries were used to isolate low-copy DNA fragments to be used as DNA probes. These DNA probes were hybridized back to restriction digests of genomic DNA to reveal allelic length variation. This is referred to as RFLP. The advantages of RFLP markers over other types, such as RAPD and AFLP markers, include their codominant nature and the ease with which map information can be transferred to a different mapping population (Beckmann and Soller 1986; Helentjaris 1987).

Although RFLP analysis requires a large amount of genomic DNA and is time-consuming and costly, an informative RFLP linkage map is useful in:

- Analyzing the structural organization of genomes (Berhan et al. 1993)
- Generating a physical map of specific chromosomes through in situ hybridization using DNA markers (Werner et al. 1992; Wanous and Gustafson 1995; King et al. 2002)
- Improving the agronomic attributes of crop plants by MAS

In addition, comparative RFLP mapping of related species holds the potential to provide important insights into the evolution of plant genomes (Ahn and Tanksley 1993; Huang and Kochert 1994). As a result, detailed linkage maps have been constructed using RFLP markers in several major crops. Among forage-grass species, the RFLP maps of perennial ryegrass, Italian ryegrass, meadow fescue, and tall fescue have been constructed (Xu et al. 1995; Chen et al. 1998; Hayward et al. 1998; Armstead et al. 2002; Jones et al. 2002a; Alm et al. 2003; Inoue et al. 2004a).

The AFLP technique is based on the amplification of subsets of genomic restriction fragments by using the polymerase chain reaction (PCR). In this method, most alleles are present in single copies in the parental lines, most AFLP markers are dominant, and the method can score markers individually on the basis of their presence or absence in the progeny. The AFLP technique provides a marker system that allows the rapid detection of a large number of polymorphic loci through relatively modest levels of experimental activity. Consequently, the technique has been widely used in linkage map construction and marker development (Vos et al. 1995; Simons et al. 1997).

The genomes of all eukaryotes contain a class of sequences termed SSRs (Tautz et al.1986) or microsatellites (Litt and Luty 1989). Microsatellites with tandem repeats of a basic motif of fewer than 6 bp have emerged as an important source of ubiquitous genetic markers for many eukaryotic genomes (Wang et al. 1994). Unlike other markers, SSR markers have the advantage of being PCR-based and multiallelic, and possessing high polymorphism. However, the development of SSR markers from genomic libraries is expensive and inefficient.

The latest of the DNA-based markers are the SNPs. Ever since DNA sequencing became feasible, alleles of one locus could be aligned and differences in a given sequence could be identified. Researchers have estimated that a SNP occurs possibly once in every few hundred or perhaps once in 1,000 nucleotides in coding DNA (Tenaillon et al. 2001; Kanazin et al. 2002;

Somers et al. 2003). SNPs could allow high through-put and automation as new detection platforms are developed. Currently, allele-specific PCR, single-base extension, and array hybridization are only some of the methods for detecting SNPs, with varying input costs and level of sophistication (Gupta et al. 2001). SNP is an excellent marker system, but SNP discovery is costly and, in outcrossing species that include most forage crops, acquiring SNPs is difficult because of the great genetic diversity among individuals.

2.2.2
Construction of First-Generation Maps

First-generation maps have been mostly constructed using RFLP, RAPD, or isozyme markers, and map densities have not been very high.

A genetic linkage map of *Lolium* has been constructed using a segregating population derived from a cross between an F_1 hybrid of *L. perenne* × *L. multiflorum* with a DH *L. perenne* (Hayward et al. 1998). A total of 106 markers comprising 17 isozymes, 48 RAPDs, and 41 RFLPs were mapped to seven linkage groups. This map has a total length of 692 cM, with linkage groups ranging from 67- to 155-cM long.

RFLP markers were also used to construct a genetic linkage map of meadow fescue and compare its genomic relationship with a closely related hexaploid species, tall fescue (Chen et al. 1998). Heterologous RFLP markers originally isolated from a tall fescue *Pst*I genomic DNA library (Xu et al. 1991, 1995) were used in the linkage map construction; 66 markers were mapped in seven linkage groups with a total length of 280.1 cM. Of those, 33 were commonly mapped in both species, and 70% were located in corresponding linkage groups in meadow fescue and tall fescue.

Five genetic linkage maps have been reported in *M. sativa*, four in diploid populations (Brummer et al. 1993; Kiss et al. 1993; Echt et al. 1994; Tavoletti et al. 1996) and one in a tetraploid population (Brouwer and Osborn 1999). Kiss et al. (1993) have constructed a genetic map by ordering the linkage values of 89 RFLP, RAPD, isozyme, and morphological markers collected from a diploid segregating population of 138 individuals. The markers span 659-cM distance, and the average of the flanking markers was 7.4 cM. Brouwer and Osborn (1999) constructed a linkage map using 82 SDRFs in two backcross autotetraploid alfalfa populations of 101 individuals each, and four homologous coupling-phase cosegregation groups were detected for seven of eight linkage groups of diploid alfalfa and

aligned probes in common. No cosegregation groups were found for linkage group 7 because of the lack of polymorphisms in this cross. A composite map was generated by integrating the four homologous cosegregation groups; it consisted of 88 loci on seven linkage groups covering 443 cM. The locus map orders and distances were in general in agreement with those found for diploid alfalfa.

2.2.3
Construction of Second-Generation Maps

Recently, many molecular markers, including AFLPs, SSRs, and SNPs, have been developed in forage crops, and high-density, high-resolution, comparative linkage maps have been constructed (Table 2).

Inoue et al. (2004a) constructed a high-density linkage map for Italian ryegrass using RFLP, AFLP, and telomeric-repeat-associated sequence markers. A two-way pseudo-testcross F_1 population consisting of 82 individuals was used to analyze the three types of markers. The final map included 385 markers, which were separated into seven major linkage groups. The total map length was 1,244.4 cM, and the average interval between markers was 3.7 cM. In the comparison of this map with the perennial ryegrass map reported by Jones et al. (2002b), linkage groups 1, 2, 3, 4, and 6 were comparable. Furthermore, the results showed some common locations of the anchor probes between this map and the homologous groups in wheat.

A molecular-marker linkage map has also been constructed for perennial ryegrass by using a one-way pseudo-testcross population based on the mating of a multiple heterozygous individual with a DH genotype (Jones et al. 2002a). RFLP, AFLP, isozyme, and EST data were combined to produce an integrated genetic map containing 240 loci covering 811 cM on seven linkage groups. The map contained 124 codominant markers, of which 109 were heterozygous anchor-RFLP probes from wheat, barley, oat, and rice, allowing inference of comparative relationships between perennial ryegrass and other Poaceae species. The genetic maps of perennial ryegrass and the Triticeae are highly conserved in synteny and colinearity. In addition, Jones et al. (2002b) have also constructed a SSR-based linkage map using the same one-way pseudo-testcross reference population (Fig. 1). Ninety-three loci have been assigned to positions on seven linkage groups. The SSR locus data have been integrated with selected data for RFLP, AFLP, and other loci mapped

Table 2. Molecular linkage maps of forage crops

Species	Population type	No. of individuals	Marker (no. of markers)	Map length (cM)	References
Meadow fescue	2-way pseudo-testcross	138	RFLP, AFLP, isozyme, SSR (466)	658.8	Alm et al. (2003)
	F_2	56	RFLP (66)	280.1	Chen et al. (1998)
Tall fescue	F_2	105	RFLP (95)	1,274	Xu et al. (1995)
	2-way pseudo-testcross		SSR, AFLP (918)	1,841	Saha et al. (2005)
Italian ryegrass	2-way pseudo-testcross F_1	82	RFLP (274), AFLP (867), TAS (85)	1,244.4	Inoue et al. (2004a)
Perennial ryegrass	1-way pseudo-testcross F_1	155	SSR, RFLP, AFLP, isozyme, STS (258)	814	Jones et al. (2002b)
	1-way pseudo-testcross F_1	165	RFLP, AFLP, isozyme, EST (240)	811	Jones et al. (2002a)
	1-way pseudo-testcross F_1	95	AFLP (463), isozyme (3), EST (5)	930	Bert et al. (1999)
	F_2, BC_1	180, 156	RFLP, AFLP, isozyme, SSR (74, 134)	446; 327	Armstead et al. (2002)
Perennial × Italian	Perennial × Italian F_1	89	Isozyme, RFLP, RAPD (106)	692	Hayward et al. (1998)
Italian × perennial	2-way pseudo-testcross F_1	91	AFLP, RAPD, RFLP, SSR, isozyme (440)	712; 537	Warnke et al. (2004)
Italian × perennial	3-generation interspecific population	156	RFLP (120)	664	Sim et al. (2005b)
Rhodesgrass	Pseudo-testcross F_1	149	AFLP (164), RFLP (25)	488.3; 443.3	Ubi et al. (2004)
Alfalfa (tetraploid)	F_1	168	SSR (107)	709	Julier et al. (2003)
Alfalfa (tetraploid)	Backcross population	101	RFLP (88)	443	Brouwer et al. (1999)
Alfalfa (diploid)	F_2	86	RFLP (108)	467.5	Brummer et al. (1993)
Alfalfa (diploid)	Self-mated progeny	138	RFLP, RAPD, isozyme, morphological (89)	659	Kiss et al. (1993)
Alfalfa (diploid)	F_2	137	RFLP, RAPD, isozyme, morphological seed protein, PCR marker (868)	754	Kaló et al. (2000)
Alfalfa (diploid)	F_1	55	RFLP (50, 55)	234; 261	Tavoletti et al. (1996)
Alfalfa (diploid)	Backcross population	87	RFLP (46), (33), RAPD (40), (28)	603; 553	Echt et al. (1994)
Medicago truncatula	F_2	124	RAPD, AFLP (289), isoenzyme	1,225	Thoquet et al. (2002)
Clover, red	Backcross population	167	RFLP (157)	535.7	Isobe et al. (2003)
Clover, white	2-way pseudo-testcross F_1	92	EST-SSR (493)	1,144	Barrett et al. (2004)
Kentucky bluegrass	2-way pseudo-testcross F_1	67	AFLP (275), SAMPL (299)	367 (P), 338.4 (M)	Porceddu et al. (2002)
Zoysia	Self-pollinated F_1	105	RFLP (115)	1,506.3	Yaneshita et al. (1999)
	F_2	78	AFLP (471)	932.5	Cai et al. (2004)

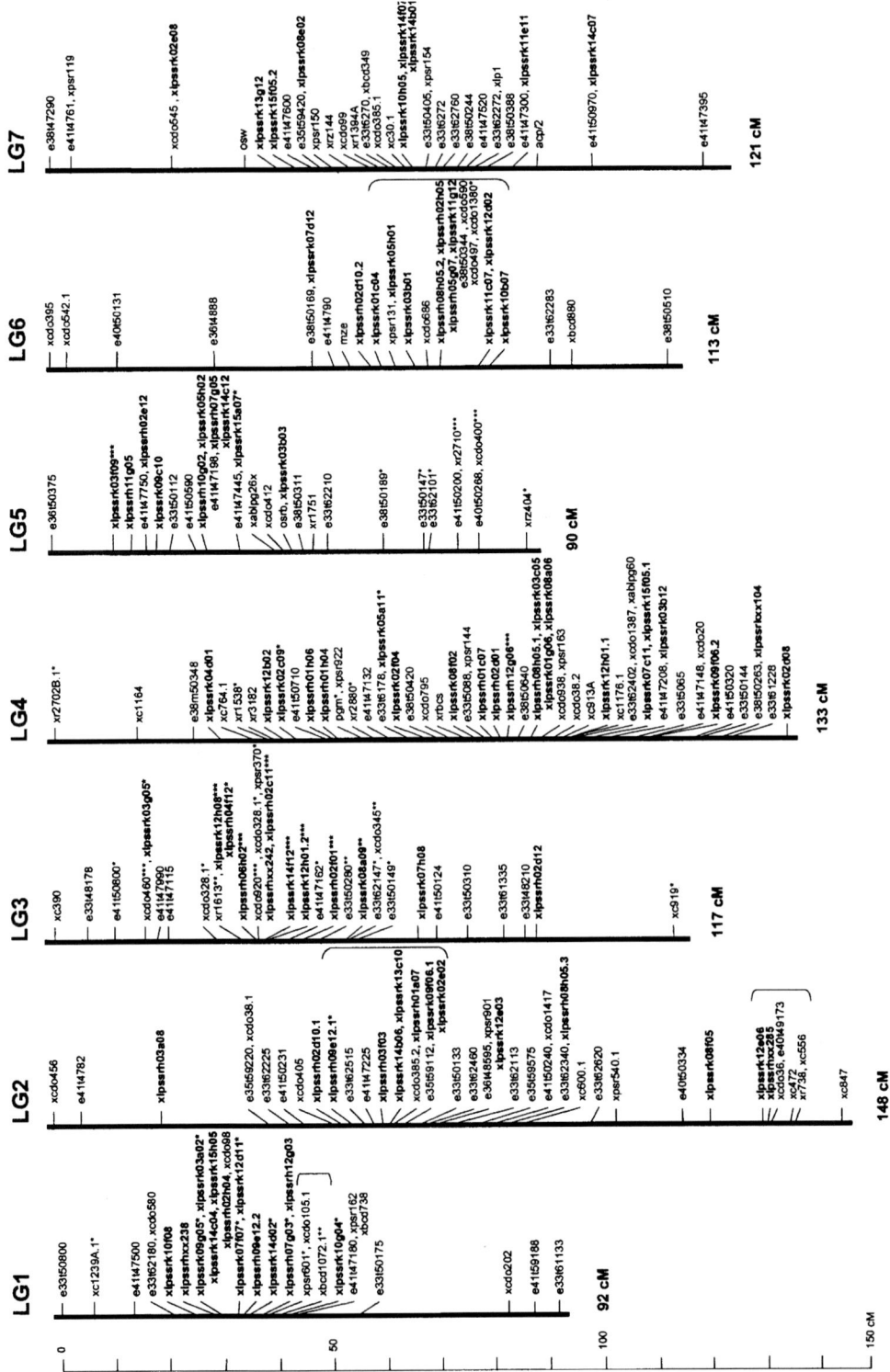

Fig. 1. Seven linkage groups (*LG*) of perennial ryegrass for the cross p150/112 based on simple sequence repeat (*SSR*) markers (LPSSR) and a selected subset of restriction fragment length polymorphism and amplified fragment length polymorphism loci. Markers showing segregation distortion are indicated as follows: significant distortion at *P < 0.05; **P < 0.01; and ***P < 0.001. The *bracketed regions* indicate markers that could not be ordered at logarithm of odds greater than 2. (From Jones et al. 2002a, with permission of Springer Science and Business Media and the authors)

in the same population to produce a composite map containing 258 loci.

Male and female molecular-marker linkage maps of an interspecific annual × perennial ryegrass mapping population were developed to determine the map location of the seedling root florescence (SRF) character and to identify additional genomic regions useful for species separation (Warnke et al. 2004). A total of 235 AFLP markers, 81 RAPD markers, 16 comparative grass RFLPs, 106 SSR markers, two isozyme loci, and two morphological characteristics (8-h flowering and SRF) were used to construct the linkage maps. The lengths of the maps differed, with the male map being 537-cM long and the female map being 712-cM long.

A genetic linkage map has been constructed for meadow fescue by using a full-sib family (Alm et al. 2003). The map consists of 466 markers (RFLPs, AFLPs, isozymes, and SSRs), and has a total length of 658.8 cM with an average marker density of 1.4 cM per marker.

Saha et al. (2004) reported the first genetic map of tall fescue constructed with PCR-based markers (AFLP and EST-SSRs). Two parental maps were initially constructed by using a two-way pseudo-testcross mapping strategy. The female map included 558 loci placed in 22 linkage groups and covered 2,013 cM of the genome. In the male map, 579 loci were grouped in 22 linkage groups with a total map length of 1,722 cM. The marker density in the two maps varied from 3.61 cM per marker (female parent) to 2.97 cM per marker (male parent). Markers that revealed polymorphism within both parents and showed 3:1 segregation ratio were used as bridging loci to integrate the two parental maps as a biparental consensus map. The integrated map covers 1,841 cM on 17 linkage groups, with an average of 54 loci per linkage group, and has an average marker density of 2.0 cM per marker. Six homeologous linkage groups of the seven predicted homeologous groups were identified.

Cultivated alfalfa (*M. sativa* L.) is an autotetraploid ($2n = 4x = 32$). Many linkage maps of alfalfa have been constructed (Brummer et al. 1993; Kiss et al. 1993; Echt et al. 1994; Tavoletti et al. 1996; Brouwer et al. 1999; Kaló et al. 2000; Thoquet et al. 2002; Julier et al. 2003; Table 2). Among them, either eight or ten linkage groups were reported. However, because of the complexity of the tetrasomic inheritance in autotetraploid alfalfa, most existing alfalfa genetic linkage maps are based on diploid rather than tetraploid alfalfa mapping populations. Another reason for using diploid populations is the complex segregating pattern of polyploidy (especially autopolyploidy) and the lack of computer programs to analyze such data. The outcrossing nature and high genetic diversity among individuals and species make the marker analysis in alfalfa more complex.

Kaló et al. (2000) constructed an improved linkage map of diploid alfalfa by using more than 868 markers distributed within an F_2 population of 137 individuals. The map covers 754 cM with an average marker density of 0.8/cM. Thoquet et al. (2002) reported the linkage map of *M. truncatula*, a model legume closely related to cultivated alfalfa. The map spans 1,225 cM and comprises 289 markers including RAPD, AFLP, known genes, and isozymes arranged in eight linkage groups. By mapping a number of common markers, they showed the eight linkage groups to be homologous to those of diploid alfalfa. Another linkage map of *M. truncatula* was also reported; it consisted of 141 ESTs, 80 bacterial artificial chromosome (BAC) end-sequence tags, and 67 resistance gene analogs, and covered 513 cM (Choi et al. 2004). Julier et al. (2003) constructed a genetic map of tetraploid alfalfa that consisted of 599 AFLP and 107 SSR markers by using an F_1 mapping population of 168 individuals produced from the cross of two heterozygous parental plants from cultivars Magali and Mercedes. For each parent, the genetic map contained eight groups of four homologous chromosomes. The lengths of the maps were 2,649 and 3,045 cM, with an average distance of 7.6 and 9.0 cM between markers, respectively. Using only the SSR markers, they built a composite map covering 709 cM (Fig. 2.2.3). Compared with diploid alfalfa genetic maps, the tetraploid alfalfa maps cover about 88–100% of the genome and are close to saturation. And except for two out of 107 SSR markers, a similar order of chromosomal markers was found between the tetraploid alfalfa and *M. truncatula* genomes, indicating a high level of colinearity between these two species.

To discuss other forage legumes, linkage maps of red clover and white clover have been reported (Isobe et al. 2003; Barrett et al. 2004). Isobe et al. (2003) constructed a genetic linkage map of red clover (*T. pratense* L., $2n = 2x = 14$) using complementary DNA (cDNA) RFLP probes in a backcross mapping population. The map contains 157 RFLP markers and one morphological marker on seven linkage groups. The total map distance was 535.7 cM and the average distance between two markers was 3.4 cM. Barrett et al. (2004) developed SSR markers for white clover

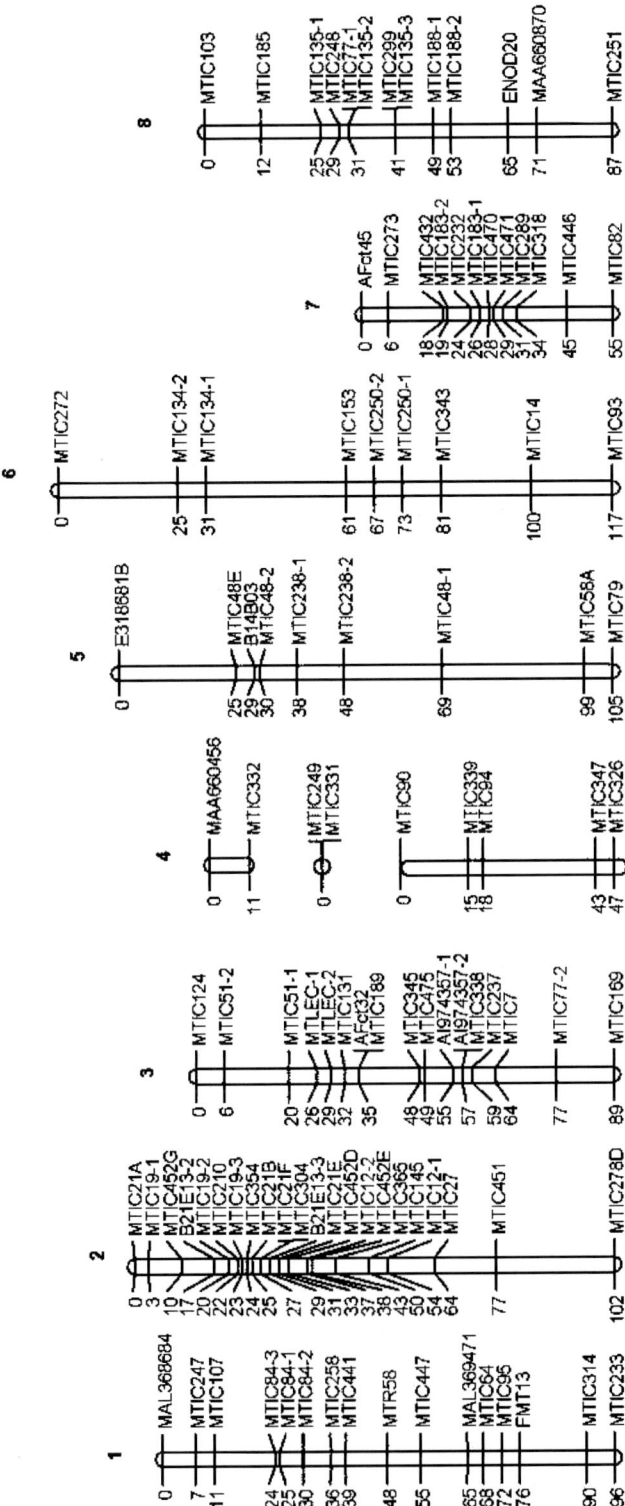

Fig. 2. Composite map with the SSR markers, for an F_1 mapping population in tetraploid alfalfa. (From Julier et al. 2003, with permission of the authors)

by mining an EST database and by isolation from enriched genomic libraries. They reported that 335 polymorphic EST-SSRs detected 493 loci in the white clover genome among 92 F_1 progeny from a pair-cross between two highly heterozygous genotypes. The total map length was 1,144 cM and the SSR markers had spanned all 16 homologs.

2.3
Gene Mapping

Recently, some major genes, including, among others, a disease resistance gene and self-incompatibility genes, were mapped and the markers closely linked to those genes were identified.

2.3.1
Crown-Rust Resistance

Crown rust, which is caused by *Puccinia coronata* f. sp. *lolii*, infests a broad host range including many forage grass genera such as *Lolium*, *Festuca*, *Agropyron*, *Agrostis*, *Paspalum*, *Phleum*, *Poa*, and *Puccinellia* (Smiley et al. 1992), and was recognized as a new disease in barley (Jin and Steffenson 2002). The disease is a very serious disease of ryegrasses (Italian and perennial) worldwide, and can cause severe losses in yield (Lancashire and Latch 1966, 1970; Plummer et al. 1990), quality (Potter et al. 1987), and palatability (Cruickshank 1957).

The studies of the genetic analysis of crown-rust resistance demonstrated that the trait is conferred by both major and minor resistance genes (Wilkins 1975, 1978a, b; Hayward 1977; Schmidt 1980; Reheul and Ghesquiere 1996; reviewed by Kimbeng 1999), and a cytoplasmic effect has been observed by Adams et al. (2000). Dumsday et al. (2003) demonstrated a major-effect locus in perennial ryegrass by using bulked segregant analysis and QTL analysis. A resistance gene locus, *LpPc1*, conferring a major effect was located in linkage group 2; and comparative genetic analysis revealed a conserved syntenic relationship between linkage group 2 of perennial ryegrass and linkage group B of *Avena*, which is the location of a cluster of genes for resistance to crown rust (Yu and Wise 2000). Roderick et al. (2002) also identified some QTLs for crown-rust resistance by using an F_2 perennial ryegrass mapping family and four single-pustule isolates. Two QTLs with moderate effect on linkage group 5 and QTLs with minor effect

on linkage groups 2 and 3 were identified by using crown-rust isolates 1 and 2. Using isolates 3 and 4, they detected two QTLs on linkage group 5 and two on linkage group 6 that had moderate effect. At least nine additional putative QTLs exhibiting minor effects were identified by using these isolates. Muylle et al. (2003) detected four QTLs for crown-rust resistance that explain 45% of the variations; these four QTLs were located in two different linkage groups. The locus with the strongest effect explained 30% of the phenotypic variation. Sim et al. (2005a) evaluated the phenotypic segregation of crown-rust resistance at two geographically and environmentally different locations. Two common QTLs were detected in linkage groups 2 and 3, and two location-specific QTLs were detected in linkage groups 6 and 7.

A major resistance gene for crown rust in Italian ryegrass has been demonstrated at the molecular level (Fujimori et al. 2003). A resistance-gene locus designated *Pc1* was detected from Yamaiku 130, a breeding line highly resistant to crown rust; three AFLP markers were found tightly linked to *Pc1* with a map distance of 0.9 cM, another three were found on the opposite side with a distance of 1.8 cM, and AFLP marker ATC-CATG153 cosegregated with *Pc1*. Hirata et al. (2003) demonstrated another major resistance gene designated *Pc2* in the cultivar Harukaze. Linkage analysis using SSR markers developed in Italian ryegrass was used to assign these resistance genes into a reference map of Italian ryegrass. The mapping data suggested that *Pc1* and *Pc2* are located in linkage groups 4 and 6, respectively (Hirata et al. 2003).

2.3.2
Blast Resistance

Blast, also called gray leaf spot (GLS), is a serious fungal disease recently reported on ryegrass. It is caused by *Magnaporthe grisea*, which also causes rice blast and many other grass diseases. Curley et al. (2004, 2005) conducted QTL analysis on the segregating population to determine the number, location, and degree of effect of the genomic regions associated with the resistance trait. A ryegrass genetic mapping population consisting of 156 progeny derived from a cross between two highly heterozygous ryegrass clones was constructed. QTL analysis using GLS reaction data and the linkage map being constructed have revealed at least two genomic regions associated with QTLs for GLS resistance, and one of the regions appears syntenic with rice linkage group 7. Miura et al. (2005)

screened EST-derived CAPS and AFLP markers linked to a gene for resistance to ryegrass blast (*Pyricularia* sp.) in Italian ryegrass. They analyzed segregation of resistance in an F_1 population derived from a cross between a resistant and a susceptible cultivar; the results indicated that the resistance was controlled by a single dominant gene (named *LmPi1*) located on linkage group 5 of Italian ryegrass. Of the 30 EST-CAPS markers screened, one marker, p56, flanking the *LmPi1* locus was identified. The restriction pattern of p56 amplification showed a unique fragment corresponding to the resistant allele at the *LmPi1* locus.

2.3.3
Incompatibility Mechanisms in Legumes and Grasses

The presence of incompatibility mechanisms in the legume and grass families has been the subject of many studies. Self-incompatibility in grass was first elucidated in rye (*Secale cereale* L.) (Lundqvist 1954). It is controlled by a gametophytic two-locus system (*S* and *Z*) (Cornish et al. 1979; Thorogood et al. 2002). In perennial ryegrass, the two loci were mapped to linkage groups 1 and 2, respectively, in accordance with the Triticeae consensus map. The *L. perenne S* and *Z* loci showed conserved synteny with the equivalent loci in rye. On the other hand, when self-compatibility (SC) was investigated in an F_2 family, distorted segregation ratios of markers on linkage groups 5 were found, indicating the possible presence of a gametophytic *SC* locus. Interval linkage analysis of pollen compatibility after selfing confirmed that this distortion was due to a locus (*T*) analogous to the *S5* locus of rye (Thorogood et al. 2004).

Self-incompatibility in *Phalaris coerulescens* is also gametophytically controlled by two unlinked multiallelic loci (*S* and *Z*). As part of a map-based cloning strategy, high-resolution maps of the *S* and *Z* regions were generated from distorted segregating populations by using RFLP probes from wheat, barley, oat, and *Phalaris*. The *S* locus was delimited to 0.26 cM with two boundary markers (Xwg811 and Xpsr168) and cosegregated with Xbm2 and Xbcd762. Xbcd266 was the closest marker linked to *Z* (0.9 cM). The *S* locus was localized to the subcentromere region of chromosome 1 and the *Z* locus to the long-arm end of chromosome 2. Several rice BAC clones orthologous to the *S* and *Z* locus regions were also identified (Bian et al. 2004).

SRF, an indicator for distinguishing perennial ryegrass and Italian ryegrass, was mapped on linkage group 1 by using an annual × perennial ryegrass (MFA × MFB) pseudo-F_2 population (Warnke et al. 2004).

2.3.4
Gene Mapping in Alfalfa

In *M. sativa*, Kiss et al. (1993) mapped genes for two phenotypic mutants, dwarf (*DWARF*) and sticky leaf (*STLF*), and two loci controlling flower color, anthocyanin, and xanthophyll (*ANT* and *XANT*). The genetic map was constructed by ordering the linkage values of 89 RFLP, RAPD, isozyme, and morphological markers collected from a segregating population of 138 individuals. *ANT* and *XANT* were mapped on linkage group 1, and *DWARF* and *STLF* on linkage groups 2 and 6, respectively. Brouwer and Osborn (1997) located the unifoliate leaf/cauliflower head mutation to linkage group 4. A nonnodulation phenotype was mapped to tetraploid alfalfa linkage group 5, and two tightly linked flanking markers were identified (Endre et al. 2002).

Tavoletti et al. (2000) mapped the jumbo-pollen trait in diploid alfalfa; homozygous recessive (*jpjp*) plants are characterized by the complete failure of postmeiotic cytokinesis during microsporogenesis, which results in 100% 4*n*-pollen formation. Two F_1 mapping populations, subsequently tested, showed Mendelian segregation for the jumbo-pollen trait and were effective in locating the *jp* gene on linkage group 6, close to RFLP marker Vg1G1b. This RFLP marker was also linked to one QTL for multinucleate-microspore formation.

Several markers linked to gene *Tne1* (2*n*-egg production) were identified (Barcaccia et al. 2000), by combined bulked segregant analysis, and RAPD, inter-simple sequence repeat (ISSR), and AFLP markers. One paternal ISSR marker located at 9.8 cM from *Tne1* was detected, and eight additional RAPD and AFLP markers were also identified.

2.4
Quantitative Trait Loci Detected

Recently, many quantitative traits, including yield, plant height, heading date, disease resistance, and stress tolerance, have been studied in several forage species. The results of QTL analysis will provide basic information for quantitative gene targeting, cloning, and gene isolation, and will also provide markers for MAS of important agronomic characters.

Humphreys et al. (2003) reported QTLs in *L. perenne* for forage quality, including water-soluble carbohydrate content (WSC), neutral detergent fiber (NDF), plant size, leaf extension rate, and regrowth rate. The results showed that QTLs were found on all seven linkage groups for morphological and growth traits and explained between 23 and 40% of the phenotypic variation in the traits. The QTLs that explained around 20–25% of trait variation for nutritive quality traits (total WSC, crude protein, and NDF) were found on four linkage groups. QTLs for individual WSC components were found on six linkage groups. The QTL for total WSC usually coincided with the QTL for high molecular weight fructan. The QTL for NDF coincided with the QTL for WSC on linkage groups 1 and 2.

Inoue et al. (2004b) reported on the chromosomal positions and the contributions of putative QTLs affecting lodging resistance and related traits in Italian ryegrass. Traits included seven quantitative characters including heading date, plant height, culm weight, culm diameter, culm strength, tiller number, and culm pushing resistance. They evaluated lodging scores in the field in a two-way pseudo-testcross F_1 population. The results revealed 17 QTLs for all traits except culm weight on six of seven linkage groups. This was accomplished by simple interval mapping using a cross-pollination algorithm. Thirty-three independent QTLs were also detected by composite interval mapping from both male and female parental linkage maps. In addition, up to 18 QTLs for lodging scores evaluated at nine different times were detected on all linkage groups.

In perennial ryegrass, Cogan et al. (2005) reported QTLs for five herbage quality traits including crude protein content, estimated in vivo dry-matter digestibility, NDF content, estimated metabolizable energy, and WSC.

They measured these through near-infrared reflectance spectroscopy analysis in the p150/112 reference genetic mapping population (183 individuals). The samples were prepared for herbage quality analyses from individual plants at six different times or locations. A total of 42 QTLs from six different sampling experiments varying by developmental stage (anthesis or vegetative growth), location, or year were detected, and some coincident QTLs were detected on linkage groups 3, 5, and 7.

Armstead et al. (2004) identified a genomic region with a major effect on heading date in *L. perenne* and identified the orthologous region in the fully sequenced rice genome. By comparing the relative position of RFLP probes CDO545, RZ144, R2869, and C764 in *L. perenne* and rice, they found that this region of *L. perenne* L7 and rice chromosome 6, which covered the region of the rice genome containing the *Hd3* locus (Yamamoto et al. 1998) or the *Hd3a* and *Hd3b* loci (Monna et al. 2002), showed synteny. QTL analysis revealed the genotype–trait association between C764 on L7 of *L. perenne* and days to heading. This major QTL represented up to 70% of the variance of heading date in *L. perenne*.

Jensen et al. (2005) mapped the QTLs for the vernalization response in perennial ryegrass. A total of five QTLs for the vernalization response, measured as days to heading, were identified and mapped to linkage groups 2, 4, 6, and 7. These QTLs explained 5.4–28.0% of the total phenotypic variation and the overall contribution of these five QTLs was 80% of the total phenotypic variation.

In alfalfa, QTLs for yield, winter hardiness, fall growth (FG), and freezing injury (FI) were detected in tetraploid alfalfa (Shah et al. 1999; Brouwer et al. 2000). The characters of FG, FI, and winter injury (WI) in 2 years of replicated field trials were studied (Brouwer et al. 2000). Two plants, B17 and P13, representing the extremes for each target trait, were crossed, and an F_1 plant was backcrossed to each parent to create two populations of 101 individuals each. Each population was scored for 82 SDRF loci and 17 or 19 two-allele loci, and evaluated for the three traits in 2 years of replicated field trials (Brouwer et al. 2000). Trait measures over the 2 years were significantly correlated (FG, $r = 0.71$; FI, $r = 0.42$; WI, $r = 0.76$). Significant correlations also existed between WI and FG ($r = 0.50$ and 0.56) and between WI and FI ($r = 0.34$ and 0.58) in each year. Within the B17 backcross population, genomic regions that contain QTLs affecting FG, FI, and WI were identified on linkage groups 5 and 8, but QTLs affecting only FG and FI were identified on linkage groups 1 and 3.

2.5
Marker-Assisted Breeding

Although many studies of gene mapping and QTL analysis directed toward MAS have been conducted, a few studies of marker-aided introgression that covered forage grasses were reported (Chen et al. 1995; Lesniewska et al. 2001; Humphreys et al. 2005; Moore et al. 2005). Humphreys et al. (2005) mapped the gene for drought tolerance transferred from *Festuca* into *Lolium*. Following the initial hybridization of a synthetic autotetraploid of *L. multiflorum* ($2n = 4x = 28$) with *F. glaucescens*, the F_1 hybrid was backcrossed twice onto diploid *L. multiflorum* ($2n = 2x = 14$) to produce a diploid *Lolium* genotype with a single *F. glaucescens* introgression located distally on the nucleolar-organizer region arm of chromosome 3. The transmission of *F. glaucescens* derived AFLPs and an STS marker, Fg71673, was monitored throughout the breeding program. RFLP markers from the *F. arundinacea* PstI genomic library were used in the determination of intergeneric hybrids of tetraploid *F. mairei* and diploid *L. perenne* (Chen et al. 1995). Intergeneric hybrids between *L. multiflorum* and *F. pratensis* (Lm/Fp) and their derivatives exhibit a unique combination of genetic and cytogenetic characteristics. This combination of characters has been used to transfer an *F. pratensis* chromosome segment carrying a mutant "stay-green" gene conferring a disrupted leaf-senescence phenotype into *L. multiflorum* and isolate it. The genetic location within the introgressed *F. pratensis* segment of the senescence gene has been mapped by using 29 AFLPs generated by 22 selected primer combinations. The final genetic distance of the *F. pratensis* chromosome segment between the terminal *F. pratensis* derived AFLP markers was estimated to be 19.8 cM, and the stay-green gene (*sid*) mutation was located at 9.8 cM. The closest flanking markers to *sid* were at 0.6 and 1.3 cM. The *F. pratensis* specific AFLP markers closely flanking the green gene have been cloned. Two STS markers showing polymorphism between the *Lolium* and *F. pratensis* parental DNA were mapped back onto the genetic map and in both cases were found to map precisely to the same position as the original AFLP markers (Moore et al. 2005).

Several authors have reported the results of genetic diversity studies on *Lolium* and *Medicago* species using molecular markers (Brummer et al. 1995; Bena et al. 1998; Cresswell et al. 2001; Skot et al. 2002; Segovia-Lerma et al. 2003). Cresswell et al. (2001) evaluated the use of AFLP markers to distinguish genotypes, populations, and species of *Lolium* collected from Portugal. Principal component analysis (PCA) of similarities between 127 plants based on 765 polymorphic bands generated from three primer pairs showed high dimensionality in the data.

Axes 1–3 were associated primarily with species differences; axes 4–14 with population differences within species; and axis 15 onward with differences within populations. Unweighted pair group method with arithmetic mean (UPGMA) analysis confirmed the groupings. Analyses of individual bands showed that every interspecific and intraspecific contrast involved a different set of bands, again confirming the high dimensionality of the data. No single band was strictly diagnostic of any population or species.

Skot et al. (2002) reported a genecological approach to identify AFLP markers that are associated with adaptation to low winter temperatures. Twenty-nine wild populations of ryegrass (*L. perenne*) representing a pan-European temperature cline in terms of their geographical origin were used. A further 18 populations from a temperature cline in Bulgaria were also analyzed. In addition, two varieties and five populations representing parents of mapping families currently in use at the Institute of Grassland and Environmental Research were included in the analysis. PCA and cluster analyses of the molecular marker data showed that the Bulgarian populations could be distinguished clearly from the other populations.

In alfalfa, Brummer et al. (1995) analyzed the variability of annual *Medicago* species using RAPD markers. A phylogeny reconstructed with the computer program Phylogenetic Analysis Using Parsimony (PAUP) showed the same relationships as traditional taxonomy. Variation was present among accessions of all species. Several accessions were considerably different from others within the species. Although the species are autogamous, the original seed collections may have been made from a number of plants in the same area. Also, some outcrossing or seed mixing may have occurred.

Finally, at least ten RAPD primers appear to be necessary in order to develop reliable estimates of relatedness among annual *Medicago* accessions. Bena et al. (1998) performed a molecular phylogenetic study based on the nuclear ribosomal internal and external transcribed spacer sequences. Thirty-one annual *Medicago* species were included in the study, representing more than half of the genus and 85% of the annuals of the genus. Major incongruence was found between phylogenetic relationships and morphological classification of the genus. Segovia-Lerma et al. (2003) reported the results of AFLP-based assessment of genetic diversity among nine alfalfa germplasms using bulked DNA templates. Bulked DNA templates were evaluated from 30 genotypes within each of nine well-recognized germplasms: African, Chilean, Flemish, Indian, Ladakh, Peruvian, Turkistani, *M. sativa* ssp. *falcata* , and *M. sativa* ssp. *varia*. These were evaluated using 34 AFLP primer combinations. A total of 3,754 fragments were identified, of which 1,541 were polymorphic. The number of polymorphic fragments detected per primer combination ranged from 20–85. UPGMA analysis of the marker data produced two main clusters: (1) *M. sativa* ssp. *sativa* and *M. sativa* ssp. *varia*; and (2) *M. sativa* ssp. *falcata*.

2.6
Map-Based Cloning

Several groups in the USA and Italy have extensively cloned and functionally analyzed some apomictic genes (Roche et al. 1999, 2002; Pupilli et al. 2001, 2004; Labombarda et al. 2002). Twelve sequence-characterized amplified region (SCAR) markers linked to pseudogamous apospory, a form of gameto-phytic apomixis, were previously isolated from *Pennisetum squamulatum* Fresen. No recombination between these markers was found in a segregating population of 397 individuals (Ozias-Akins et al. 1998). Roche et al. (1999) tested whether these markers were also linked to the aposporous mode of reproduction in two small segregating populations of *Cenchrus ciliaris* (= *Pennisetum ciliare* L. Link), another apomictic grass species. Among 12 markers, six were scored as dominant markers between apporous and sexual *C. ciliaris* genotypes. Five were always linked to apospory, and one showed a low level of recombination in 84 progeny. RFLPs were observed between sexual and apomictic phenotypes for three of the six remaining SCARs from *P. squamulatum* when used as probes. No recombination was observed in the F$_1$ progeny. Preliminary data from megabase DNA analysis and sequencing in both species indicate that an apospory-specific genomic region (ASGR) is highly conserved between the two species. Subsequently, Roche et al. (2002) constructed BAC libraries from two grass species that reproduce by apospory, a form of gametophytic apomixis. Both libraries were screened by hybridization with six SCARs, whose tight linkage in a single ASGR had been previously demonstrated in both species. Analysis of these BAC clones indicated that some of the SCAR markers actually amplify duplicated regions linked in the coupling phase in both genomes.

A mapping population generated by backcross-ing a sexual genotype with an apomictic genotype of *Paspalum simplex*, both at the tetraploid level, was used to find markers cosegregating with apomixis (Pupilli et al. 2001). Five rice markers, mapped in the telomeric region of the long arm of rice chromosome 12, showed tight linkage with apomixis. Four AFLPs linked to apomixis in the coupling phase were also found in the same population (Labombarda et al. 2002). Integrating the AFLP data with those obtained previously with rice RFLP anchor markers, the researchers drew a map for the chromosome region of *P. simplex* encompassing apomixis.

Meanwhile, Molinari et al. (2003) established a regeneration protocol in apomictic and sexual lines of *P. simplex* with the dual perspective of developing a transformation system to screen candidate genes for apomixis and for investigating whether tissue culture could induce rearrangements of several kinds at the apomixis-controlling locus. A common set of RFLP markers, including five rice anchor markers previously shown to be linked to apomixis in *P. simplex*, were used to detect linkage with apomixis in *P. notatum* and *P. malacophyllum* (Pupilli et al. 2004). A comparative map of the region around the apomixis locus was drawn for the three *Paspalum* species and compared to the rice map. The locus that controls apomixis in *P. simplex* was almost completely conserved in the closely related species *P. malacophyllum*, whereas it was only partially represented in the distantly related species *P. notatum*. Although strong synteny of markers was noted between this locus and a portion of rice chromosome 12 in both *P. simplex* and *P. malacophyllum*, the same locus in *P. notatum* was localized to a hybrid chromosome that carried markers that map to rice chromosomes 2 and 12.

Thoquet et al. (2002) established the first genetic map of *Medicago truncatula* for comparative legume genomics and the isolation of agronomically important genes such as a symbiosis gene (*Mtsym6*) and the sense of pod-coiling (*SPC*) gene, responsible for the direction the pod coils. This map spans 1,225 cM (average 470 kb/cM) and comprises 289 markers, including RAPD, AFLP, known genes, and isozymes arranged in eight linkage groups (2*n* = 16). It used an F$_2$ segregating population of 124 individuals. The agronomically important legume *M. sativa* is taxonomically very close to *M. truncatula*. The relative orientation of linkage groups 1, 3, 4, 5, 6, and 7 between *M. sativa* and *M. truncatula* was determined. *Mtsym6* and *SPC* were mapped on linkage groups 8 and 7, respectively. The map-based cloning of *Mtsym6* and *SPC* is currently under way.

Until now, no ryegrass gene has been isolated through map-based cloning; however, the basic tools for map-based cloning in ryegrass are complete. BAC libraries have been constructed for both Italian ryegrass and perennial ryegrass (Fujimori et al. 2004; Farrar et al. 2005). Many RFLP and SSR markers are mapped on the seven linkage groups (Jones et al. 2001; Inoue et al. 2004a; Hirata et al. 2006), and more than 50,000 ESTs have been developed by various institutes (Sawbridge et al. 2003; Ikeda et al. 2004).

Sawbridge et al. (2003) have reported the generation and analysis of an EST and cDNA microarray resource for perennial ryegrass. From 29 cDNA libraries of perennial ryegrass representing a range of plant organs and developmental stages, over 44,000 ESTs were produced, analyzed by Basic Local Alignment Search Tool (BLAST) searches, categorized functionally, and subjected to cluster analysis, leading to the identification of a unigene set corresponding to 14,767 genes. This unigene set, representing approximately half or one third of all expressed sequences in ryegrass, was compared with the *Arabidopsis* and rice proteomes and used to develop a unigene cDNA microarray for genomewide gene expression analysis in grasses. The study of the isolation of some resistance genes such as the crown-rust gene *Pc1* is under way (Fujimori et al., unpublished data).

2.7
Future Scope of Works

2.7.1
Synteny Between Forage Crops and Other Well-Studied Grass Species

Anchor probes that can hybridize between various species have become an efficient, powerful tool for comparative genome mapping or study of synteny. Van Deynze et al. (1998) screened a RFLP marker set for the Poaceae family selected from cDNA libraries developed from rice, oat, and barley. A total of 1,800 probes were screened on garden blots containing the DNA of rice, maize, sorghum, sugarcane, wheat, barley, and oat, and 152 of them were selected as "anchors" because (1) they hybridized to the majority of target grass species in Southern blot analysis, (2) they appeared to be low-copy or single-copy sequences in rice, and (3) they helped provide reasonably good genome coverage of all species. The 152 probes were then screened for polymorphism in mapping parents, and polymorphic markers were mapped in rice, oat, maize, and wheat. The use of anchor probes for comparative mapping is an efficient way of establishing genetic relationships for comparisons among all the species and genera being studied (Ahn and Tanksley 1993; Ahn et al. 1993; Van Deynze et al. 1995a–c).

A genetic linkage map has been constructed for meadow fescue by using a full-sib family (Alm et al. 2003). The map consists of 466 markers (RFLPs, AFLPs, isozymes, and SSRs). The RFLP probes used were derived from the original genomic library and many kinds of anchor probes such as wheat (PSR, WG), barley (ABG, BCD), oat (CDO), rice (RZ, RGC, RGG, RGR), maize (CSU), sorghum (SbRPG), pearl millet (PSM), and meadow fescue (IBF). The results showed conserved syntenic relationships between the meadow fescue linkage groups and the maps of *Lolium*, the Triticeae, oats, rice, maize, and sorghum. For *Lolium*, the number of loci on the meadow fescue map with equivalent map locations in perennial ryegrass was 46, representing 62% genome coverage. The fescue maps contain 117 loci with known map locations in the Triticeae. Of the 72% of the fescue genome that was covered by Triticeae markers, 94% was orthologous, and a high degree of orthology was observed. On the basis of the known relationships between the Triticeae and oats (Van Deynze et al. 1995c), the relationship between fescue and oats was guessed. The 48 loci with map locations in oats represented genome coverage of only 48%, of which 90% was orthologous between oats and fescue.

Sim et al. (2005b) constructed a RFLP-based genetic map of ryegrass based on an interspecific population that was derived by crossing perennial ryegrass and Italian ryegrass, for comparative mapping with other Poaceae species using heterologous anchor probes. First, a genetic map containing 235 AFLP, 81 RAPD, 160 SSR, two isozyme, 16 RFLP, and two morphological markers was constructed (Warnke et al. 2004). Next, they reconstructed a second linkage map using only RFLP markers for comparative mapping. This map covers a total distance of 573 cM with a total of 123 loci, including 16 loci previously reported in Warnke et al. (2004). Of the 123 loci, 112 were common with the Triticeae consensus linkage map, 82 with the oat linkage map, and 108 with the rice linkage map. Seven linkage groups were represented by ten syntenic segments of Triticeae chromosomes, 12 syntenic segments of oat chromosomes, or 16 syntenic segments of rice chromosomes, suggesting that the ryegrass genome has a high degree of genome conservation in relation to the Triticeae, oat, and rice.

ESTs are important resources for gene discovery and molecular-marker development. The partial sequencing of anonymous cDNA clones (ESTs) has become the method of choice for the rapid, cost-effective generation of data on the coding capacity of the genome.

EST-SSR has also been an efficient tool for comparative mapping. Saha et al. (2004) designed 157 EST-SSR primer pairs from tall fescue ESTs and tested them on 11 genotypes representing seven grass species. Nearly 92% of the primer pairs produced characteristic SSR bands in at least one species. A high level of marker polymorphism was observed in the outcrossing ryegrass and tall fescue species (66%). Sequencing of selected PCR bands revealed that the nucleotide sequences of the forage grass genotypes were highly conserved.

2.7.2
Medicago truncatula as a Model Plant for Legumes

A legume species, *M. truncatula*, which is closely related to alfalfa, has emerged as a model plant for the molecular and genetic dissection of various plant processes involved in rhizobial, mycorrhizal, and plant–pathogen interactions. Thoquet et al. (2002) established the first map of this species. They mapped 18 gene or isoenzyme markers that tag the eight linkage groups of the diploid *M. sativa* genetic map constructed by Kaló et al. (2000). All these markers were similarly linked in the two species with the exception of the ribosomal DNA, which was found on linkage group 5 of *M. truncatula* and linkage group 6 of *M. sativa*. The relative orientation of linkage groups 1, 3, 4, 5, 6, and 7 between *M. sativa* and *M. truncatula* could be determined because at least two common markers in each linkage group had been mapped.

Eujayl et al. (2004) identified 4,384 ESTs containing perfect SSRs (EST-SSR) from over 147,000 ESTs of *M. truncatula*. Six hundred and sixteen primer pairs were designed and screened over a panel of eight genotypes representing six *Medicago* species and subspecies. Nearly 74% (455) of the primer pairs produced characteristic SSR bands of expected size and length in at least one *Medicago* species. Four hundred and six (89%) of these 455 primer pairs produced SSR bands in all eight genotypes tested. Only

17 primer pairs were specific to *M. truncatula*. High levels of polymorphism (more than 70%) were detected for these markers in alfalfa, *M. truncatula*, and other annual medics (*Medicago* species). Gutierrez et al. (2005) screened 209 EST-based or BAC-based SSR markers from *M. truncatula* in the three most important European legumes, pea, fava bean, and chickpea, and the results showed significant transferability of *M. truncatula* SSRs to the three species.

Kaló et al. (2004) compared the genomes of alfalfa (*M. sativa*) and pea (*Pisum sativum*), which represent two closely related tribes of the subfamily Papilionoideae, but with different basic chromosome numbers. The positions of genes on the most recent linkage map of diploid alfalfa were compared with those of homologous loci on the combined genetic map of pea to analyze the degree of colinearity between their linkage groups. Besides using unique genes, analysis of the map position of multicopy (homologous) genes identified syntenic homologs and pinpointed the positions of nonsyntenic homologs. The results of the comparison revealed extensive conservation of the gene order between alfalfa and pea. The high degree of synteny observed between pea and *Medicago* loci has generated a promising strategy for further map-based cloning of pea genes based on the genome resources now available for *M. truncatula*.

2.7.3
Genetic Transformation

Genetic transformation is a new technique for plant breeding. Several methods are used to develop transgenic plants, including use of *Agrobacterium tumefaciens*, particle bombardment, and electroporation. Among forage crops, tall fescue, pearl millet, Italian ryegrass, blue grama grass, and bermudagrass were transformed by the particle bombardment method. Alfalfa, white clover, and Chinese milk vetch were transformed by the *Agrobacterium* method (Tabe et al. 1995; Ye et al. 1997; Cho et al. 2000; Lee et al. 2001; Aguado-Santacruz et al. 2002; Takahashi et al. 2002; Goldman et al. 2003; Ray et al. 2003; Wang et al. 2003; Li and Qu 2004).

Takahashi et al. (2004) introduced the rice chitinase (Cht-2; *RCC2*) gene into calluses of Italian ryegrass by particle bombardment. Chitinase is one of the most representative pathogenesis-related proteins, and has led to many achievements in transgenic research. Higher chitinase activity was detected in all investigated transgenic plants expressing the *RCC2* gene than in the nontransgenic plants. Two plants that were the most resistant to the development of uredospore colonies had approximately 8.7 times the chitinase activity of the nontransgenic plants.

Transgenic plants of tall fescue were obtained by the biolistic transformation of embryogenic suspension cells (Wang et al. 2003). Cell-suspension-derived and seed-derived transgenic tall fescue plants were transferred to the field. The aim of the study was to comparatively evaluate the agronomic performance of transgenic and nontransgenic control plants as well as that of their half-sib families of this important outcrossing forage species. The agronomic performance of the primary transgenics and regenerants was generally inferior to that of the seed-derived plants, with primary transgenics having fewer tillers and a lower seed yield. However, no major differences between the progeny of transgenics and the progeny of seed-derived plants were found in the agronomic traits evaluated. Primary transgenics and regenerants from the same genotype were more uniform than were plants from seeds. Progeny of transgenics performed similarly to progeny of the regenerants. Those results indicate that outcrossing grass plants generated through transgenic approaches can be incorporated into forage breeding programs.

Tabe et al. (1995) reported the establishment of an efficient protocol for transfer of foreign genes into two Australian commercial cultivars of alfalfa, and also presented an analysis of the expression in transgenic plants of four different genes encoding the SFA8 (sunflower albumin 8) protein. The level of accumulation of the foreign protein would be predicted to supply an extra 40 mg of sulfur amino acids daily to sheep fed the modified forage.

2.7.4
Endophytic Fungi

Endophytes have been reported in perennial ryegrass and meadow fescue and are present in many other cool-season grass species (Christensen et al. 1993). An endophyte (*endo* meaning "within" + *phyte* meaning "plant") is a plant or fungus living within another plant (or animal), usually as a parasite. Infected plants show no external symptoms, even though the hyphae of the fungus may be found within the seeds, pith, leaf sheath, or other tissues. The endophytic fungus is seed-transmitted.

The presence of an endophyte in a grass plant is not completely undesirable. There are also beneficial

aspects: infected plants are relatively free of disease and insect pests (Prestidge and Gallagher 1988; Breen 1993). Nevertheless, some alkaloids, notably ergot, are synthesized in infected plants, and can be a serious problem for cattle grazing on infected plants (Gallagher et al. 1984; Yates et al. 1985). However, the existence of an endophytic fungus that causes no adverse effects and restricts damage from disease and insect pests was recently discovered (Prestidge and Gallagher 1988; Breen 1993). An attempt is under way to bring the endophytic fungus into use in the forage breeding system.

References

Adams E, Roldan-Ruiz I, Depicker A, van Bockstaele E, de Loose M (2000) A maternal factor conferring resistance to crown rust in *Lolium multiflorum* cv. "Axis." Plant Breed 119:182–184

Aguado-Santacruz GA, Rascon-Cruz Q, Cabrera-Ponce JL, Martinez-Hernandez A, Olalde-Portugal V, Herrera-Estrella L (2002) Transgenic plants of blue grama grass, *Bouteloua gracilis* (H. B. K) Lag. Ex Steud., from microprojectile bombardment of highly chlorophyllous embryogenic cells. Theor Appl Genet 104:763–771

Ahloowalia BS (1972) Trisomics and aneuploids of ryegrass. Theor Appl Genet 42:363–367

Ahn S, Anderson JA, Sorrells ME, Tanksley SD (1993) Homoeologous relationships of rice, wheat and maize chromosomes. Mol Gen Genet 241:483–490

Ahn S, Tanksley SD (1993) Comparative linkage maps of the rice and maize genomes. Proc Natl Acad Sci USA 90:7980–7984

Alm V, Fang C, Busso CS, Devos KM, Vollan K, Grieg Z, Rognli OA (2003) A linkage map of meadow fescue (*Festuca pratensis* Huds.) and comparative mapping with other Poaceae species. Theor Appl Genet 108:25–40

Armstead IP, Turner LB, Farrell M, Skot L, Gomez P, Montoya T, Donnison IS, King IP, Humphreys MO (2004) Synteny between a major heading-date QTL in perennial ryegrass (*Lolium perenne* L.) and the *Hd3* heading-date locus in rice. Theor Appl Genet 108:822–828

Armstead IP, Turner LB, King IP, Cairns AJ, Humphreys MO (2002) Comparison and integration of genetic maps generated from F_2 and BC_1-type mapping populations in perennial ryegrass. Plant Breed 121:501–507

Barcaccia G, Albertini E, Rosellini D, Tavoletti S, Veronesi F (2000) Inheritance and mapping of 2n-egg production on diploid alfalfa. Genome 43:528–537

Barrett B, Griffiths A, Schreiber M, Ellison N, Mercer C, Bouton J, Ong B, Forster J, Sawbridge T, Spangenberg G, Bryan G, Woodfield D (2004) A microsatellite map of white clover. Theor Appl Genet 109:596–608

Baumann U, Juttner J, Bian XY, Langridge P (2000) Self-incompatibility in the grasses. Ann Bot 85 (Suppl A) 203–209

Beckmann JS, Soller M (1986) Restriction fragment length polymorphisms and genetic improvement of agricultural species. Euphytica 35:111–124

Bena G, Prosperi JM, Lejeune B, Olivieri I (1998) Evolution of annual species of the genus *Medicago*: a molecular phylogenetic approach. Mol Phylogenet Evol 3:552–559

Berhan AM, Hulbert SH, Butler LG, Bennetzen JL (1993) Structure and evolution of the genomes of *Sorghum bicolor* and *Zea mays*. Theor Appl Genet 86:598–604

Bert PF, Charmet G, Sourdille P, Hayward MD, Balfourier F (1999) A high-density molecular map for ryegrass (*Lolium perenne*) using AFLP markers. Theor Appl Genet 99:445–452

Bian XY, Friedrich A, Bai JR, Baumann U, Hayman DL, Barker SJ, Langridge P (2004) High-resolution mapping of the S and Z loci of *Phalaris coerulescens*. Genome 47:918–930

Bolton JL (1962) Alfalfa: Botany, Cultivation, and Utilization. Interscience, New York, USA

Breen JP (1993) Enhanced resistance to three species of aphids (Homoptera, Aphididae) in *Acremonium* endophyte-infected turfgrasses. J Econ Entomol 86:1279–1286

Brouwer DJ, Duke SH, Osborn TC (2000) Mapping genetic factors associated with winter hardiness, fall growth, and freezing injury in autotetraploid alfalfa. Crop Sci 40:1387–1396

Brouwer DJ, Osborn TC (1997) Identification of RFLP markers linked to the unifoliate leaf, cauliflower head mutation of alfalfa. J Hered 88:150–152

Brouwer DJ, Osborn TC (1999) A molecular marker linkage map of tetraploid alfalfa (*Medicago sativa* L.). Theor Appl Genet 99:1194–1200

Brummer EC, Bouton JH, Kochert G (1993) Development of an RFLP map in diploid alfalfa. Theor Appl Genet 86:329–332

Brummer EC, Bouton JH, Kochert G (1995) Analysis of annual *Medicago* species using RAPD markers. Genome 2:362–367

Cai H, Inoue M, Yuyama N, Nakayama S (2004) An AFLP-based linkage map of Zoysiagrass (*Zoysia japonica*). Plant Breed 123:543–548

Chen C, Sleper DA, West CP (1995) RFLP and cytogenetic analyses of hybrid between *Festuca mairei* and *Lolium perenne*. Crop Sci 35:720–725

Chen C, Sleper DA, Johal GS (1998) Comparative RFLP mapping of meadow and tall fescue. Theor Appl Genet 97:255–260

Cho HJ, Brotherton JE, Song HS, Widholm JM (2000) Increasing tryptophan synthesis in a forage legume *Astragalus sinicus* by expressing the tobacco feedback-insensitive anthranilate synthase (*ASA2*) gene. Plant Physiol 123:1069–1076

Choi HK, Kim DJ, Uhm T, Limpens E, Lim HJ, Mun JH, Kaló P, Penmetsa RV, Seres A, Kulikova O, Roe BA, Bisseling T, Kiss GB, Cook DR (2004) A sequence-based genetic map of *Medicago truncatula* and comparison of marker colinearity with *M. sativa*. Genetics 166:1463–1502

Christensen MJ, Leuchtmann A, Rowan DD, Tapper BA (1993) Taxonomy of *Acremonium* endophytes of tall fescue (*Festuca arundinacea*), meadow fescue (*F. pratensis*) and perennial ryegrass (*Lolium perenne*). Mycol Res 97:1083–1092

Cogan NOI, Smith KF, Yamada T, Francki MG, Vecchies AC, Jones ES, Spangenberg GC, Forster JW (2005) QTL analysis and comparative genomics of herbage quality traits in perennial ryegrass (*Lolium perenne* L.). Theor Appl Genet 110:364–380

Cornish MA, Hayward MD, Lawrence MJ (1979) Self incompatibility in ryegrass. 1. Genetic control in diploid *Lolium perenne* L. Heredity 43:95–106

Cresswell A, Sackville Hamilton NR, Roy AK, Viegas BM (2001) Use of amplified fragment length polymorphism markers to assess genetic diversity of *Lolium* species from Portugal. Mol Ecol 1:229–241

Cruickshank IAM (1957) Crown rust of ryegrass. NZ J Sci Technol 38:539–543

Curley J, Jung G (2005) QTL mapping of resistance to gray leaf spot in ryegrass: consistency of QTL between two mapping populations. In: Plant and Animal Genome XIII Conf, San Diego, CA, USA, p 374

Curley J, Sim SC, Jung G, Leong S, Warnke S, Barker RE (2004) QTL mapping of ryegrass blast resistance in ryegrass, and synteny-based comparison with rice blast resistance genes in rice. In: Hopkins A, Wang ZY, Mian R, Sledge M, Barkers RE (eds) Molecular Breeding of Forage and Turf. Kluwer Academic Publ, Dordrecht, The Netherlands, pp 37–46

Diwan N, Bhagwat AA, Bauchan GR, Cregan PB (1997) Simple sequence repeat (SSR) DNA markers in alfalfa and perennial and annual *Medicago* species. Genome 40:887–895

Diwan N, Bouton JH, Kochert G, Cregan PB (2000) Mapping of simple sequence repeat (SSR) DNA markers in diploid and tetraploid alfalfa. Theor Appl Genet 101:165–172

Dumsday J, Trigg P, Jones E, Batley J, Smith K, Forster J (2003) SSR-based genetic linkage analysis of resistance to crown rust (*Puccinia coronata Corda* f. sp. *lolii*) in perennial ryegrass (*Lolium perenne* L.). In: Plant and Animal Genome XI Conf, San Diego, CA, USA, p 709

Echt CS, Kidwell KK, Knapp SJ, Osborn TC, McCoy TJ (1994) Linkage mapping in diploid alfalfa (*Medicago sativa*). Genome 37:61–71

Endre G, Kaló P, Kevei Z, Kiss P, Mihacea S, Szakal B, Kereszt A, Kiss GB (2002) Genetic mapping of the non-nodulation phenotype of the mutant MN-1008 in tetraploid alfalfa (*Medicago sativa*). Mol Genet Genom 266:1012–1019

Eujayl I, Sledge MK, Wang L, May GD, Chekhovskiy K, Zwonitzer JC, Mian MAR (2004) *Medicago truncatula* EST-SSRs reveal cross-species genetic markers for *Medicago* spp. Theor Appl Genet 108:414–422

Farrar K, Thomas A, Humphreys M, Donnison I (2005) Construction of a BAC library for *Lolium perenne* (perennial ryegrass). In: Plant and Animal Genome XIII Conf, San Diego, CA, USA

Forster JW, Jones ES, Kolliler R, Drayton MC, DumsdayJ, Dupal MP, Guthridge KM, Mahoney NL, van Zijll de Jong E, Smith KF (2001) Development and implementation of molecular markers for forage crop improvement. In: Spangenberg G (ed) Molecular Breeding of Forage Crops. Kluwer, Dordrecht, The Netherlands, pp 101–133

Fujimori M, Hayashi K, Hirata M, Ikeda S, Takahashi Y, Mano Y, Sato H, Takamizo T, Mizuno K, Fujiwara T, Sugita S (2004) Molecular breeding and functional genomics for tolerance to biotic stress. In: Hopkins A, Wang ZY, Mian R, Sledge M, Barker RE (eds) Molecular Breeding of Forage and Turf. Kluwer, Dordrecht, The Netherlands, Boston London, UK, pp 21–36

Fujimori M, Hayashi K, Hirata M, Mizuno K, Fujiwara T, Akiyama F, Mano Y, Komatsu T, Takamizo T (2003) Linkage analysis of crown rust resistance gene in Italian ryegrass (*Lolium multiflorum* Lam.). In: Plant and Animal Genome XI Conf, San Diego, CA, USA, p 203

Gallagher RT, Hawkes AD, Steyn PS, Vleggaar R (1984) Tremorgenic neurotoxins from perennial ryegrass causing ryegrass staggers disorder of livestock: structure elucidation of lolitrem B. J Chem Soc Chem Commun 9:614–616

Goldman JJ, Hanna WW, Fleming G, Ozias-Akins P (2003) Fertile transgenic pearl millet (*Pennisetum glaucum* [L.] R. Br.) plants recovered through microprojectile bombardment and phosphinothricin selection of apical meristem-, inflorescence-, and immature embryo-derived embryogenic tissues. Plant Cell Rep 21:999–1009

Grattapaglia D, Sederoff R (1994) Genetic linkage maps of *Eucalyptus grandis* and *Eucalyptus uraphylla* using a pseudo-testcross mapping strategy and RAPD markers. Genetics 137:1121–1137

Gupta PK, Roy JK, Prasad M (2001) Single nucleotide polymorphisms (SNPs): a new paradigm in molecular marker technology and DNA polymorphism detection with emphasis on their use in plants. Curr Sci 80:524–536

Gutierrez MV, Vaz Patto MC, Hugurt T, Cubero JI, Moreno MT, Torres AM (2005) Cross-species amplification of *Medicago truncatula* microsatellites across three major pulse crops. Theor Appl Genet 110:1210–1217

Hayward MD (1977) Genetic control of resistance to crown rust (*Puccinia coronata* Corda) in *Lolium perenne* L. and its implications in breeding. Theor Appl Genet 51:49–53

Hayward MD, McAdam NJ (1988) The effect of isozyme selection on yield and flowering time in *Lolium perenne*. Plant Breed 101:24–29

Hayward MD, McAdam NJ, Jones JG, Evans C, Evans GM, Forster JW, Ustin A, Hossain KG, Quader B, Stammers M, Will JK (1994) Genetic markers and the selection of quantitative traits in forage grasses. Euphytica 77:269–275

Hayward MD, Degenaars GH, Balfourier F, Eickmeyer F (1995) Isozyme procedure for the characterization of germplasm, exemplified by the collection of *Lolium perenne* L. Genet Resour Crop Evol 42:327–337

Hayward MD, Forster JW, Jones JG, Dolstra O, Evans C, McAdam NJ, Hossain KG, Stammers M, Will J, Humphreys MO, Evans GM (1998) Genetic analysis of Lolium. I. Identification of linkage groups and the establishment of a genetic map. Plant Breed 117:451–455

Helentjaris T (1987) A genetic linkage map for maize based on RFLPs. Trends Genet 3:217–221

Hirata M, Fujimori M, Inoue M, Miura Y, Cai H, Satoh H, Mano Y, Takamizo T (2003) Mapping of a new crown rust resistance gene, *Pc2*, in Italian ryegrass cultivar "Harukaze". In: Proceeding of Molecular Breeding of Forage and Turf 2003, 3rd Intl Symp, Dallas, USA, p 15

Hirata M, Cai H, Inoue M, Miura Y, Komatsu T, Takamizo T, Fujimori M (2006) Development of simple sequence repeat (SSR) markers and construction of an SSR-based linkage map in Italian ryegrass (*Lolium multiflorum* Lam.). Theor Appl Genet 113:270–279

Huang H, Kochert G (1994) Comparative RFLP mapping of an allotetraploid wild rice species (*Oryza latifolia*) and cultivated rice (*O. sativa*). Plant Mol Biol 25:633–648

Humphreys J, Harper JA, Armstead IP, Humphreys MW (2005) Introgression-mapping of genes for drought resistance transferred from *Festuca arundinacea* var. *glaucescens* into *Lolium multiflorum*. Theor Appl Genet 110:579–587

Humphreys M, Turner L, Armstead I (2003) QTL mapping in *Lolium perenne*. In: Plant and Animal Genome XI Conf, San Diego, CA, USA

Hutchinson J, Rees H, Seal AG (1979) An assay of the activity of supplementary DNA in *Lolium*. Heredity 43:411–421

Ikeda S, Takahashi W, Oishi M (2004) Generation of expressed sequence tags from cDNA libraries of Italian ryegrass (*Lolium multiflorum* Lam.). Grassland Sci 49:593–598

Inoue M, Gao Z, Hirata M, Fujimori M, Cai H (2004a) Construction of a high-density linkage map of Italian ryegrass (*Lolium multiflorum* Lam.) using restriction fragment length polymorphism, amplified fragment length polymorphism, and telomeric repeat associated sequence markers. Genome 47:57–65

Inoue M, Gao Z, Cai H (2004b) QTL analysis of lodging resistance and related traits in Italian ryegrass (*Lolium multiflorum* Lam.). Theor Appl Genet 109:1576–1585

Isobe S, Klimenko I, Ivashuta S, Gau M, Kozlov NN (2003) First RFLP linkage map of red clover (*Trifolium pretense* L.) based on cDNA probes and its transferability to other red clover germplasm. Theor Appl Genet 108:105–112

Jensen LB, Andersen JR, Frei U, Xing Y, Taylor C, Holm PB, Lubberstedt T (2005) QTL mapping of vernalization response in perennial ryegrass (*Lolium perenne* L.) reveals co-location with an orthologue of wheat *VRN1*. Theor Appl Genet 110:527–536

Jin Y, Steffenson BJ (2002) Sources and genetics of crown rust resistance in barley. Phytopathology 92:1064–1067

Jones ES, Dupal MP, Kolliler R, Drayton MC, Forster JW (2001) Development and characterization of simple sequence repeat (SSR) markers for perennial ryegrass (*Lolium perenne* L.). Theor Appl Genet 102:405–415

Jones ES, Mahoney NL, Hayward MD, Armstead IP, Jones JG, Humphreys MO, King IP, Kishida T, Yamada T, Balfourier F, Chrmet G, Forster JW (2002a) An enhanced molecular marker based genetic map of perennial ryegrass (*Lolium perenne*) reveals comparative relationships with other Poaceae genomes. Genome 45:282–295

Jones ES, Dupal MP, Dumsday JL, Hughes LJ, Forster JW (2002b) An SSR-based genetic linkage map for perennial ryegrass (*Lolium perenne* L.). Theor Appl Genet 105:577–584

Julier B, Flajoulot S, Barre P, Cardinet G, Santoni S, Huguet T, Huyghe C (2003) Construction of two genetic linkage maps in cultivated tetraploid alfalfa (*Medicago sativa*) using microsatellite and AFLP markers. BMC Plant Biol 3(1):9

Kaló P, Endre G, Zimanyi L, Csanadi G, Kiss GB (2000) Construction of an improved linkage map of diploid alfalfa (*Medicago sativa*). Theor Appl Genet 100:641–657

Kaló P, Seres A, Taylor SA, Jakab J, Kevei Z, Kereszt A, Endre G, Ellis TH, Kiss GB (2004) Comparative mapping between *Medicago sativa* and *Pisum sativum*. Mol Genet Genom 272:235–246

Kanazin V, Talbert H, See D, DeCamp P, Nevo E, Blake T (2002) Discovery and assay of single nucleotide polymorphisms in barley (*Hordeum vulgare*). Plant Mol Biol 48:529–537

Kasha KJ, McLennan HA (1967) Trisomics in diploid alfalfa. I. Production, fertility, and transmission. Chromosoma 21:232–242

Kimbeng CA (1999) Genetic basis of crown rust resistance in perennial ryegrass, breeding strategies, and genetic variation among pathogen populations: a review. Aust J Exp Agric 39:361–378

King J, Armstead IP, Donnison IS, Thomas HM, Jones RN, Kearsey MJ, Roberts LA, Thomas A, Morgan WG, King IP (2002) Physical and genetic mapping in the grasses *Lolium perenne* and *Festuca pratensis*. Genetics 161:315–324

Kiss GB, Csanadi G, Kalman K, Kaló P, Okresz L (1993) Construction of a basic genetic map for alfalfa using RFLP, RAPD, isozyme, and morphological markers. Mol Gen Genet 238(1–2):129–137

Labombarda P, Busti A, Caceres ME, Pupilli F, Arcioni S (2002) An AFLP marker tightly linked to apomixis reveals hemizygosity in a portion of the apomixis-controlling locus in *Paspalum simplex*. Genome 45:513–519

Lancashire JA, Latch GCM (1966) Some effects of crown rust (*Puccinia coronata* Corda) on the growth of two ryegrass varieties in New Zealand. N Z J Agric Res 9:628–640

Lancashire JA, Latch GCM (1970) The effect of crown rust (*Puccinia coronata* Corda) on the yield and botanical composition of two ryegrass/white clover pastures. N Z J Agric Res 13:279–281

Lee RWH, Strommer J, Hodgins D, Shewen PE, Niu YQ, Lo RYC (2001) Towards development of an edible vaccine against bovine pneumonic pasteurellosis using transgenic white

clover expressing a *Mannheimia haemolytica* A1 leukotoxin 50 fusion protein. Infect Immun 69:5786–5793

Lesniewska A, Ponitka A, Slusarkiewicz-Jarzina A, Zwierzykowska E, Zwierzykowski Z, James AR, Thomas H, Humphreys MW (2001) Androgenesis from *Festuca pratensis* × *Lolium multiflorum* amphidiploid cultivars in order to select and stabilize rare gene combinations for grass breeding. Heredity 86:167–176

Li L, Qu R (2004) Development of highly regenerable callus lines and biolistic transformation of turf-type common bermudagrass (*Cynodon dactylon* [L.] Pers.). Plant Cell Rep 22:403–407

Litt M, Luty JA (1989) A hypervariable microsatellite revealed by *in vitro* amplification of a dinucleotide repeat within the cardiac muscle actin gene. Am J Hum Genet 44:397–401

Lundqvist A (1954) Studies on self-sterility in rye, *Secale cereale*. L. Hereditas 40:278–294

Maliepaard C, Alston FH, van Arkel G, Brown LM, Chevreau E, Dunemann F, Evans KM, Gardiner S, Guilford P, van Heusden AW, Janse J, Laurens F, Lynn JR, Manganaris AG, den Nijs APM, Periam N, Rikkerink E, Roche P, Ryder C, Sansavini S, Schmidt H, Tartarini S, Verhaegh JJ, Vrielink-van Ginkel M, King GJ (1998) Aligning male and female linkage maps of apple (*Malus primula* Mill) using multi-allelic markers. Theor Appl Genet 97:60–73

Meijer EGM, Ahloowalia BS (1981) Trisomics of ryegrass and their transmission. Theor Appl Genet 60:135–140

Michaud R, Lehman WF, Rumbaugh MD (1988) World distribution and historical development. In: Hanson AA, Barnes DK, Hill Jr RR (eds) Alfalfa and Alfalfa Improvement. ASA–CSSA–SSSA, Madison, WI, USA, pp 25–91

Miura Y, Ding C, Ozaki R, Hirata M, Fujimori M, Takahashi W, Cai HW, Mizuno K (2005) Development of EST-derived CAPS and AFLP markers linked to a gene for resistance to ryegrass blast (*Pyricularia* sp.) in Italian ryegrass (*Lolium multiflorum* Lam.) Theor Appl Genet 111:811–818

Molinari L, Busti A, Calderini O, Arcioni S, Pupilli F (2003) Plant regeneration from callus of apomictic and sexual lines of *Paspalum simplex* and RFLP analysis of regenerated plants. Plant Cell Rep 21:1040–1046

Monna L, Lin HX, Kojima S, Sasaki T, Yano M (2002) Genetic dissection of a genomic region for a quantitative trait locus, *Hd3*, into two loci, *Hd3a* and *Hd3b*, controlling heading date in rice. Theor Appl Genet 104:772–778

Moore BJ, Donnison IS, Harper JA, Armstead IP, King J, Thomas H, Jones RN, Jones TH, Thomas HM, Morgan WG, Thomas A, Ougham HJ, Huang L, Fentem T, Roberts LA, King IP (2005) Molecular tagging of a senescence gene by introgression mapping of a stay-green mutation from *Festuca pratensis*. New Phytol 165:801–806

Muntzing A (1956) Chromosome in relation to species differentiation and plant breeding. Conf Chrom Lect 6:161–197

Muylle H, Bockstaele EV Rolandan-Ruiz I (2003) Identification of four genomic regions involved in crown rust resistance in a *L. perenne* population. In: Proc Mol Breed of Forage and Turf 2003, 3rd Intl Symp, Dallas, USA

OECD (2005) List of varieties eligible for seed certification. http://www.oecd.org/agr/seed, accessed 23 April 2005. United Nations Organization for Economic Co-operation and Development, Paris, France

Ozias-Akins P, Roche D, Hanna WW (1998) Tight clustering and hemizygosity of apomixis-linked molecular markers in *Pennisetum squamulatum* implies genetic control of apospory by a divergent locus that may have no allelic form in sexual genotypes. Proc Natl Acad Sci USA 95:5127–5132

Plummer RM, Hall RL, Watt TA (1990) The influence of crown rust (*Puccinia coronata*) on tiller production and survival of perennial ryegrass (*Lolium perenne*) plants in simulated swards. Grass Forage Sci 45:9–16

Poehlman JM, Sleper DA (1995) Breeding Field Crops, 4th edn. Iowa State University Press, Ames

Porceddu A, Albertini E, Barcaccia G, Falistocco E, Falcinelli M (2002) Linkage mapping in apomictic and sexual Kentucky bluegrass (*Poa pratensis* L.) genotypes using a two way pseudo-testcross strategy based on AFLP and SAMPL markers. Theor Appl Genet 104:273–280

Potter LR (1987) Effect of crown rust on regrowth, competitive ability and nutritional quality of perennial and Italian ryegrasses. Plant Pathol 36:455–461

Prestidge RA, Gallagher RT (1988) Endophyte fungus confers resistance to ryegrass: Argentine stem weevil larval studies. Ecol Entomol 13:429–435

Pupilli F, Labombarda P, Caceres ME, Quarin CL, Arcioni S (2001) The chromosome segment related to apomixis in *Paspalum simplex* is homoeologous to the telomeric region of the long arm of rice chromosome 12. Mol Breed 8:53–61

Pupilli F, Martinez EJ, Busti A, Calderini O, Quarin CL, Arcioni S (2004) Comparative mapping reveals partial conservation of synteny at the apomixis locus in *Paspalum* spp. Mol Genet Genom 270:539–548

Quiros CF (1982) Tetrasomic segregation for multiple alleles in alfalfa. Genetics 101:117–127

Quiros CF, Morgan K (1981) Peroxidase and leucine-aminopeptidase in diploid *Medicago* species closely related to alfalfa: Multiple gene loci. Multiple allelism and linkage. Theor Appl Genet 60:221–228

Ray H, Yu M, Auser P, Blahut-Beatty L, McKersie B, Bowley S, Westcott N, Coulman B, Lloyd A, Gruber MY (2003) Expression of anthocyanins and proanthocyanidins after transformation of alfalfa with maize *Lc*. Plant Physiol 132:1448–1463

Reheul D and Ghesquiere A (1996) Breeding perennial ryegrass with better crown rust resistance. Plant Breed 115:465–469

Ritter E, Gebhardt C, Salamini F (1990) Estimation of recombination frequencies and construction of RFLP linkage maps in plants from crosses between heterozygous parents. Genetics 125:645–654

Roche D, Cong P, Chen Z, Hanna WW, Gustine DL, Sherwood RT, Ozias-Akins P (1999) Short Communication: An apospory-specific genomic region is conserved between Buffelgrass (*Cenchrus ciliaris* L.) and *Pennisetum squamulatum* Fresen. Plant J 19:203–208

Roche D, Conner JA, Budiman MA, Frisch D, Wing R, Hanna WW, Ozias-Akins P (2002) Construction of BAC libraries from two apomictic grasses to study the microcolinearity of their apospory-specific genomic regions. Theor Appl Genet 104:804–812

Roderick HW, Humphreys MO, Turner L, Armstead I, Thorogood D (2002) Isolate specific quantitative trait loci for resistance to crown rust in perennial ryegrass. In: Proceedings of 24th EUCARPIA Fodder Crops and Amenity Grasses Section Meeting, Braunschweig, Germany, pp 22–26109:783–791

Saha MC, Mian MAR, Eujayl I, Zwonitzer JC, Wang L, May GD (2004) Tall fescue EST-SSR markers with transferability across several grass species. Theor Appl Genet 109:783–791

Saha MC, Mian R, Zwonitzer JC, Chekhovskiy K, Hopkins AA (2005) An SSR- and AFLP- based genetic linkage map of tall fescue (*Festuca arundinacea* Schreb.). Theor Appl Genet 110:323–336

Sawbridge T, Ong EK, Binnion C, Emmerling M, McInnes R, et al. (2003) Generation and analysis of expressed sequence tags in perennial ryegrass (*Lolium perenne* L.). Plant Sci 165:1089–1100

Schmidt D (1980) La selection du ray-grass d'Italie pour la resistance a la rouille couronnee. Rech Agron Suisse 1971–1984

Segovia-Lerma A, Cantrell RG, Conway JM, Ray IM (2003) AFLP-based assessment of genetic diversity among nine alfalfa germplasms using bulk DNA templates. Genome 1:51–58

Shah MM, Luth D, Brummer EC, Council CL, Kunz RC (1999) Molecular mapping of QTLs for yield heterosis in tetraploid alfalfa. In: Plant and Animal Genome VII Conf, San Diego, CA, USA

Sim SC, Chang T, Curley J, Diesburg K, Nelson L, Jung G (2005a) Genetic Dissection of Genes Controlling Resistance to Crown Rust (*Puccinia coronata* f. sp. *lolii*) in Ryegrass. In: Plant, and Animal Genome XIII Conf, San Diego, CA, USA, P 334

Sim S, Chang T, Curley J, Warnke SE, Barker RE, Jung G (2005b) Chromosomal rearrangements differentiating the ryegrass genome from the Triticeae, oat, and rice genomes using common heterologous RFLP probes. Theor Appl Genet 110:1011–1019

Simons G, van der Lee T, Diergarde P, van Daelen R, Groendijk J, Frijters A, Buschges R, Hollicher K, Topsch S, Schulze-Lefert P, Salamini F, Zabeau M, Vos P (1997) AFLP-based fine mapping of the *Mlo* gene to a 30-kb DNA segment of the barley genome. Genomics 44:61–70

Skot L, Sackville Hamilton NR, Mizen S, Chorlton KH, Thomas ID (2002) Molecular genecology of temperature response in *Lolium perenne*: 2. association of AFLP markers with ecogeography. Mol Ecol 9:1865–1867

Smiley RW, Dernoeden PH, Clarke BB (1992) Compendium of Turfgrass Disease. 2nd ed. The American Phytopathological Society Press, St Paul, MN, USA

Somers DJ, Kirkpatrick R, Moniwa M, Walsh A (2003) Mining single nucleotide polymorphisms from hexaploid wheat ESTs. Genome 49:431–437

Stam P (1993) Construction of integrated genetic linkage maps by means of a new computer package, JoinMap. Plant J 3:739–744

Stanford EH (1959) The use of chromosome deficient plants in cytogenetic analyses of alfalfa. Agron J 51:470–472

Tabe LM, Wardley-Rich T, Ceriotti A, Aryan A, McNabb W, Moore A, Higgins TJV (1995) A biotechnological approach to improving the nutritive value of Alfalfa. J Anim Sci 73:2752–2759

Takahashi W, Oishi H, Ebina M, Takamizo T, Komatsu T (2002) Production of transgenic Italian ryegrass (*Lolium multiflorum* Lam.) via microprojectile bombardment of embryogenic calli. Plant Biotechnol 19(4):241–249

Tautz D, Trick M, Dover GA (1986) Cryptic simplicity in DNA is a major source of genetic variation. Nature 322:652–656

Tavoletti S, Veronesi F, Osborn TC (1996) RFLP linkage map of an Alfalfa meiotic mutant based on an F_1 population. J Hered 87:167–170

Tavoletti S, Pesaresi P, Barcaccia G, Albertini E, Veronesi F (2000) Mapping the *jp* (jumbo pollen) gene and QTLs involved in multinucleate microspore formation in diploid alfalfa. Theor Appl Genet 101:372–378

Tenaillon MI, Swakins MC, Long AD, Gaut RL, Doebley JF, Gaut BS (2001) Patterns of DNA sequence polymorphism along chromosome 1 of maize (*Zea mays* ssp. *mays* L.). Proc Natl Acad Sci USA 98:9161–9166

Thoquet P, Ghérardi M, Journet EP, Kereszt A, Ané JM, Prosperi JM, Huguet T (2002) The molecular genetic linkage map of the model legume *Medicago truncatula*: an essential tool for comparative legume genomics and the isolation of agronomically important genes. BMC Plant Biol 2(1):1

Thorogood D, Armstead I, Turner LB, Humphreys MO, Hayward MD (2004) Identification and mode of action of self-compatibility loci in *Lolium perenne* L. Heredity 94:356–363

Thorogood D, Kaiser WJ, Jones JG, Armstead I (2002) Self-incompatibility in ryegrass 12. Genotyping and mapping the S and Z loci of *Lolium perenne* L. Heredity 88:385–390

Ubi BE, Fujimori M, Mano Y, Komatsu T (2004) A genetic linkage map of rhodesgrass based on an F_1 pseudo-testcross population. Plant Breed 123:247–253

Van Deynze AE, Dubcovsky J, Gill KS, Nelson JC, Sorrells ME, Dvorak J, Gill BS, Lagudah ES, McCouch SR, Appels R (1995a) Molecular-genetic maps for chromosome 1 in Triticeae species and their relation to chromosomes in rice and oats. Genome 38:45–59

Van Deynze AE, Nelson JC, O'Donoughue LS, Ahn SN, Siripoonwiwat W, Harrington SE, Yglesias ES, Braga DP, McCouch SR, Sorrells ME (1995b) Comparative mapping in grasses. Oat relationships. Mol Gen Genet 249:349–356

Van Deynze AE, Nelson JC, Yglesias ES, Harrington SE, Braga DP, McCouch SR, Sorrells ME (1995c) Comparative mapping in grasses. Wheat relationships. Mol Gen Genet 248:744–754

Van Deynze AE, Sorrells ME, Park WD, Ayres NM, Fu H, Cartinhour SW, Paul E, McCouch SR (1998) Anchor probes for comparative mapping of grass genera. Theor Appl Genet 97:356–369

Viruel MA, Messeguer R, de Vicente MC, Mas JG, Puigdomenech P, Vargas F, Arús P (1995) A linkage map with RFLP and isozyme markers for almond. Theor Appl Genet 91:964–971

Vos P, Hogers R, Bleekers M, Reijans M, van der Lee T, Hornes M, Frijters A, Pot J, Peleman J, Kuiper M, Zabeau M (1995) AFLP: a new technique for DNA fingerprinting. Nucl Acids Res 23:4407–4414

Wang Z, Weber JL, Zhong G, Tanksley SD (1994) Survey of plant short tandem DNA repeats. Theor Appl Genet 88:1–6

Wang ZY, Scott M, Bell J, Hopkins A, Lehmann D (2003) Field performance of transgenic tall fescue (*Festuca arundinacea* Schreb.) plants and their progenies. Theor Appl Genet 107:406–412

Wanous MK, Gustafson JP (1995) A genetic map of rye chromosome 1R integrating RFLP and cytogenetic loci. Theor Appl Genet 91:720–726

Warnke SE, Barker RE, Jung G, Sim SC, RoufMian MA, Saha MC, Brilman LA, Dupal MP, Forster JW (2004) Genetic linkage mapping of an annual × perennial ryegrass population. Theor Appl Genet 109:294–304

Werner JE, Endo TR, Gill BS (1992) Toward a cytogenetically based physical map of the wheat genome. Proc Natl Acad Sci USA 89:11307–11311

Wilkins PW (1975) Inheritance of resistance to *Puccinia coronata* Corda and *Rhynchosporium orthosporum* Caldwell in Italian ryegrass. Euphytica 24:191-196

Wilkins PW (1978a) Specialisation of crown rust on highly and moderately resistant plants of perennial ryegrass. Ann Appl Biol 88:179–184

Wilkins PW (1978b) Specialization of crown rust (*Puccinia coronata* Corda) on clones of Italian ryegrass (*Lolium multiflorum* Lam.). Euphytica 27:837–841

Williams JGK, Kubelik AR, Livak KJ, Rafalski JA, Tingey SV (1990) DNA polymorphisms amplified by arbitrary primers are useful as genetic markers. Nucl Acids Res 18:6531–6535

Wu KK, Burnquist W, Sorrells ME, Tew TL, Moore PH, Tanksley SD (1992) The detection and estimation of linkage in polyploidy using single-dose restriction fragments. Theor Appl Genet 83:294–300

Xu WW, Sleper DA, Chao S (1995) Genome mapping of polyploid tall fescue (*Festuca arundinacea* Schreb.) with RFLP markers. Theor Appl Genet 91:947–955

Yamamoto T, Kuboki Y, Lin SY, Sasaki T, Yano M (1998) Fine mapping of quantitative trait loci *Hd-1*, *Hd-2* and *Hd-3*, controlling heading date of rice, as single Mendelian factors. Theor Appl Genet 97:37–44

Yaneshita M, Kaneko S, Sasakuma T (1999) Allotetraploidy of *Zoysia* species with $2n = 40$ based on a RFLP genetic map. Theor Appl Genet 98:751–756

Yates SG, Plattner RD, Garner GB (1985) Detection of ergopeptine alkaloids in endophyte infected, toxic Ky-31 tall fescue by mass spectrometry. J Agric Food Chem 33:719

Ye X, Wang ZY, Wu X, Potrykus I, Spangenberg G (1997) Transgenic Italian ryegrass (*Lolium multiflorum*) plants from microprojectile bombardment of embryogenic suspension cells. Plant Cell Rep 16:379–384

Yu GX, Wise RP (2000) An anchored AFLP and retrotransposon-based map of diploid *Avena*. Genome 43:736–749

3 Ornamentals

Marcus Linde[1], Zifu Yan[2], and Thomas Debener[1]

[1] Hannover Univerity, Institute of Plant Genetics, Department of Molecular Breeding, Herrenhäuser Str. 2, 30419 Hannover, Germany, *e-mail*: debener@genetik.uni-hannover.de

[2] Plant Research International, Wageningen University and Research Centre, P.O. Box 16, 6700 AA, Wageningen, The Netherlands

3.1 Introduction

Ornamentals are an important economical factor worldwide. Often underestimated, ornamental plant production even exceeds vegetable and fruit production in many countries of the world. However, scientific research on genetics and molecular biology has only been conducted on a few species. There are several reasons for this situation: The group of ornamental plants is very large, comprising hundreds of different species from various taxonomic groups. Therefore, even the most important species, for example, roses, carnations, and gerberas, have a smaller individual market share compared with the major agricultural crops, for example, corn, wheat, and soybean. Furthermore, many ornamental species are polyploids with characters that hamper genetic char-

acterization, for example, high heterozygosity, low seed set, and crossing barriers. As a consequence, the application of biotechnological strategies is mainly restricted to in vitro techniques, for example, in vitro propagation and molecular fingerprinting techniques distinguishing varieties and analyzing genetic diversity. Only in a few economically important species like roses, *Alstroemeria*, *Lilium*, carnations, and *Rhododendron* as well as in *Petunia* and *Antirrhinum* as model organisms, mapping experiments have been conducted (Debener 2002; Table 1).

In the following only the most extensive mapping activities carried out in roses as well as the important genetic models *Petunia* and *Antirrhinum* will be considered for a description of the work on marker maps followed by a general overview on the use of molecular markers for the description of genetic diversity in ornamentals.

Table 1. Ornamental crops for which mapping experiments have been conducted

Species	Maps constructed	Number of genes mapped	References
Rosa species	5	7 single genes 15 QTLs	Debener and Mattiesch (1999), Rajapakse et al. (2001), Crespel et al. (2002), Yan et al. (2005), Linde and Debener (2004), von Malek et al. (2000), Kaufmann et al. (2003), Dugo et al. (2005)
Rhododendron	1	2 QTLs for leaf chlorosis 2 QTLs for flower color	Dunemann et al. (1999)
Alstroemeria	1	1 single gene for anthocyanin pigmentation 2 QTLs for tepal spot numbers	Han et al. (2000, 2002a, b)
Petunia	3		Strommer et al. (2002)
Antirrhinum	1	164	Schwarz-Sommer et al. (2003)
Lilium	1	1 single gene and 2 QTLs	Abe et al. (2002)
Carnation	–	1 single gene	Scovel et al. (1998)

QTL quantitative trait locus

Genome Mapping and Molecular Breeding in Plants, Volume 6
Technical Crops
C. Kole (Ed.)
© Springer-Verlag Berlin Heidelberg 2007

3.2
Maps and Mapped Genes in Selected Ornamentals

3.2.1
Roses

Roses are among the economically most important and also oldest ornamental crops. The genus *Rosa* comprises more than 150 different species distributed over the temperate regions of the northern hemisphere (Wissemann 2003). Rose cultivation dates back to a time period around 3000 BC in China and Egypt (Gudin 2000). Modern cultivars are mostly tetraploids; ploidy levels of wild species range from diploid to octoploid (Gudin 2000) with a basic chromosome number of 7. About ten different wild rose species contributed to the genome of present cultivars, leading to a wide diversity among cultivated roses.

Nuclear DNA amounts vary from 0.78 pg/2C in some diploid species to 2.91 pg/2C in some pentaploid species (Ben-Sade and Samach 2003) with about the threefold to four fold size compared with the *Arabidopsis* genome. Considering modern rose breeding, one has to differentiate major rose groups (garden roses, cut roses, pot roses, and rootstocks) which differ significantly in respect to the breeding strategy, the breeding goals and the gene pools which are utilized for the selection of parents. Goals for past breeding programs have been mainly focused on ornamental characters, for example, flower color, scent and morphology, recurrent blooming, and plant habit. In recent years, goals like disease resistance against the major pathogens and pests and stress tolerance (mainly frost tolerance) in garden roses as well as productivity and vase life for cut roses and shelf life and plant habit for pot roses have become increasingly important. Owing to a high heterozygosity rose cultivars are vegetatively propagated either as scions on rootstocks or on their own roots. As breeders were able to select new varieties easily from first-generation hybrids and owing to the tetraploid nature, vegetative propagation, and heterozygosity of most cultivars genetic analyses are scarce in the genus *Rosa* (Debener and Mattiesch 1996; Debener 1999, 2003). Only a few characters like flower colors, flower morphology, the presence of prickles, recurrent blooming, disease resistance against black spot and powdery mildew have been characterized as single Mendelian genes. In ad-

dition, petal numbers, prickle numbers, and mildew resistance have been described as quantitative trait loci (QTLs; summarized in Debener 2003). Therefore, the first genetic maps were published only after diploid segregating populations and molecular markers were available (Debener and Mattiesch 1999).

First-Generation Maps

The first rose map was published in 1999 and was based on a population of 60 individuals (population 94/1) derived from an intercross between diploid heterozygous half sibs from an introgression program (Debener and Mattiesch 1996, 1999; Tables 1, 2). This was the first diploid recurrent flowering rose population resulting from a controlled cross between polymorphic parents that was utilized for genetic analyses. The parental plants (93/1-117, parent A, and 93/1-119, parent B) are recurrent flowering plants and carry several other characters from cultivated roses introgressed into the genetic background of a variety of the wild species *Rosa multiflora*. The parents were derived from the same female grandparent by open pollination and marker analyses showed that both were the result of pollinations by different fathers. Segregation analysis and map construction used the so-called double-pseudo-testcross strategy which is widely used for many highly heterozygous crops for which true F_2 populations cannot be obtained easily. Linkage groups 1, 2, 3, 4, 5, and 7 were linked by at least one marker segregating in both parents. However, the reliability of this approach was low as this type of marker was scored dominantly with a 3:1 segregation in a population of only 60 individuals. Clustering of markers was observed for almost all linkage groups and might be enhanced by the small population size. As both parents were shown to be self-incompatible (data not shown) it cannot be excluded that some chromosomal regions around the self-incompatibility locus might be underrepresented in the map. Along with the first map for roses two morphological characters, double flowers and pink flower color, also segregated in the mapping population and could be assigned to one linkage group each.

The second map published for roses was constructed on the basis of the segregation of 52 tetraploid F_2 plants (Rajapakse et al. 2001; Table 2). These were derived from a cross between a tetraploid, black spot resistant amphidiploid genotype (86-7, parent A) and a susceptible tetraploid genotype

Table 2. Marker maps constructed for roses

Publication	Mapping population	Types of markers	Mapped markers	Segregating markers with skewed segregation	Marker coverage	Average distance between markers
Debener and Mattiesch (1999)	60 diploid F$_1$ plants pseudo-double-testcross	AFLP: 126 RAPD: 179	Parent A: 119 markers on 7 linkage groups Parent B: 127 markers on 7 linkage groups	23% parent A 11% parent B 16% from both parents	326 cM parent A 370 cM parent B	2.4 cM parent A 2.6 cM parent B
Rajapakse et al. (2001)	52 tetraploid F$_2$ plants	AFLP: 675 SSR: 6 Isozyme: 1	Parent A: 171 markers on 15 linkage groups Parent B: 167 markers on 14 linkage groups	10% parent A 8% parent B	902 cM parent A 682 cM parent B	7.6–9.3 cM not further specified
Crespel et al. (2002)	91 diploid F$_1$ plants pseudo-double-testcross	AFLP: 288	Parent A: 68 markers on 8 linkage groups Parent B: 108 markers on 6 linkage groups	50% parent A 28.9% parent B	238 cM parent A 287 cM parent B	3.7 cM parent A 3.1 cM parent B
Yan et al. (2005)	88 diploid F$_1$ plants as in Debener and Mattiesch 1999	AFLP: 320[a] SSR: 74[a] PK[b]: 24[a] RGA[c]: 51[a]	Parent A: 271 Parent B: 273	Average of 15%	490 cM parent A 487 cM parent B	2.0 cM parent A 1.9 cM parent B
Dugo et al. (2005)	96 diploid F$_1$ plants pseudo-double-testcross	RAPD: 130 SSR: 6	Parent A: 50 markers on 7 linkage groups Parent B: 87 markers on 7 linkage groups	Average of 14.8%	260 cM parent A 388 cM parent B	5.8 cM parent A 5.6 cM parent B

AFLP amplified fragment length polymorphism, RAPD randomly amplified polymorphic DNA, SSR simple sequence repeat, PK protein kinase, RGA resistance gene analog

[a] New markers added to the existing data set

[b] Markers derived from degenerate polymerase chain reaction (PCR) with primers specific for PKs

[c] Markers derived from degenerate PCR with primers for nucleotide binding site leucine-rich repeat RGAs

(82-1134, parent B). The resulting hybrid was open-pollinated and among the 115 seedlings those being the result of true self-pollination were selected as the F_2 mapping population. A total of 675 amplified fragment length polymorphism (AFLP) markers, six simple sequence repeat (SSR) markers, and one isozyme marker were analyzed in the progeny and resulted in a total of 338 markers that could be placed onto the map. Only single-dose markers were used for mapping and could be located on 15 and 14 linkage groups for parents A and B, respectively. Owing to the small population size and the tetraploid nature of the plants no attempt was made to use markers segregating from both parents to link some of the linkage groups. The isozyme marker for malate dehydrogenase 2 could be scored as a codominant marker and was located on linkage group 4 of parent A. In addition, the authors tested 21 SSR primers from peach, apple, and sour cherry, eight of which allowed the amplification of distinct marker fragments. From these only two primer combinations from peach could be placed onto the map. However, these primers are very interesting candidate markers for comparative genomics approaches. In addition to molecular and isozyme markers, a gene controlling the presence of prickles on the petiole could be located on linkage group 7 of parent A.

A third map for roses was published by Crespel et al. (2002), who used 91 F_1 plants from a double-pseudo-testcross design. It is worth mentioning that one parent, the diploid line H190 (parent A), is a di-haploid genotype that resulted from the tetraploid cultivar Zambra by the irradiated pollen strategy (Meynet et al. 1994). The second parent is a genotype of the diploid species *R. wichurana* (parent B). Only AFLP markers were used and 64 primer combinations produced a total of 288 segregating markers (Table 2). From these, 68 could be mapped for parent A onto eight linkage groups and 108 were placed onto six linkage groups for parent B. Distorted segregation was observed for 50% of the markers from parent A and 29% of the markers from parent B. Especially the high distortion of markers from parent A exceeds the values from other rose-mapping projects and could be the result of disturbed gamete formation in the dihaploid plant which was observed in a number of plants from this project (Meynet, personal communication). The total map lengths are 238 and 287 cM for parents A and B, respectively (Table 2). Average distances between markers are 3.7 cM for parent A and 3.1 cM for parent B. In addition to the molecular markers,

two genes controlling double flowers and recurrent flowering as well as a QTL for the density of prickles on the stems were also placed onto the map.

Recently Dugo et al. (2005) published a map based on 96 diploid F_1 plants by a pseudo-double-testcross from the maternal parent Blush Noisette (D10) and the paternal *R. wichurana* (E15). The linkage maps were constructed separately for both parents using 14 isoenzymes, 169 randomly amplified polymorphic DNA (RAPD) markers, two SSRs, and two morphological markers. From these, about 15% showed a skewed segregation and were excluded from mapping. Finally 133 of the 148 markers could be mapped to 14 linkage groups using a logarithm of odds (LOD) threshold of 5.0. The authors observed a distorted segregation for 14.8% of all markers on an average. A total map length of 388 cM was calculated for the parent Blush Noisette, whereas the map for the *R. wichurana* parent spans 260 cM. Because only 4.5% of all 133 markers were shared between both parents no integrated map could be calculated. Four of these linkage groups contained only five or fewer markers. The QTL analysis of the four quantitative traits (flower size, days to flowering, leaf size, and resistance to natural powdery mildew infections in the field) yielded 13 putative loci with LOD>3.0. Two significant QTLs, one from each parent, were detected for the powdery mildew resistance and five for the leaf size. Also two loci influencing the days to flowering were mapped and four QTLs for flower size.

Second-Generation Map

On the basis of the mapping information of Debener and Mattiesch (1999), Debener et al. (2001) and Yan et al. (2005) extended the existing linkage map by mapping 469 new markers in a population of 88 F_1 plants of the population 94/1 (Fig. 1, Table 2). In addition to the already mapped markers this work produced a total of 520 markers, of which 271 were placed onto the map of parent A and 273 onto the map of parent B. With total map lengths of 490 and 487 cM for parents A and B and average marker distances of 2.0 and 1.9 cM, this map significantly improves the map from Debener and Mattiesch (1999) and shows the highest marker density of all rose maps. Codominantly scored AFLP and SSR markers were used to align the parental maps as well as integrated map with the map from Debener and Mattiesch (1999). The integrated map has a total

94-1 B3

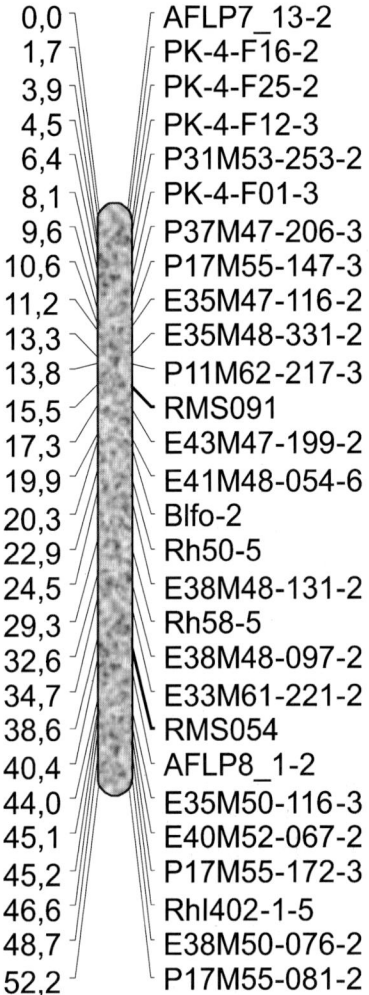

0,0	AFLP7_13-2
1,7	PK-4-F16-2
3,9	PK-4-F25-2
4,5	PK-4-F12-3
6,4	P31M53-253-2
8,1	PK-4-F01-3
9,6	P37M47-206-3
10,6	P17M55-147-3
11,2	E35M47-116-2
13,3	E35M48-331-2
13,8	P11M62-217-3
15,5	RMS091
17,3	E43M47-199-2
19,9	E41M48-054-6
20,3	Blfo-2
22,9	Rh50-5
24,5	E38M48-131-2
29,3	Rh58-5
32,6	E38M48-097-2
34,7	E33M61-221-2
38,6	RMS054
40,4	AFLP8_1-2
44,0	E35M50-116-3
45,1	E40M52-067-2
45,2	P17M55-172-3
46,6	Rhl402-1-5
48,7	E38M50-076-2
52,2	P17M55-081-2

Fig. 1. Group B3 of the linkage map constructed by Yan et al. (2005) containing the locus controlling double flowers (*Blfo*) by Debener and Mattiesch (1999) enriched with additional simple sequence repeats

map length of 545 cM. As estimated by the authors, this map covers about 90% of the rose genome and is, therefore, the most advanced rose map currently available. In addition, other characteristics make this map an interesting general resource for rose genetics:

1. On the map, 74 polymorphic SSR markers are placed, 26 of which could be scored codominantly. These markers are very interesting for an alignment of different rose maps. A set of 21 SSR markers is well spread across the seven chromosomes of the maps, which can be used as anchor points independently of the type of population.

2. Polymerase chain reaction (PCR) products resulting from amplifications with primers specific for protein kinases and nucleotide binding site leucine-rich repeat sequences were placed on the map and are interesting candidate sequences for resistance gene loci. Some of these markers were also scored codominantly and contribute to the set of markers of general interest.

3. The mapping population 94/1 has been extended to a total of 400 individuals and partially screened for morphological markers. It has already been provided to other research laboratories and could be used as a reference population for rose genetics and marker localization. Furthermore, work to map additional SSRs and other loci of interest is in progress, which is continuously increasing the number of markers on the reference map.

Mapped Single Mendelian Genes and QTLs in Roses

A total of eight genes (Table 3) and 17 QTLs have been mapped by different research groups with the application of molecular markers in rose genetic studies.

The gene for double flowers (*Blfo*) was characterized in the mapping population by Debener and Mattiesch (1999). It is a dominant gene that causes the developmental transition of stamens to petals so that double flowers have more than 15 petals in contrast to five petals found in most wild roses. It was mapped in 60 F_1 plants between markers AFLP1_1 (3.5 cM) and AFLP10_5 (15.3 cM) and without any recombination to marker P483/493_2 (0 cM).

The gene *Blfa* controls pink flower color and is inherited codominantly in the same population as *Blfo* (Debener and Mattiesch 1999). It is linked to markers P507/513_3 (1.9 cM), AFLP12_14 (5.8 cM), and P511/513, (0 cM) on linkage group 2 of parent A. In addition, it was linked to markers OPD 3, AFLP1_3, AFLP2_2, AFLP2_3, AFLP5_11, OPD 18_1, and P469/471 all at 0 cM and to markers P507/513_3 (1.9 cM) and AFLP12_14 (5.8 cM) from linkage group 2 of parent B.

Rdr1 is a dominant gene conferring resistance to rose black spot and was first characterized in tetraploid populations (von Malek and Debener 1998). It was mapped in three tetraploid populations with AFLP markers at distances of 1.1 and 7.6 cM.

Table 3. Gene loci mapped in roses

Gene	Function	Closest markers	Closest marker	References
Blfo	Transition of stamens to petals, double flowers	P483/493_2 (0 cM)	AFLP 1_1 (3.5 cM)	Debener and Mattiesch (1999)
Blfa	Pink flower color	P511/513 (0 cM)	OPD 3, AFLP1_3 (0 cM)	Debener and Mattiesch (1999)
Rdr1	Resistance to rose black spot	Hind-AAT/Mse-GGC (0.18 cM)	20T, 55T (0.18 cM)	Von Malek et al. (2000), Kaufmann et al. (2003)
Rpp1	Resistance to rose powdery mildew	H-CAg/M-ATT (1.7 cM)	H-CAT/M-ggC1 (3.4 cM)	Linde et al. (2004)
prickles	Prickles on the petioles	EACAMCAA16 (17.6 cM)	–	Rajapakse et al. (2001)
r4	Recurrent blooming	E4M1.21, E2M1.02 (0 cM)	E2M1.05, E2M5.05 (0 cM)	Crespel et al. (2002)
d6	Double flowers	E3M4.07 (10.9 cM)	E2M5.0 (4.7 cM)	Crespel et al. (2002)
Blfo	Double flowers	OPA06_826 (4.0 cM)	OPA11_1175 (9.0 cM)	Dugo et al. (2005)

The closest marker was converted into a sequence-characterized amplified region (SCAR) marker that mapped at 0.75 cM from the cloned fragment. The cloned fragment was used as a restriction fragment length polymorphism (RFLP) probe on the mapping population 94/1 to locate the gene on linkage group 1 of the rose map.

With use of this cloned marker, bacterial artificial chromosome clones from a library made of *R. rugosa* nuclear DNA and three additional diploid segregating populations with a total number of 538 diploid plants, *Rdr1* was mapped to several markers without any recombinants and between the markers 20T, *Hind*-AAT/*Mse*-GGC (0.18 cM), and 55T (0.18 cM) (Kaufmann et al. 2003). This fine map was then used to construct a contig spanning the whole locus which is now used to attempt the positional cloning of *Rdr1*.

Rpp1 is a gene conferring resistance to powdery mildew of roses (Linde et al. 2004). This gene was mapped using 117 diploid roses of the population 97/9. This population originated from a cross of the diploid line 88/124-46 (resistant against the powdery mildew isolate 9) and the susceptible diploid male parent 82/78-1 (Linde and Debener 2003). Both genotypes are open-pollinated seedlings from a breeding program aimed at the introgression of genes from tetraploid garden roses into *R. multiflora* (Reimann-Philipp 1981). The resulting F_1-hybrid 95/13-90 was backcrossed to 82/78-1, resulting in a population of 117 plants segregating for resistance to the physiological race 9. Use of a total of 260 AFLP primer combinations in a bulked segregant analysis allowed the putative resistance gene *Rpp1* to be located at an interval of 5 cM (Fig. 2) between the AFLP markers H-CAT/M-ggC1 (3.4 cM) and H-CAg/M-ATT (1.7 cM) which was converted into a SCAR marker with the same segregation pattern. Indirect mapping of *Rpp1* in relation to the black spot resistance gene *Rdr1* using the AFLP marker H-CAT/M-ggC1 which segregated in both populations (97/9 and 94/1) with fragments of identical size on the same gel revealed no linkage between the two R-genes. Use of JoinMap 2.0 (Stam and Van Ooijen 1995) allowed the fragment to be located on linkage group A3 on the map from Debener and Mattiesch (1999), whereas the black spot resistance gene *Rdr1* is located on linkage group B1 (Fig. 2).

A single recessive gene for the absence of prickles on the petioles (simply called "prickles") was characterized (Rajapakse et al. 2001), and was located at the end of linkage group 7 of parent A at a distance of 17.6 cM to marker EACAMCAA16.

The gene *r4* controls recurrent blooming and is recessive as shown by various research groups. Wild roses flower only once a year around springtime for a period of about 8–10 weeks. Recurrent blooming cultivars flower throughout the whole season. How this gene has been introgressed into all modern rose cultivars is also of scientific and practical interest. Crespel et al. (2002) observed single Mendelian segregation in their mapping population and could map this gene in relation to several AFLP markers (E4M1.21, E2M1.02, E2M1.05, E2M5.05) without recombinants and several at distances of 1.1–3.1 cM on linkage group B4.

The gene *d6* controls double flowers in the mapping population of Crespel et al. (2002) as a dominant

Fig. 2. Linkage maps for *Rpp1* in the BC$_1$ population 97/9 segregating for PM race 9 (*left*) and indirect mapping of this gene in the population 94/1 using amplified fragment length polymorphism H-CAT/M-ggC-1 (*right*)

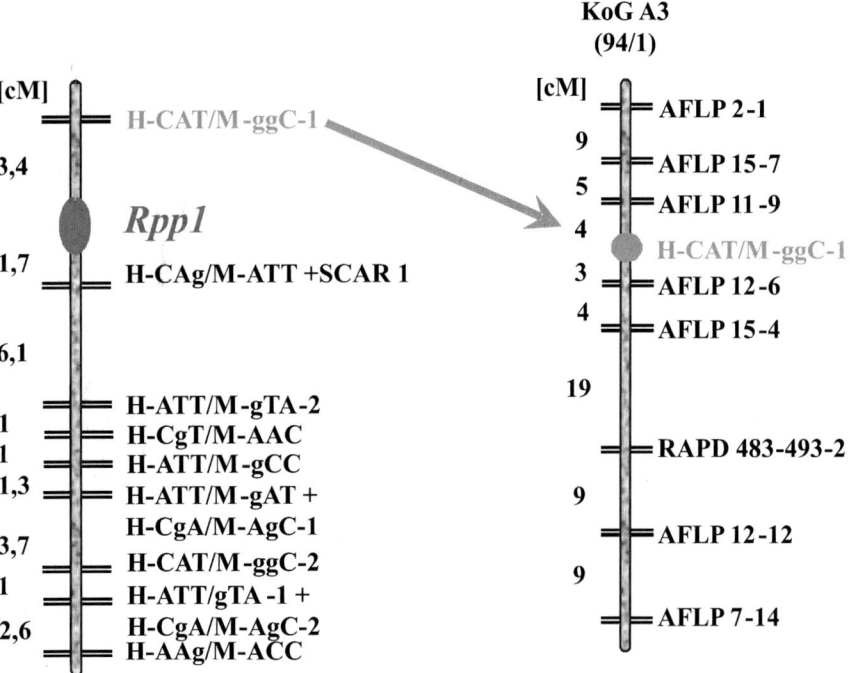

trait. It was mapped on group A6 between markers E3M4.07 (10.9 cM) and E2M5.0 (4.7 cM). Dugo et al. (2005) also mapped a qualitative locus controlling simple flowers with five petals to double flowers in a population of 96 F$_1$ plants. This locus was located on linkage group D10-2 (Blush Noisette) between the RAPD markers OPA06_826 (4.0 cM) and OPA11_1175 (9.0 cM). However, the question of whether these genes are identical to the gene characterized by Debener and Mattiesch (1999) is still open as markers have not been exchanged between these populations yet.

The gene *t4* is a QTL controlling the number of prickles on the stems. It was characterized in the mapping population of Crespel et al. (2002) and is responsible for 66.4% of the observed phenotypic variation. The peak of the LOD score (LOD = 36.4) is located between the markers D4M1.01 and E5M8.02 on linkage group B4 of the map of parent B. In addition, a minor QTL *t4b* explains 13.8% of the phenotypic variation (LOD = 2.9) and is located on marker E5M7.11 at a distance of 39.7 and 3.2 cM from *t4* and *r4*, respectively.

Altogether 13 QTLs were detected in the work of Dugo et al. (2005). Using a threshold of LOD>3.0, they mapped two QTLs for powdery mildew resistance on linkage groups D10-3 and E15-4, explaining together around 80% of phenotypic variation. For the trait leaf size they calculated two QTLs coming from the D10 parent and three from the E15 parent. The flower-

ing time (days to flower) was controlled by two QTLs on the D10 map, the early flowering genotype. The size of the flowers was inherited by two QTLs, each on linkage groups 1 and 2, on both parental linkage maps.

3.2.2
Snapdragon

Antirrhinum május L. is used as one of the most important model systems in developmental plant genetics. Recently a molecular map was established using an F$_2$ hybrid population (Schwarz-Sommer et al. 2003).

A. május L. is one of the 40–50 species of the genus *Antirrhinum* (Veronicaceae; Olmstead et al. 2001), which are distributed over the temperate zones of the northern hemisphere. Originating from the western Mediterranean peninsula (Sutton 1988; Horn 1996), the perennial species were cultivated already in ancient times. Within the genus a considerable variation in the phenotype and habitat of these small shrubs and herbs is found. All species behave as diploids with a chromosome number of $2n = 16$. The establishment of the cultivars was based on the selection on flower color and plant growth mutants. The first F$_1$ hybrid cultivars were introduced into the market in the late 1940s. Mainly used as a cut flower, the economic im-

Table 4. Maps constructed for *Antirrhinum* and *Petunia*

Species	Population type	2n	Population size	Marker number	Map length (cM)	Marker types	Marker interval (cM)	Largest gap (cM)	References
Antirrhinum	F₁	$2x = 16$	–[a]	57	420	Morphological	7.3	–	Stubbe (1966)
Antirrhinum	F₂	$2x = 16$	92	243	613	AFLP, CAPS, ISTR, ISSR, RFLP	2.5	19	Schwarz-Sommer et al. (2003)
Petunia	BC₁	$2x = 14$	100	35	263	RAPD, morphological	8.2	27	Peltier et al. (1994)
Petunia	BC₁	$2x = 14$	32, 41, 79	38	368	RFLP, biochemical, morphological	9.7	63	Strommer et al. (2000)
Petunia	BC₁	$2x = 14$	175	163	332	RFLP, AFLP	2.0	23	Strommer et al. (2002)
Petunia	BC₁	$2x = 14$	61	228	261	RFLP, AFLP	1.1	24	Strommer et al. (2002)

CAPS cleaved amplified polymorphism sequence, *ISTR* inverse sequence-tagged repeat, *ISSR* intersimple sequence repeat, *RFLP* restriction fragment length polymorphism

[a] Population size could not be given because the map resulted from a large number of crossing experiments and linkage studies beginning in 1911 and ending with a map combining all these results in 1966.

portance with about 20 million stems sold yearly only in the Netherlands is substantial (Horn 1996). Used as a model plant for genetics, more than 40 genes and the corresponding mutations have been characterized up to now, including floral meristem and organ identity genes as well as factors affecting the asymmetry of lateral organs and flowers which are dependent on two closely related genes, *CYCLOIDEA* (*CYC*) and *DICHOTOMA* (*DICH*) (Coen et al. 1990; Waites et al. 1998; Schwarz-Sommer et al. 2003). The majority of these mutations have been caused by two families of mutagenic transposons (Sommer et al. 1990; Kunze et al. 1997).

A classical map of *A. majus* with 57 morphological markers in eight linkage groups spanning about 420 cM was developed by Stubbe in 1966, resulting from the experiments of Kuckuck, Schick, Sampson and others in the 1930s to 1950s with crosses of mutants (Stubbe 1966; Table 4). Use of morphological characters of the chromosomes and analyses of deletions allowed two linkage groups to be assigned to the chromosomes (Pohlendt 1942a, b).

The recent map of Schwarz-Sommer et al. (2003) was produced using a segregating F₂ population of *A. majus* and *A. molle* (Table 4). The female parent was derived by repeated self-pollination over more then ten generations. The self-incompatible *A. molle* parent was derived from an accession from Catalonia (Spain)

and maintained by cross-pollination. The *A. molle* parent was found to be heterozygous at many loci. Ninety-two genotypes from the F₂s were randomly selected for the analysis.

The molecular map was constructed using 418 markers defining a minimum of 243 different loci. The 108 codominant and 36 dominant RFLP markers were mostly detected with expressed sequence tags (ESTs) selected at random from a complementary DNA (cDNA) library in λ-phage. Twenty-two of the RFLP markers represent the coding or 5′regions of genes with known mutant phenotypes and six transposon insertion sites. Codominant cleaved amplified polymorphism sequences were detected on agarose gels at 18 loci and dominant ones at 2 loci. Copia/Ty-like retrotransposons were amplified as inverse sequence-tagged repeats with two primer combinations, generating 25 dominant fragments. With three AFLP primer combinations 99 dominant markers could be produced. An additional four markers were obtained using a single intersimple sequence repeat primer. Significant ($\alpha = 0.05$; χ^2 tests) deviations from Mendelian segregation ratios were observed for all marker types. Distorted ratios were found for 60% of the codominant markers and 38% of the dominant ones.

The map was constructed first using only the codominant markers. From these data a core map was built using the MAPRF program (Ritter and Salamini 1996) implementing the EM algorithm and multipoint linkage analysis as described by Ritter et al. (1990). This map with 16 linkage groups was used in fixed order to integrate the remaining dominant markers. Each pair of homologous linkage groups was projected onto one combined group using allelic bridges (Ritter and Salamini 1996). Altogether 50 markers showing ambiguous associations to mapped loci or representing identical loci were excluded from the map. The resulting map of Schwarz-Sommer et al. (2003) contains 243 markers (125 codominant) with a total map length of 613 cM, which is similar to the *A. thaliana* map of Lister and Dean (1993) with 690 cM and a similar number of markers. The average distance between two markers is 2.5 cM, with no interval larger than 20 cM. Except for linkage groups 5 and 8, all other groups contain at least one MADS-box gene mapped as RFLP. Closely located to all of these mapped genes are other, mostly codominant, RFLP markers which could be useful in marker-assisted selection. On the presence of common markers with the classical map by Stubbe (1966) the authors could align LG1 to the LUV chromosome, LG3 with *PAL* to the UNI chromosome, LG4 with *DEL* and *CEN* to the GRAM chromosome. LG5 with *DEF*, *NIV*, and *PHAN* could represent the AUR chromosome and LG6 with *DEF* the DEFICIENS chromosome.

In contrast to the classical map, *GLO* and *CEN* map on LG1 and LG4, respectively, and not together on the GRAM chromosome as assumed so far. Whereas *DEF* and *RAD* map together on LG6 in the map of Schwarz-Sommer et al., they are located on different chromosomes of the classical map. This first molecular map of *Antirrhinum* should facilitate the further mapping of mutations, QTLs and other coding sequences in this model genus. It should also be a starting point for further studies on the syntheny in gene order among dicotyl model species such as *Antirrhinum, Arabidopsis*, and others.

3.2.3
Petunia

Petunia hybrida Vilm is a horticulturally important dicot Solanaceae with seven chromosomes in the haploid complement. The origin of the artificial hybrid is unclear; it is thought to be a hybrid between a white flowered species (*P. axillaries* or *P. parodii*) and a pigmented one (*P. violaceae, P. integrifolia*, or *P. inflate*) (Röber et al. 1996; Strommer et al. 2002). Most of the 25 perennial petunia species originated in South America. It is a relatively fast breeding hybrid which is primarily used as an annual bedding plant but can be maintained indefinitely in the greenhouse. Commercial breeding has produced hundreds of varieties with defined genes, owing to the straightforward crossing strategies leading to hundreds of seeds per cross. Being a close relative to tobacco and tomato, petunias are also an attractive subject for transgenic studies and transformation (Strommer et al. 2000, 2002). The moderate genome size of 1.2×10^6 kb (Arumuganathan and Earle 1991) and only seven basic chromosomes make petunia an interesting system for physical and genetic mapping.

A classical genetic map of *P. hybrida* has existed since 1993 with about 120 phenotypic markers and altogether 134 loci (Gerats et al. 1993; Table 4), in which the 74 petunia genes described by Wiering et al. (1979) were integrated. Also some years earlier, an RFLP map was used to map seven complete members of the chalcone synthase multigene family which could be placed into two subfamilies on chromosomes 2 and 5 in 1987 (Koes et al. 1987). McLean et al. (1990) presented an RFLP map with members of the actin gene family. In 1994, Peltier et al. presented an integrated RAPD map from a BC$_1$ population of 100 individuals with 35 markers from potential ancestors of *P. hybrida*. Using Mapmaker v2.0, they constructed a linkage map in which they could assign six linkage groups to chromosomes with phenotypic markers that had already been mapped by Cornu et al. (1990) (in Peltier et al. 1994). One large linkage group with eight RAPD markers could not be assigned to chromosomes, owing to no significant linkage to any morphological marker.

Strommer et al. (2000) presented a gene-based RFLP map of petunia on three segregating populations, in which they merged 27 gene loci with 11 previously mapped morphological and biochemical markers. Three BC$_1$ populations with 32, 41, and 79 individuals were made by crossing a hybrid of V23×R51 with two varieties, V23 and R51, and the doubled haploid *P. mitchel*. Twenty-seven genes were used as probes in RFLP mapping and merged with the 11 markers from McLean et al. (1990) with three-point analysis. In this mapping project, Strommer et al. (2000) included members of six gene families: actin, chalcone synthase, alcohol dehydrogenase, aminocyclopropane-1-carboxylate oxidase, aminocyclopropane-1-carboxylic acid synthase, and ribosomal DNA (rDNA). The total map length is 368 cM, showing no general correlation between the number of loci per chromosome or total length and physical chromosome length. The average crossover frequency was calculated to be only about one quarter of the expected one from cytological examinations. In this map the previous positions for the actin and chalcone synthase loci could be verified with minor changes owing to the larger mapping population. Variation in the position for an rDNA locus on chromosome 2 was recorded, which was previously mapped to chromosome 2 by fluorescent in situ hybridization (Franzs et al. 1996). A shift in map position, possibly due to different mapping populations, was also recorded for the *adh2* locus on chromosome 2 near *prx*A with six recombinants among 66 individuals. This locus was assigned previously to chromosome 4 near the *blind* locus by Gregerson et al. (1993) with eight recombinants out of 85 individuals.

As a future scope of their work the authors stated the generation of an AFLP linkage map with a high marker number, which was published 2 years later by Strommer et al. (2002). The authors used the same BC$_1$ populations and the same DNA samples as in Strommer et al. (2000). The two VR backcrosses with V23 and R51 consisting of 48 individuals were merged with the 79 individuals from the backcross of the doubled haploid *P. mitchel* to the VRM population for linkage mapping. An additional map was created for a fourth backcross population (W137) with 61 individuals. Both maps were constructed using AFLPs and RFLPs from the maps of McLean et al. (1990) and Strommer et al. (2000) for the six gene families. Also integrated were RFLPs for *apetala-2* from Maes (in Strommer et al. 2002) and *pFlo* from Sauer and Gerats (unpublished data).

With the minor exception of the inversion of an RFLP cluster on chromosome 4 and the repositioning of two rDNA loci on chromosome 3, the data of the VRM map are in agreement with the map of Strommer et al. (2000). A high degree of locus clustering was detected in the VRM and the W137 maps. On chromosome 5 of the VRM map, 26 of 31 markers were clustered within 15 cM and on chromosome 2, 28 out of 36 markers within 17 cM. In the W137 population, clustering of markers was even higher.

3.3
Variety and Genotype Identification

Variety identification is an important area of applied research in ornamentals owing to the relatively high value of individual plants compared with agricultural crops. The conventional identification of varieties based on their morphological and physiological characteristics is becoming insufficient owing to the lack of precision and subjectivity of the observations, the increase of the number of varieties to be tested and the limitation of the genetic distances between varieties. Molecular markers provide information about the difference between individuals of unknown pedigree and have clear advantages both in accuracy and in speed for genotypic identification in ornamental breeding, variety registration and protection. The studies on variety identification in ornamentals have been performed in more than 40 species by using different molecular marker systems. The methods of variety identification are different depending on the varying propagations of crops.

3.3.1
Vegetatively Propagated Crops

The annual release of newly developed varieties of vegetatively propagated crops like rose, carnation, tulip, and chrysanthemum can be hundreds to thousands. At the same time, infringements of protected varieties with vegetative multiplication frequently happen. Compared with the identification of sexually propagated crops, the identification of the vegetatively propagated varieties with molecular markers should be easier since the plants belonging to the same clone are expected to display the same marker profile. For example, Esselink et al. (2003) differentiated 46 rose hybrid tea varieties and 30 rootstock varieties with as few as three sequence-tagged microsatellite site markers.

3.3.2
Sexually Propagated Crops

For sexually propagated crops the picture is more complicated since each variety consists of a population of genetically distinct individuals. In these cases markers can be used to differentiate varieties via distinctive marker allele combinations (Iqbal and Rayburn 1994) or the differentiation of populations can be based on allele frequencies (Huff et al. 1993). This, however, cannot rule out misinterpretations for varieties of small sample sizes. In each case the discriminative power of molecular markers is much smaller than in vegetatively propagated plants.

3.3.3
Sports

Sports are the result of spontaneous and artificially induced mutation. Owing to small changes in chromosomes due to mutation, an interesting question connected to the problem of variety identification is whether the sports could be distinguished from their ancestral variety at the DNA level. Recent changes in breeder's rights within the EU define sport varieties as "essentially derived" varieties and concede part of the royalties from these to the breeders of the ancestral variety (EU regulation no. 2100/94). Therefore, the clear assignment of sports to their ancestral varieties is of immense economic importance for breeders of vegetatively propagated ornamentals.

Sports and their ancestral varieties have been investigated, for example, in *Alstroemeria* (Anastassopolous and Keil 1996), chrysanthemum (Trigiano et al. 1998), *Pelargonium* (Becher et al. 2000), poinsettia (Starman et al. 1999), and roses (Debener et al. 2000; Zhang et al. 2000; Esselink et al. 2003). It is not easy to distinguish sports from their ancestral varieties. In most cases, however, the low level of polymorphisms in the sports still allowed the assignment to the ancestral variety compared with the case for sexually propagated progenies (Becher et al. 2000; Debener et al. 2000; Esselink et al. 2003). This demonstrates the value of fingerprinting techniques with molecular markers as a new strategy for breeders to claim their rights to these "derived varieties".

3.4
Analysis of Genetic Relatedness

Knowledge of genetic relationships in crops is important for genetic resource conservation, plant breeding, variety protection, and genetic evaluation. Morpho-logical and pedigree estimates of genetic relatedness among genotypes are not precise owing to use of agronomic traits as parameters and the effects of the environments in which plants are cultivated. Molecular markers not only are used to distinguish genotypes but also provide information about the genetic relatedness between genotypes and even phylogenetic relationships between species and related taxa. Marker profiling gives an indication that the closer the genetic relationship between a given pair of genotypes is, the larger will be the number of shared markers.

3.4.1
Methods of Classification

The best numerical classification strategy is the one that produces the most compact and well-separated groups; therefore, the degree of association is strong between members of the same group and weak between members of different groups. Different strategies may be employed to infer the genetic similarities or phylogenetic relationships (Staub et al. 1996). In most cases markers are treated as individual characters and their presence or absence is recorded without weighing the different marker fragments under the assumption that markers of identical size (e.g., RFLP or AFLP fragments) represent homologous DNA fragments and that the genetic relationships of the taxa under investigation is not disturbed by processes summarized under the term "reticulate evolution" (e.g., sexual hybridization of taxa or "horizontal gene flow").

Two basic strategies are widely applied. The first one is the so-called maximum parsimony method (Felsenstein 1988) that uses the marker fragment matrix directly to compute a dendrogram with a minimum of character differences between the individual fragment patterns. The second is fragment matrices that are first converted into a matrix of pairwise values for genetic distance or similarity. This can be done by a variety of different algorithms (Rohlf 1989). From the similarity matrix a dendrogram can be constructed. Different ways may be employed, for example, unweighted pair group method with arithmetic mean (UPGMA; Felsenstein 1988) and the neighbor-joining method (Saitou and Nei 1987). As an alternative to dendrograms the genetic distance between genotypes may also be displayed in three-dimensional representations using principal component analyses (Rohlf 1989).

3.4.2
Studies of Genetic Relatedness of Genotypes

For both general strategies and numerous available methods, the conditions are rarely met with most real sets of data (Avise 1994). A major violation of the underlying conditions for all methods is the presence of hybrid genotypes in the datasets. However, pairwise distances inferred from fragment data and dendrogram computations with UPGMA clustering turned out to be relatively robust against inconsistencies in the datasets and are therefore the strategy most widely applied for the inference of genetic relationships between ornamental plant varieties and species. A selection of the latest studies on genetic distances between ornamental plant varieties and species is listed in Table 5.

The information about the genetic distance between ornamental plant genotypes or species may be applied in several ways. In the case of phyloge-

netic information about the relationships between different ornamental taxa, the data may be used to infer the evolutionary history of the cultivated varieties. Apart from the increase in basic knowledge, this information may be useful for the exploitation of exotic germplasm in which either hybrid ornamental species are resynthesized or most closely related species are selected for the introgression of horticulturally interesting traits. Information on the genetic diversity within the gene pool of cultivated varieties may be used to select the most distantly related parents in order to minimize inbreeding depression. This may be of particular importance for groups of ornamental varieties like roses, which, over the last decades, have been bred only within a limited gene pool.

Some computer programs are available for processing of marker data and the computation of dendrograms and phylogenetic trees (Rohlf 1989; Felsenstein 1988). A useful overview of available soft-

Table 5. Selection of ornamentals for which genetic distances or phylogenetic relationships were investigated with molecular markers

Genus	Marker type	Method	References
Aglaonema	AFLP	UPGMA	Chen et al. (2004a)
Alstroemeria	AFLP	NJM	Han et al. (2000)
Aster	AFLP	UPGMA	Cammareri et al. (2004)
Calladium	AFLP	UPGMA	Loh et al. (1999)
Campanula	RAPD	NJM	Joung et al. (2001)
Dieffenbachia	AFLP	UPGMA	Chen et al. (2004b)
Echinacea	AFLP	NJM	Kim et al. (2004)
Euphorbia (Poinsettis)	RAPD	UPGMA, PCA	Ling et al. (1997)
Freesia	RAPD	UPGMA	Wongchaochant et al. (2002)
Gladiolus	RAPD	UPGMA	Takatsu et al. (2001)
Iris	RAPD	NJM	Kim et al. (2003)
Juniperus	RAPD	PCA	Le Duc et al. (1999)
Lachenalia	RAPD	UPGMA	Kleynhans and Spies (2000)
Lilium	RAPD	NJM	Kim and Ahn (2002)
Limonnium	RAPD, AFLP	NJM	Palacios et al. (1999)
Narcissus	RAPD	UPGMA	Jiao et al. (2003)
Osteospermum	RAPD; AFLP	UPGMA	Berio et al. (2001)
Penstemon	SSR	PCA	Wolfe et al. (1998)
Petunia	DAF	MPM	Cerny et al. (1996)
Phalaenopsis	RAPD	UPGMA	Been et al. (2002)
Rhododendron	AFLP	UPGMA	Riek et al. (2000)
Rosa	RAPD; AFLP	UPGMA	Debener et al. (1996), Leus et al. (2004)
Viola	RAPD; SSR	MPM, NJM	Ko et al. (1998), Yockteng et al. (2003)

DAF DNA amplification fingerprinting, *UPGMA* unweighted paired group method with arithmetic mean, *NJM* neighbour-joining method, *PCA* principal component analysis, *MPM* maximum parsimony method

ware packages can be found via the Internet: http://evolution.genetics.washington.edu/phylip.html; http://corba.ebi.ac.uk/Biocatalog/Phylogeny.html; http://linkage.rockefeller.edu/soft/list.html

3.5
Future Prospects

To a large extent, marker analyses in ornamental crops still rely on the application of dominant marker systems, for example, RAPDs and AFLPs for which no information about target sequences and locations of individual markers are required. However, several trends have already become visible which parallel those in the more advanced agricultural crops:

- Dominant markers will be increasingly complemented and even replaced by locus-specific and codominant marker types. To date this is almost exclusively done by means of microsatellite markers (Yan et al. 2005). However, as the number of annotated DNA sequences deposited in databases is also increasing at exponential rates for ornamentals, locus-specific markers based on EST and full-length cDNA sequences are being used more and more in candidate gene approaches supplementing map information of first- and second-generation maps.
- Once sufficient highly polymorphic locus-specific markers are available, multiplex analyses will be possible by the combination of several markers in a single PCR reaction, therefore increasing the informational content of the single PCR reaction while retaining accuracy and reliability of the respective marker systems. These multiplex marker systems will be used both for discrimination of genotypes as well as for genetic analyses.
- The availability of so-called conserved orthologous sets of markers makes it possible to investigate genome synteny between different taxa by comparative mapping. Within the Rosaceae family this has led to valuable information about the conservation between the genomes of fruit crops (Dirlewanger et al. 2004) and is currently being extended to ornamental Rosaceae, for example, roses (Rousseau et al. 2006). In the near future, this information will allow the exploitation of data gathered on species for which a large body of information is available (e.g., peach) for crops with minor scientific input (e.g., roses). Furthermore, this research will shed

light on the evolutionary process of important crop species.
- In the near future, with sequencing technologies becoming cheaper and faster and with many of the major food crops completely sequenced it might be possible that sequencing of whole genomes of ornamentals will come into reach of the scientific community.

In summary, ornamental crops will finally greatly profit from recent technical developments of genomics tools and therefore will provide additional information both for basic research and for research aiming at the genetic improvement of economically relevant traits.

References

Abe H, Nakano M, Nakatsuka A, Nakayama M, Koshioka M, Yamagishi M (2002) Genetic analysis of floral anthocyanin pigmentation traits in Asiatic hybrid lily using molecular linkage maps. Theor Appl Genet 105:1175–1182

Anastassopoulos E, Keil M (1996) Assessment of natural and induced genetic variation in *Alstroemeria* using random amplified polymorphic DNA (RAPD) markers. Euphytica 90:235–244

Arumuganathan K, Earle ED (1991) Nuclear DNA content of some important plant species. Plant Mol Biol Rep 9:208–218

Avise JC (1994) Molecular Markers, Natural History and Evolution. Chapman & Hall, New York, USA, London, UK

Becher SA, Steinmetz K, Weising K, Boury S, Peltier D, Renou JP, Kahl G, Wolff K (2000) Microsatellites for cultivar identification in *Pelargonium*. Theor Appl Genet 101:643–651

Been CG, Na AS, Kim JB, Kim HY (2002) Random amplified polymorphic DNA (RAPD) for genetic analysis of *Phalaenopsis* species. J Kor Soc Hort Sci 43(4):338–391

Ben-Sade H, Samach A (2003) Nuclear DNA Amounts In: Roberts A, Debener T, Gudin S (eds) Encyclopedia of Rose Sciences. Elsevier Science, Oxford, UK, pp 279–285

Berio T, Morreale G, Giovannini A, Allavena A, Arru L, Faccioli P, Terzi V, Pecchioni N (2001) RAPD and AFLP genetic markers for the characterisation of *Osteospermum* germplasm. Acta Hort 546:171–176

Cammareri M, Errico A, Sebastiano A, Conicella C (2004) Genetic relationships among *Aster* species by multivariate analysis and AFLP markers. Hered Lund 140(3):193–200

Cerny TA, Caetano AG, Trigiano RN, Starman TW (1996) Molecular phylogeny and DNA amplification fingerprinting of *Petunia* taxa. Theor Appl Genet 92:1009–1016

Chen JJ, Devanand PS, Norman DJ, Henny RJ, Chao CC (2004a) Genetic relationships of *Aglaonema* species and cultivars inferred from AFLP markers. Ann Bot 93(2):157–166

Chen JJ, Henny RJ, Norman DJ, Devanand PS, Chao CC (2004b) Analysis of genetic relatedness of *Dieffenbachia* cultivars using AFLP markers. J Am Soc Hort Sci 129(1):81–87

Coen ES, Romero JM, Doyle S, Eliot R, Murphy G, Carpenter RC (1990) *Floricaula*: a homeotic gene required for flower development in *Antirrhinum majus*. Cell 63:1311–1322

Crespel L, Chirollet M, Durel E, Zhang D, Meynet J, Gudin S (2002) Mapping of qualitative and quantitative phenotypic traits in *Rosa* using AFLP markers. Theor Appl Genet 105:1207–1214

Debener T (1999) Genetic analyses of important morphological and physiological characters in diploid roses. Gartenbauwissenschaft 64:14–20

Debener T (2002) Molecular markers as a tool for analyses of genetic relatedness and selection in ornamentals. In: Vainstein A (ed) Breeding for Ornamentals: Classical and Molecular Approaches. Kluwer Academic Publ, Dordrecht, The Netherlands, pp 329–346

Debener T (2003) Inheritance of characters. In: Roberts A, Debener T, Gudin S (eds) Encyclopedia of Rose Sciences. Elsevier Science, Oxford, UK, pp 286–292

Debener T, Mattiesch L (1996) Genetic analysis of molecular markers in crosses between diploid roses. Acta Hort 424:249–253

Debener T, Mattiesch L (1999) Construction of a genetic linkage map for roses using RAPD and AFLP markers. Theor Appl Genet 99:891–899

Debener T, Bartels C, Mattiesch L (1996) RAPD analysis of genetic variation between a group of rose cultivars and selected wild rose species. Mol Breed 2:321–327

Debener T, Janakiram T, Mattiesch L (2000) Sports and seedlings of rose varieties analysed with molecular markers. Plant Breed 119:71–74

Debener T, Mattiesch L, Vosman B (2001) A molecular marker map for roses. Acta Hort 547:283–287

Dirlewanger E, Graziano E, Joobeur T, Garriga-Caldere F, Cosson P, Howad W, Arus P (2004) Comparative mapping and marker-assisted selection in Rosaceae fruit crops. Proc Natl Acad Sci USA 101:9891–9896

Dugo ML, Satovic Z, Millán T, Cubero JI, Rubiales D, Cabrera A, Torres A (2005) Genetic mapping of QTLs controlling horticultural traits in diploid roses. Theor Appl Genet 111:511–520

Dunemann F, Kahnau R, Stange I (1999) Analysis of complex leaf and flower characters in *Rhododendron* using a molecular linkage map. Theor Appl Genet 98:1146–1155

Esselink GD, Smulders MM, Vosman B (2003) Identification of cut rose (*Rosa hybrida*) and rootstock varieties using robust sequence tagged microsatellite site markers. Theor Appl Genet 106(2):277–286

Felsenstein J (1988) Phylogenies from molecular sequences: inference and reliability. Annu Rev Genet 22:521–565

Fransz PF, Stam M, Montign B, ten Hoopen R, Wiegant J, Kooter K, Oud O, Nanninga N (1996) Detection of single copy genes and chromosome rearrangements in *Petunia hybrida* by fluorescent in situ hybridization. Plant J 9:767–774

Gerats AG, Souver E, Kroon J, McLean M, Farcy E, Maizonnier D (1993) *Petunia hybrida*. In: O'Brien S (ed) Genetic Maps: Locus Maps of Complex Genomes, 6th ed. Cold Spring Harbor Laboratory Press, New York, USA, pp 6.13–6.23

Gregerson RG, Cameron L, McLean M, Dennis P, Strommer J (1993) Structure, expression, chromosomal location and product of the gene encoding ADH2 in *Petunia*. Genetics 133:999–1007

Gudin S (2000) Rose: genetics and breeding. Plant Breed Rev 17:59–189

Han TH, van Eck H, de Jeu MJ, Jacobsen E (2000) Genetic diversity of Chilean and Brazilian *Alstroemeria* species assessed by AFLP analysis. Heredity 84(5):564–569

Han TH, van Eck H, de Jeu M, Jacobsen E (2002a) The construction of a linkage map of *Alstroemeria aurea* by AFLP markers. Euphytica 128:153–164

Han TH, van Eck H, de Jeu M, Jacobsen E (2002b) Mapping of quantitative trait loci involved in ornamental traits in *Alstroemeria*. HortScience 37:585–592

Horn W (1996) Scrophulariaceae. In: Horn W (ed) Zierpflanzenbau. Blackwell, Vienna, Austria, pp 464–466

Huff DR, Peakall R, Smouse PE (1993) RAPD variation within and among natural populations of outcrossing buffalograss [*Buchloe dactyloides* (Nutt.) Engelm.]. Theor Appl Genet 86:927–934

Iqbal MJ, Rayburn AL (1994) Stability of RAPD markers for determining cultivar specific DNA profiles in rye (*Secale cereale*). Euphytica 75:215–220

Jiao CL, Tian HQ, Wu J (2003) The study on RAPD fingerprints of *Narcissus* in China and Europe. J Trop Subtrop Bot 11(2):177–180

Joung YH, Roh MS, Kim TI, Song JS (2001) Forcing and molecular characterization of *Campanula*. Acta Hort 546:421–425

Kaufmann H, Mattiesch L, Lorz H, Debener T (2003) Construction of a BAC library of *Rosa rugosa* Thunb and assembly of a contig spanning *Rdr1*, a gene that confers resistance to blackspot. Mol Genet Genom 267:666–674

Kim DH, Heber D, Still DW (2004) Genetic diversity of *Echinacea* species based upon amplified fragment length polymorphism markers. Genome 47(1):102–111

Kim SC, Jung YH, Jang KC, Song EY, Koh SC (2003) Phylogenetic analysis and the bulbous specific band identification using random amplified polymorphic DNA in the genus *Iris*. J Kor Soc Hort Sci 44(2):228–232

Kim SK, Ahn BJ (2002) RAPD mediated determination of genetic relatedness in oriental lilies. Kor J Hort Sci Technol 20(3):246–251

Kleynhans R, Spies JJ (2000) Evaluation of genetic variation in *Lachenalia bulbifera* (Hyacinthaceae) using RAPDs. Euphytica 115(2):141–147

Ko MK, Yang J, Jin YH, Lee CH, Oh BJ (1998) Genetic relationships of *Viola* species evaluated by random amplified polymorphic DNA analysis. J Hort Sci Biotechnol 73:601–605

Koes RE, Spelt CE, Mol JNM, Gerats AGM (1987) The chalcone synthase multigene family of *Petunia hybrida* (V30): sequence homology, chromosomal location and evolutionary aspects. Plant Mol Biol 10:159–169

Kunze R, Saedler H, Lönning WE (1997) Plant transposable elements. Adv Bot Res 27:331–470

Le Duc A, Adams RP, Zhong M (1999) Using random amplification of polymorphic DNA for a taxonomic reevaluation of Pfitzer junipers. HortScience 34:1123–1125

Leus L, Jeanneteau F, Van Huylenbroeck J, Van Bockstaele E, De Riek J (2004) Molecular evaluation of a collection of rose species and cultivars by AFLP, ITS, *rbc*L and *mat*K. Acta Hort 651:141–147

Linde M, Debener T (2003) Isolation and identification of eight races of powdery mildew of roses (*Podosphaera pannosa* (Wallr.: Fr.) de Bary) and the genetic analysis of the resistance gene *Rpp1*. Theor Appl Genet 107:256–262

Linde M, Mattiesch L, Debener T (2004) *Rpp1*, a dominant gene providing race specific resistance to rose powdery mildew (*Podosphaera pannosa*): molecular mapping, SCAR development and confirmation of disease resistance data. Theor Appl Genet 109:1261–1266

Ling JT, Sauve R, Gawel N (1997) Identification of poinsettia cultivars using RAPD markers. HortScience 32(1):122–124

Lister C, Dean,C (1993) Recombinant inbred lines for mapping RFLP and phenotypic markers in *Arabidopsis thaliana*. Plant J 4:745–750

Loh JP, Kiew R, Keet A, Gan LH, Gan YY (1999) Amplified fragment length polymorphism (AFLP) provides molecular markers for the identification of *Caladium bicolor* cultivars. Ann Bot 84:155–161

McLean M, Gerats AGM, Baird WV, Meagher RB (1990) Six actin gene subfamilies map to five chromosomes of *Petunia hybrida*. J Hered 81:341–346

Meynet J, Barrade R, Duclos A, Siadous R (1994) Dihaploid plants of roses (*Rosa* × *hybrida* cv Sonia) obtained by parthenogenesis induced using irradiated pollen and in vitro culture of immature seeds. Agronomie 2:169–175

Olmstead RG, De Pamphelis CE, Wolfe AD, Young ND, Elisions WJ (2001) Disintegration of the Scrophulariaceae. Am J Bot 88:348–361

Palacios C, Gonzalez CF (1999) AFLP analysis of the critically endangered *Limonium cavanillesii* (Plumbaginaceae). J Hered 90(4):485–489

Peltier D, Farcy E, Dulieu H, Bervillé B (1994) Origin, distribution and mapping of RAPD markers from wild *Petunia* species in *Petunia hybrida* Hort lines. Theor Appl Genet 88:637–645

Pohlendt G (1942a) Cytologische Untersuchungen an Mutanten von *Antirrhinum majus* L. I. Deletionen im uni-Chromosom. Z Indukt Abstamm Vererbungsl 80:281–288

Pohlendt G (1942b) Cytologische Untersuchungen an Mutanten von *Antirrhinum majus* L. II. Mosaikpflanzen mit reziproken Translokationen. Chromosoma 2:388–406

Rajapakse S, Byrne DH, Zhang L, Anderson N, Arumuganathan K, Ballard RE (2001) Two genetic linkage maps of tetraploid roses. Theor Appl Genet 103:575–583

Reimann-Philipp (1981) Cytogenetics and breeding in diploid roses from the triploid hybrid R. multiflora × garden cultivars. In: EUCARPIA Section Ornamentals (ed) Meeting on Rose Breeding. Proc 11th Meeting of the Section Ornamentals, Ahrensburg, Germany, 7–8 Sept 1981, pp 27–29

Riek J, Mertens M, Dendauw J, Van Bockstaele E, Loose M, Heursel J (2000) The use of fluorescent AFLP to assess genetic conformity of a breeders collection of *R. simsii* hybrids. Acta Hort 508:99–104

Ritter E, Gebhardt C, Salamini F (1990) Estimation of recombination frequencies and construction of RFLP linkage maps in plants from crosses between heterozygous parents. Genetics 125:645–654

Ritter E, Salamini F (1996) The calculation of recombination frequencies in crosses of allogamous plant species with application to linkage mapping. Genet Res 67:55–65

Röber R (1996) Solanaceae. In: Horn W (ed) Zierpflanzenbau. Blackwell, Berlin, Germany, pp 464–466

Rohlf FJ (1989) NTSYS-pc numerical taxonomy and multivariate analysis system. ExeterPubl, New York, USA

Rousseau M, Hibrand Saint Oyant L, Foucher F, Barrot L, Lalanne D, Sargent D, Simpson D, Laigret F, Denoyes-Rothan B. (2006) Comparative mapping in the Rosoideae tribe: *Rosa* and *Fragaria*. (Abstr). 3rd Intl Rosaceae conf, Napier, New Zealand

Saitou N, Nei M (1987) The neighbour joining method: a new method for reconstructing phylogenetic trees. Mol Biol Evol 4:406–425

Schwarz-Sommer Z, Silva ED, Berndtgen R, Lönning WE, Muller A, Nindl I, Stuber K, Wunder J, Saedler H, Gubitz T, Borking A, Golz JF, Ritter E, Hudson A (2003) A linkage map of an F₂ hybrid population of *Antirrhinum majus* and *A. molle*. Genetics 163(2):699–710

Scovel G, Ben Meir H, Ovadis M, Itzhaki H, Vainstein A (1998) RAPD and RFLP markers tightly linked to the locus controlling carnation (*Dianthus caryophyllus*) flower type. Theor Appl Genet 96:117–122

Sommer H, Beltran JP, Huijser P, Pape, Lönning WE (1990) *Deficiens*, a homeotic gene involved in the control of flower morphogenesis in *Antirrhinum majus*: the protein shows homology to transcription factors. EMBO J 9:605–613

Stam P, Van Ooijen JW (1995) JoinMap version 2.0: Software for the calculation of genetic linkage maps. CPRO-DLO, Wageningen, The Netherlands

Starman TW, Duan X, Abbit S (1999) Nucleic acid scanning techniques distinguish closely related cultivars of *Poinsettia*. HortScience 34:119–122

Staub JE, Serquen FC, Gupta M (1996) Genetic markers, map construction, and their application in plant breeding. HortScience 31:729–741

Strommer J, Gerats AGM, Sanago M, Molnar SJ (2000) A gene-based RFLP map of petunia. Theor Appl Genet 100:899–905

Strommer J, Peters J, Zethof J, de Keukeleire P, Gerats T (2002) AFLP maps of *Petunia hybrida*: building maps when markers cluster. Theor Appl Genet 105:1000–1009

Stubbe H (1966) Genetik und Zytologie von *Antirrhinum* L. sect. Antirrhinum.VEB Fischer Verlag, Jena

Sutton DA (1988) A Revision of the Tribe Antirhineae. Oxford Univ Press, London, New York, USA, Oxford, UK

Takatsu Y, Miyamoto M, Inoue E, Yamada T, Manabe T, Kasumi M, Hayashi M, Sakuma F, Marubashi W, Niwa M (2001) Interspecific hybridization among wild *Gladiolus* species of southern Africa based on randomly amplified polymorphic DNA markers. Sci Hort 91(3/4):339–348

Trigiano RN, Scott MC, Caetano-Anolles G (1998) Genetic signatures from amplification profiles characterise DNA mutation in somatic and radiation induced sports of chrysanthemum. J Am Soc Hort Sci 123:642–646

Von Malek B, Debener T (1998). Genetic analysis of resistance to blackspot (*Diplocarpon rosae*) in tetraploid roses. Theor Appl Genet 96:228–231

Von Malek B, Weber WE, Debener T (2000) Identification of molecular markers linked to *Rdr1*, a gene conferring resistance to blackspot in roses. Theor Appl Genet 101:977–983

Waites R, Selvadurai HRN, Oliver IR, Hudson A (1998) The *PHANTASTICA* gene encodes a MYB transcription factor involved in growth and dorsiventrality of lateral organs in *Antirrhinum*. Cell 93:779–789

Wiering H, deValming P, Cornu A, Maizonnier D (1979) *Petunia* genetics. I. List of genes. Ann Amél Plant 29:611–622

Wissemann V (2003) Conventional Taxonomy (Wild Roses). In: Roberts A, Debener T, Gudin S (eds) Encyclopedia of Rose Sciences. Elsevier Science, Oxford, UK, pp 326–334

Wolfe AD, Xiang QY, Kephart SR (1998) Assessing hybridization in natural populations of *Penstemon* (Scrophulariaceae) using hypervariable intersimple sequence repeat (ISSR) bands. Mol Ecol 7(9):1107–1125

Wongchaochant S, Doi M, Inamoto K, Imanishi H (2002) Phylogenetic classification of *Freesia* spp. by morphological and physiological characteristics and RAPD markers. J Jpn Soc Hort Sci 71(6):758–764

Yan Z, Denneboom C, Hattendorf A, Dolstra O, Debener T, Stam P, Visser PB (2005) Construction of an integrated map of rose with AFLP, SSR, PK, RGA, RFLP, SCAR and morphological markers, Theor Appl Genet 110:766–777

Yockteng R, Ballard HE, Mansion G, Dajoz I, Nadot S (2003) Relationships among pansies (*Viola* section Melanium) investigated using ITS and ISSR markers. Plant Syst Evol 241(3/4):153–170

Zhang D, Germain E, Reynders-Aloisi S, Gandelin MH (2000) Development of amplified fragment length polymorphism markers for variety identification in rose. Acta Hort 508:113–120

4 Oil Palm

Zuzana Price[1], Sean Mayes[2], Norbert Billotte[3], Farah Hafeez[1], Frederic Dumortier[4], and Don MacDonald[1]

[1] Department of Genetics, Cambridge University, Downig Site, Downing Street, Cambridge, CB2 2EH, UK
[2] Division of Biosciences, Nottingham University, Sutton Boninghton Campus, Loughborough, Leicester, LE12 5RD, UK
  e-mail: sean.mayes@nottingham.ac.uk
[3] CIRAD (CIRAD-CP) TA 80/03, Avenue Agropolis, Montpellier Cedex, 34398, France
[4] DAMI, OPRS, New Britain Palm Oil Ltd., P.O. Box 165, Kimbe, West New Britain Province , Papua New Guinea

4.1
Introduction

4.1.1
Botanical Classification and Phylogeny

Palms

Palms are woody monocotyledons in the family Arecaceae (an alternative name to *Palmae*) which is placed in the order Arecales (Jones 1994). They are a natural group of plants with fossil records dating from the late Cretaceous, and with a characteristic appearance that enables them to be recognized easily, despite occasional confusion with cycads or Cordylines. The present evidence also suggests that palms probably evolved very early in the history of the monocotyledons. Phylogenetic analysis of monocot relationships based on plastid *atp*B, *rbc*L, *mat*K, and *ndh*F, mitochondrial *atp*A and nuclear 18S and 26S ribosomal DNA (rDNA) by Chase et al. (2004) placed Arecales as a sister group to the rest of Commelinids, which includes orders Commeliniales, Zingiberales, and Poales. However, Arecales (single family Arecaceae, the palms) were not included in the Commelinids although Dahlgren et al. (1985) noted that the palms shared characteristics with those families and the connection of Arecaceae to Pandanaceae and Cyclantacae based on similarities in habit and inflorescence indicates parallelism rather than close phylogenetic relationship.

There are no morphological synapomorphies (shared, derived state) for the palms overall. However, phylogenetic studies of *rbc*L (the large subunit of ribulose-1,5-bisphosphate carboxylase/oxygenase; Duval et al. 1993; Chase et al. 1995), of 18S rRNA (Hahn et al. 1996; Soltis et al. 1997), and of the chloroplast gene *rps*4 (Nadot et al. 1995) all support monophyly of the palm family (Uhl et al. 1995).

The latest phylogenetic analysis of Arecaceae (Hahn 2002a) added sequence data of 51 genera of the Arecoid line for plastid genes (*atp*B, *rbc*L, and *ndh*F) and the plastid intergenic spacers (*trn*T and *trn*Q-*rps*16). Furthermore, Hahn (2002b) added the nuclear DNA (18S rDNA) and chloroplast DNA (cpDNA; *atp*B and *rbc*L) sequence for 65 palm genera. The detailed extensive array of molecular phylogenetic studies of palms has identified four major groups corresponding to (1) Calamoideae, (2) Nypoideae, (3) Coryphoideae + Caryoteae, and (4) Arecoideae. The Arecoid line is the largest of the four groups, with approximately 60% of the genera in the family. The analysis of biogeographic patterns present in the Arecoid line suggests that the group is of Gondwana origin (Hahn 2002a).

Palms are evolving slowly at the sequence level. Studies of *rbc*L and comparative studies of chloroplast DNA restriction fragment length polymorphisms (RFLPs) revealed that palm plastid genomes have an approximately fivefold slower rate of synonymous substitution compared with other monocots, particularly grasses (Wilson et al. 1990; Gaut 1992). This difference in rates of divergence has been confirmed with the nuclear *Adh* gene (Gaut et al. 1996; Morton et al. 1996). Studies of variation in palm mitochondrial genes have been focused on the gene *atp*A (Eyre-Walker and Gaut 1997). Results showed that the nuclear genes evolve faster than the chloroplast genes, and the chloroplast genes evolve faster than the mitochondrial genes as shown previously by Wolfe et al. (1987). The relative rates of divergence within groups seem consistent, but palms evolve at synonymous sites more slowly than do grasses.

The patterns of non-synonymous substitutions re different; only *rbc*L and *Adh*1 evolve significantly faster in grasses compared with palms (Morton et al. 1996). The rates of nucleotide substitutions in the nuclear ribosomal small subunit (18S rDNA) are significantly lower than that seen in other monocots and are comparable to that seen in with the plastid gene *rbc*L (Soltis et al. 1997).

In the palm family there is little variation in chromosome number but the genome size can vary significantly. Differences in chromosome numbers are unusual and polyploidy (multiple presence of whole chromosome sets) is also rare in palms (Uhl and Dransfield 1987).

The Elaeidinae

The oil palm (*Elaeis guineensis* Jacq.) belongs to the subfamily Arecoideae, tribe Cocoeae and subtribe Elaeidinae. The analysis of biogeographic patterns present in tribe Cocoeae suggests that the tribe is of Gondwana origin and primary diversification in this group might have coincided with continental breakup (Hahn 2002a). The subtribe Elaeidinae includes only the genus *Elaeis* (from the greek *elaia*, for the olive tree) and *Barcella* and is always recovered as monophyletic (Hahn 2002a). The genus *Barcella* has no commercial use at present. The genus *Elaeis* consists of only two species: the African oil palm – *E. guineensis* Jacq. – and the Latin American oil palm – *E. oleifera* Cortez (Corley and Tinker 2003).

Mitochondrial DNA has been used to assess the phylogeny of the subtribe Elaeidinae to which the African and Latin American oil palm and the genus *Barcella* belong (Barcellos et al. 1999). The authors analyzed 288 accessions representative of *E. oleifera* and 38 of *E. guineensis* by performing RFLP with four mitochondrial probes. The RFLP analysis identified more mitotypes in *E. oleifera* compared to *E. guineensis* and also confirmed that the divergence between the two species was very low.

Elaeis guineensis – the African Oil Palm

There are no subspecies in *E. guineensis* Jacq. However, there are a range of breeding populations of restricted origin (BPRO) such as Pobe, Yangambi, Deli *dura*, Algemene Verneiging Rubber Planters Oostkust, Sumatra (AVROS), and others, documented by Rosenquist (1985), which play an important role in many breeding programs. One of the most important ones is the Deli *dura*, which is believed to be descended from four palms which were planted in 1848 in the Bogor Botanical Gardens, Indonesia. Concerns

have already been raised about the limited within-population genetic diversity for some BPRO such as Deli *dura*. Marker studies have investigated this possibility (Sect. 4.1.6).

Elaeis species have 16 pairs of chromosomes (Madon and Clyde 1995). On the basis of their length, the chromosomes of *E. guineensis* were divided into three groups (Madon and Clyde 1995). The size of the haploid *E. guineensis* genome has been estimated to be 1.7×10^9bp ($2C = 3.7$ pg) (Rival et al. 1997).

4.1.2
Geographical Distribution

Although the two *Elaeis* species occur on separate continents and have different growth habits, they are very similar. However, while they can hybridize to produce some fertile offspring, the differences between them are sufficient to treat them as separate species. Barcellos et al. (1999) have proposed the center of origin of the genus *Elaeis* to be Latin America on the basis of their results which revealed more variability in *E. oleifera* than in *E. guineensis* accessions they examined, and botanical evidence.

However, the fossil, historical, and linguistic evidence (particularly from Brazil) for the African origin of *E. guineensis* is strong. Zeven (1964) reported finding fossil pollen similar to *E. guinensis* from Miocene and earlier layers in the Niger delta and there similar results have been obtained (Elenga et al. 1994; Raynaud et al. 1996; Ergo 1997). Zeven suggested that *Elaeis* sp. originated in Africa and spread to South America via the Tertiary bridges which are believed to have connected Africa and America. Both results from Africa and South America are consistent with a initial *Elaeis* species which underwent geographical speciation with the breakup of Gondwana, some 60 million years ago, without the need for a specific mechanism to prevent cross-fertilization of the derived species (although fertility is a major issue with the interspecific F_1) (Heywood 1993).

The first historical evidence of oil palm cultivation in Africa comes from the Portuguese explorer Ca' Da Mosto (1434–1460; Crone 1937) although the species was recorded in detail by Jacquin (1793).

It is generally accepted that the present geographical distribution of oil palm is the reflection of favorable climate and of human farming activities, with palms almost certainly being spread by the migration of humans (Corley and Tinker 2003).

4.1.3
Biology of Oil Palm

Separate male and female inflorescences arise in the leaf axils among the leaves and the infructescence is large and densely packed with fruit primordia. Approximately two inflorescences are initiated per month and take up to 3 years to develop into a mature male or female inflorescence. The oil palm is monoecious, alternately producing male and female inflorescences in a cycle of around 6 months and is thus naturally outcrossing. Detailed studies of the flowers have shown, however, that each flower primordium is a potential producer of both female and male organs, though one or the other almost always remains rudimentary (Hartley 1977). The female inflorescence is a panicle which consists of a variable number of rachillae that carry five to 30 floral triads, each composed of one female flower accompanied by two nonfunctional male flowers. Although each female flower has a tricarpellate ovary, only one carpel usually develops to give rise to a seed. The ovary is accompanied by two rudimentary androecia. In male inflorescences each male rachilla is composed of 400–1,500 staminate flowers. In very rare cases (tissue culture stress, or very young plants) both androecium and gynoecium develop to give rise to a hermaphrodite flower. Inflorescence abnormalities are by no means uncommon in oil palm. The sex of the inflorescence is influenced by the external conditions around 2–3 years before anthesis and also by a genetic component (Corley et al. 1995). The cycle between male and female inflorescences can be biased toward male inflorescences under harsh external conditions, such as drought, and toward female inflorescences under favorable external conditions.

It was originally believed that oil palm, because of its abundant pollen and a reduced flower structure, was wind-pollinated. In fact, the introduction into Malaysia of the pollinating insect *Elaiedobius kamerunicus* from Africa showed that insect pollination plays an essential part in the fruit set, especially in wet conditions (Hartley 1988).

The seed (kernel) and the pulp (mesocarp) of the fruit are very rich in oil. In the internal fruit structure the most important differences are to be found in the thickness of the shell (endocarp). There are three fruit forms known: *dura*, *pisifera*, and *tenera*. These forms were determined by Beinaret and Vanderveryen (1941) to be due to a single gene. The importance of this gene is well understood, because only plants of intermediate type (*tenera*) are grown commercially.

In the *pisifera* form (*shsh*) there is no shell (endocarp) as such, only a fibrous ring. In the dura form (*ShSh*) there is a thick endocarp and in the intermediate *tenera* form (*Shsh*) the endocarp is thinner and the fibrous ring is present too. The thickness of the endocarp varies considerably in the *dura* and *tenera* forms; the distributions even overlap, so the ultimate criterion for distinguishing *dura* from *tenera* is the presence of the fibrous ring in the *tenera* form. There are no fruit types as such recognized in *E. oleifera*; all fruits appear to be of *dura* form.

The external appearance of a fruit varies when ripened. There are four major fruit types known: *nigrescens*, *virescens*, *albescens*, and *poissoni* (which may have more than one seed in the fruit). The normal fruit is dark and is called *nigrescens*. A relatively uncommon type is *virescens*. Its green color is due to the absence of anthocyanin in the exocarp of the fruit and the color changes at maturity to orange. The white color of the mesocarp (*albescens*) is caused by the absence of, or low level of, carotenoids in the mesocarp. *Poissoni* is an abnormal fruit type which is often referred to as "mantled," or as "a fruit with supplementary carpels." This is somewhat misleading because in fact it is the rudimentary androecia which develop into carpel-like structures, and it is not a tricarpellate ovary developing the two extra seeds.

4.1.4
Economy

Oil palm plays a vital role in the economy of many developing countries in Southeast Asia, particularly Malaysia and Indonesia. Oil palm is a very important crop often used for replacement of primary and secondary tropical forests in many developing countries. During the last century the oil palm industry developed from an economy based on the wild palm groves of West Africa (before the 1920s) to become one of the most successful perennial plantation crop industries in human history. The change was connected with the establishment of plantations, where the major product was palm mesocarp oil. By the end of the colonial era, a well-developed oil palm industry, supported by research activities, existed in Africa and Southeast Asia. In 1985, Malaysia was producing most of the palm oil in the world (Fig. 1). Palm oil was expected to overtake soybean oil as the world's leading vegetable oil in 2005 (Corley and Tinker 2003; Fig. 2), with Europe and China at present the major markets for palm oil (Fig. 3).

Fig. 1. World major producers of oil palm: 1994–2003 (1,000 t). (Source: *Oil World Annual* 1999–2003 and *Oil World Weekly* December 12, 2003; Malaysian Palm Oil Board)

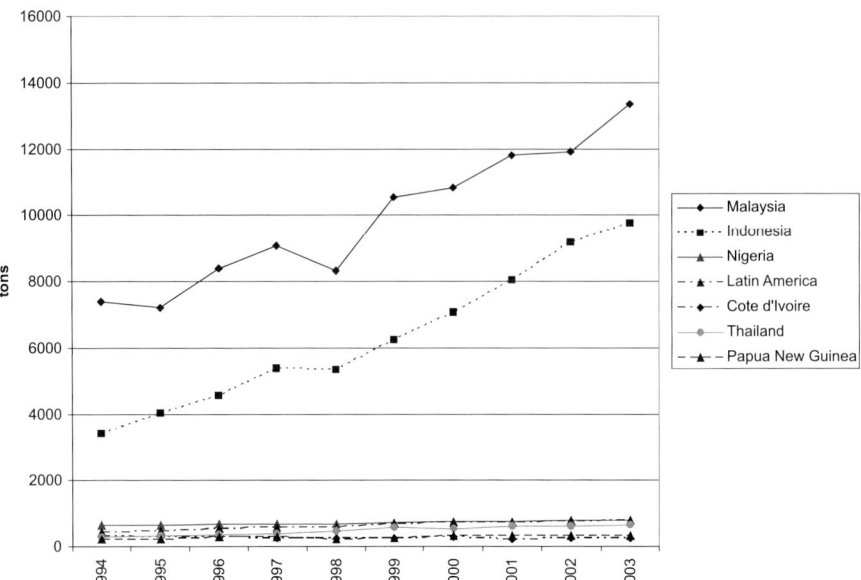

Fig. 2. World oil production of vegetable oils: 1994–2003 (1,000 t). (Source: *Oil World Annual* 1999–2003 and *Oil World Weekly* December 12, 2003; Malaysian Palm Oil Board)

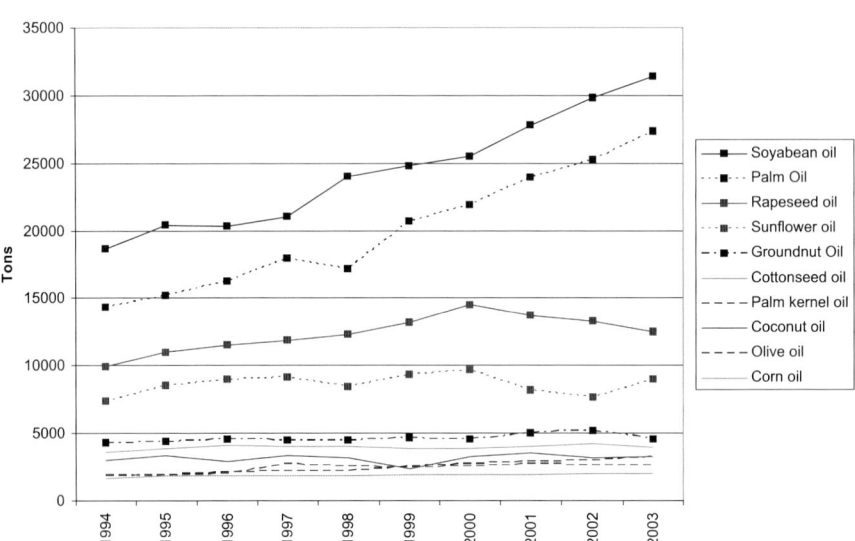

The 1980s campaign against "tropical oils" launched by the soybean interests in the USA produced strong negative publicity for palm oil (Enig 1998). The basis for this campaign was the claim that unsaturated fats were much healthier than saturated fats. Moreover, the oil palm industry in general was heavily criticized because of the indiscriminate and wasteful approach of some Indonesian companies to opening up land for oil palm cultivation, causing the loss of large areas of natural forests (Casson 2000). However, very significant increases in the area planted with oil palm continue to occur, some at the expense of other traditional plantation crops, such as rubber and coconut (Corley and Tinker 2003).

At present, production costs for palm oil are below those of other vegetable oils even though the crop is manually harvested. This is mainly due to the availability of cheap labor (Corley and Tinker 2003) and the roughly five-fold yield per hectare per year advantage that oil palm has over temperate oil crops. As labor costs are likely to rise, mechanization will become more important. Increasing environmental concerns about loss of tropical rain forest will also create pressure for increasing oil yields per hectare on existing plantations. The average yield in Malaysia ranges from 3–4.3 t/ha/year (Corley and Tinker 2003), which is substantially below the predicted physiological maximum of 17 t/ha/year (Corley 1983).

Fig. 3. World major consumers of oils: 1994–2003 (1,000 t).
(Source: *Oil World Annual* 1999–2003 and *Oil World Weekly* December 12, 2003; Malaysian Palm Oil Board)

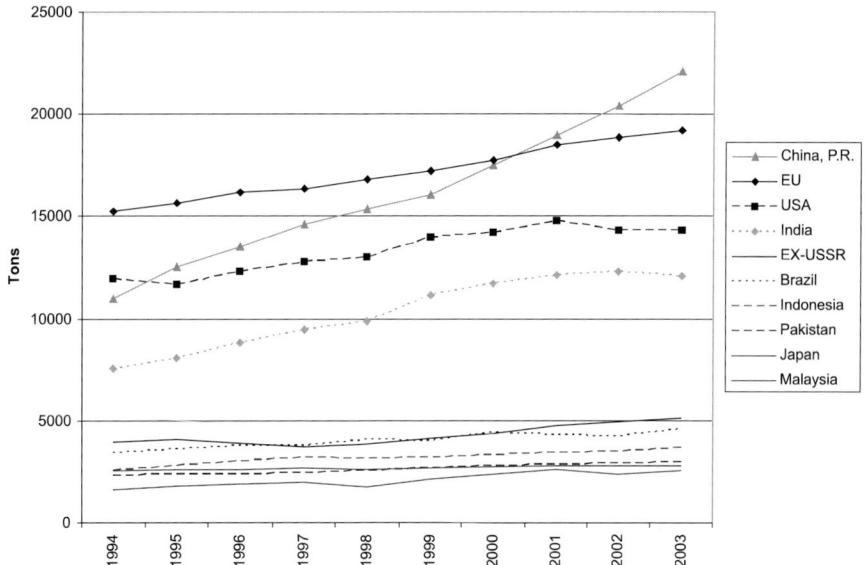

There is also still an immense gap in the productivity of the African and the Asian palm oil production systems. In many African countries oil palm plays an integral part of sustainable agriculture because of its versatile uses. Increasingly agricultural sustainability is likely to require both the maintenance of diversity in areas where traditional varieties continue to be important and its increased use for high-productivity areas. In Latin America, the oil palm industry faces immense challenges connected with developing their own breeding programs based on the hybrids between African and Latin American oil palms. These show an increased resistance to some oil palm diseases although good-quality hybrid-derived material is becoming available from seed companies such as ASD.

4.1.5
Oil Palm Products and Their Uses

Oil is the main commercial product of the oil palm. The oil is extracted from the fruit mesocarp (palm oil – by far the most important product) and nut kernels (kernel oil). Palm oil, extracted from the fibrous flesh of fruits (mesocarp) after they have been hot-squeezed, has an oil content from 40–70%. Prime oil, commercially known as palm kernel oil, is extracted from the seeds, which are firstly shelled and ground, then hot-squeezed; it can be occasionally extracted by means of chemical solvents. Oil content per seed can range from 43–51%. Palm kernel oil is very similar to coconut oil, from which it differs in its higher content of oleic acid.

Human Consumption
About 90% of world's palm oil is used for edible purposes (Sambanthamurthi et al. 2000). Numerous studies have shown the association between diet and the incidence of coronary heart disease. The fatty acid composition of palm oil has thus been the focus of attention in determining its influence in relation to coronary heart disease risk factors (Enig 1998). Palmitic acid (44%) is the major saturated fatty acid in palm oil and this is balanced by almost 39% monounsaturated oleic acid and 11% polyunsaturated linoleic acid. The remainder is largely stearic (5%) and myristic (1%) acids. This composition is significantly different from palm kernel oil (obtained as a co-product during the processing of oil palm fruits), which is almost 85% saturated, short-chain (C_{12}–C_{14}), lipids. The nutritional studies showed that the diet with a high proportion of palm oil as the fat component is as healthy as any other (Corley and Tinker 2003).

The fatty acid composition of palm oil (approximately 1:1 saturated to unsaturated fatty acids) is such that the oil is semisolid at normal room temperature (22 °C), which favors its use as the solid-fat component for margarine. More liquid oils need to be treated by hydrogenation to make them solid, leading to trans-fatty acids (van Duijn 2000). Palm oil is particularly suitable for deep frying because of its low content of polyunsaturated linoleic acid and a higher level of saturated fatty acids (Sambanthamurthi et al. 2000), which are less susceptible to oxidation. In addition, palm oil contains natural antioxidants such as tocopherols and carotenoids. In recent years there has also been interest in red palm oil as a source of vitamin A in human nutrition (Sundram et al. 2003; Zagre et al. 2003), with the current process of bleaching palm oil to make it light in color now being questioned.

Palm Oil Based Oleochemicals, Diesel, and Biodegradable Plastics

Only about 10% of palm oil is used for nonfood products such as oleochemicals, for example, sodium salts (soaps) and glycerol. Furthermore, fatty acid methyl esters from palm oil can be used as substitutes for diesel (Choo and Cheah 2000). In addition to this, alcohol can also be produced by fermentation of carbohydrates (Corley and Tinker 2003) and burned. Biodegradable plastics such as polyhydroxybutyrate (PHB) can be produced from palm oil. PHB can be synthesized from acetylcoenzyme A, the precursor for fatty acid synthesis by transforming palm oil (Houmiel et al. 1999; Masani Mat Usus et al. 2001).

Other Products from Oil Palm

One of the major "waste" products from the oil palm mill is the empty fruit bunches which are left after the fruit is steam-stripped from the fresh fruit bunches. One of the major ways in which these are used is as a fuel, to produce electricity for the plantations needs. They are also sometimes used as mulch on young field planted palms (Corley and Tinker 2003). After oil extraction, the remaining palm kernel material can be pressed to form cattle cake. While relatively poor in protein compared with other oil seed cake, kernel cake from screw press machines can still have 8–13% residual oil, which can make a useful contribution to animal diets (Corley and Tinker 2003). The wood from palm trunks is relatively soft and is only really useful for forming pressed wooden objects. Perhaps the most promising use of this is to produce a plywood substitute, which has the potential to replace some of the hardwood boards used by construction industries (Kamarudin et al. 1999; Koh et al. 1999). The stripping of rain forest for hardwood trees by multinational companies for this purpose represents a major blight on conservation efforts, particularly in developing countries, such as Papua New Guinea and Brazil.

Oil Palm Crop Improvement

A more scientific approach to crop improvement began toward the middle of the twentieth century, when Beinaret and Vanderveryen proved that the *tenera* form was generated by crossing *dura* and *pisifera* forms. This has been the single most important step in the genetic improvement of oil palm yields. The thick-shelled *dura* form is homozygous for one allele (*ShSh*) but has a yield disadvantage of about 25% compared with the thin-shelled heterozygous tenera form (*Shsh*) (commercially grown type), while the shell-less *pisifera* is homozygous for the alternative allele (*shsh*) but is often female-sterile, and cannot be grown as a crop.

The oil palm is monoecious and is naturally cross-pollinated. Although both parents have male and female flowers, for seed production the *dura* form is used as a female parent and *pisifera* has taken the role of the male parent.

Progress to date in oil palm breeding to increase oil yields has been spectacular, with a four-fold increase in yield over the last century (Corley and Lee 1992). Half of this has been ascribed to improvement in the genetic material. Within Deli *dura*, a comparison of unselected material (grown under the best agronomic practices) derived from botanic gardens and material after four generations of selection suggested a 50% increase in yield, largely from an increase in mesocarp/fruit (Corley and Lee 1992). Hardon et al. (1987) estimated that improvement in yield per generation has been 10–15%, although this only equates to approximately 1% per year. Interestingly, despite the major difference in generation time, the increase in wheat yields over the last century or so has been four-fold (half from genetic improvement and half from agronomic improvement), which is roughly 1% per year.

In general, two approaches have been adopted for oil palm breeding. The basic approaches are family and individual selection (FIS) and reciprocal recurrent selection (RRS). FIS identifies the best families and then selects the next generation of palm parents from within these, using mainly phenotypic values. Such an approach resembles animal breeding (Falconer 1981). RRS aims to develop separate and complementary pools of *pisifera* and *dura* which exploit hybrid vigor when crossed.

These are progeny-tested against each other before further development, allowing the generation of breeding values. This is the method favored by maize breeders.

Rosenquist (1991) suggested that some of the disadvantages of RRS were the tendency to produce inbreeding within the maternal and the paternal pools of material, as well as the more limited numbers of palms in the base population which could be worked with (approximately one third the number that could be used in FIS). In practice, many programs are a combination of the two approaches, as oil palm has the advantage over maize of being perennial, with palms with good breeding values making repeated contributions to breeding material, and the female sterility of many *pisifera* sources (such as AVROS) makes progeny testing necessary.

However, recent results from Dami *dura* suggest that inbreeding within the maternal pool is no longer an overriding concern (Dumortier 2000). This might argue that many of the deleterious recessive genes have already been eliminated through the RRS program.

It is difficult to see how such impressive increases in oil yield can be maintained in future generations, without a major contribution from biotechnology.

4.2
Molecular Markers in Oil Palm Breeding

The primary objective of oil palm research is to increase profit per hectare from plantations; this applies as much to the oil palm breeder as to the agronomist (Corley 1983). Breeding and selection of *E. guineensis* began in the early 1920s and since then considerable improvements have been made both in yield and in quality characters (Rosenquist 1985; Hartley 1988). The potential yield of the crop may be as high as 17 t of oil per hectare (Corley 1983). The long breeding cycle (over 10 years) and the variation still encountered (despite some concerns in Southeast Asia) suggest that there exists considerable scope for improvement in yield (Corley 1983), disease resistance (Flood et al. 1989), and oil composition. Molecular markers represent one way in which it may be possible select within material earlier in the breeding cycle to reduce the generational times.

4.2.1
Fingerprinting and Linkage Studies

Linkage mapping was initially performed using polymorphic DNA markers such as isozymes (a molecular marker system based on the staining of proteins with identical function, but different electrophoretic mobilities; Ghesquiere 1984, 1985) and RFLPs (Jack et al. 1993). There are two ways RFLPs may be generated: gain or loss of restriction sites and/or indels between restriction sites. Jack et al. (1995) and Mayes et al. (1996) reported the potential in oil palm for marker identification and application and the use of highly informative oil palm RFLP markers for genotype characterization. Mayes et al. (1997) and Jack et al. (1998) reported construction of a RFLP map for oil palm and subsequent identification of a marker linked to shell thickness. There were 24 linkage groups identified (resolved to 21 by Rance et al. 2001) although oil palm has only 16 chromosomes. Forty RFLP markers have also been used for estimating genetic similarity within oil palm breeding parents such as Deli, AVROS, etc. (Mayes et al. 2000).

RFLPs are gradually becoming replaced by less laborious and more polymorphic marker systems which are based on polymerase chain reacton, such as simple sequence repeats (SSRs; Tautz and Renz 1984; Billotte et al. 1999) and amplified fragment length polymorphisms (AFLPs; Vos et al. 1995). These systems are generally used for saturating already existing RFLP maps. Kulartne et al. (2000) used AFLP markers for studying the diversity within different populations collected by the Malaysian Palm Oil Board (MPOB). Purba et al. (2000) investigated the genetic relationships between genotypes from different *E. guineensis* populations used in the Indonesian Oil Palm Research Institute breeding program with the help of AFLP markers. The authors showed that the crosses among African subpopulations showed more potential for breeding than the crosses between the African and the Deli populations currently used in the reciprocal recurrent selection. An AFLP map with 20 linkage groups was reported by Chua et al. (2001). Microsatellites (or SSRs) are small arrays of tandem repeats that are simple in sequence, for example, $(CA)_{10}$. SSRs are thought to have been produced by mutation, unequal crossing over, and DNA slippage. This last process occurs when DNA strands mispair during replication so that short stretches of sequence slip against each other creating loops of DNA, which, when the second strand is synthesized, results in the loss or gain of motifs. Microsatellites are neutrally evolving, co-dominant (heterozygotes can be distinguished from homozygotes) markers with high levels of genetic diversity and show Mendelian inheritance.

About 400 microsatellite markers (SSR) were recently developed in the *E. guineensis* species by Centre de Coopération Internationale en Recherche Agronomique pour le Développement (CIRAD), using a microsatellite-enriched library building procedure based on a hybridization-based capture methodology using biotin-labeled microsatellite oligoprobes and streptavidin-coated magnetic beads (Billotte et al. 1999). The SSR polymorphism was characterized in *E. guineensis* and in the closely related species *E. oleifera*, in which the utility of the SSR markers was observed, and in a subset of 16 other palm species some oil palm SSRs were potentially transferable. Billotte et al. 2001 reported also a transferability of some date palm and peach palm SSRs to the *E. guineensis* species was also observed (Billotte et al. 2004a, b). A reference linkage map of oil palm was developed in the control cross LM2T × DA10D, using 944 loci (255 SSRs, 688 AFLPs, locus *Sh*) distributed on 16 linkage groups representing the 16 chromosome pairs of the oil palm (Billotte et al. 2005; http://tropgenedb.cirad.fr/oilpalm/publications.html). Two AFLP markers were located on this map at 7 and at 11 cM on each side of the *Sh* locus controlling the variety type of the fruit in oil palm, using bulked segregant analysis and linkage mapping methods. A further 103 SSRs were also developed by Price et al. (2003).

There are a range of other marker systems such as randomly amplified polymorphic DNAs (RAPDs) (Williams et al. 1990) and inter-simple sequence repeats (ISSRs) available (Zietkiewicz et al. 1994). RAPDs use a random sequence of ten nucleotides as primers and the amplification of products depends upon the presence of the complementary nucleotide sequence, in the opposite orientation, on each DNA stretch. This is generally within a stretch of no more than around 3 kbp. They are fast, simple, but are often not reproducible between laboratories. In ISSR anchored primers that anneal to microsatellites are used. Shah (1994) assessed the utility of RAPD markers for determination of genetic variation in oil palm and Moretzsohn (2000) produced a RAPD linkage map of the shell thickness locus in oil palm. Rajanaidu et al. (2000) used RAPDs and RFLPs to

estimate genetic diversity and compare different populations collected by the MPOB.

There have been a number of long-terminal-repeat retrotransposon (LTR-RTN) based marker systems such as sequence-specific amplified polymorphism, inter-retrotransposon amplified polymorphism (IRAP), and retrotransposon microsatellite amplified polymorphism (REMAP) developed in plants (Price et al. 2004). In the case of LTR-RTN-based markers, polymorphisms are generated by the unique biological process of retrotransposition, which is an irreversible process resulting in insertions of retrotransposons into new sites without the loss of the parental copies. The consequences of retrotransposition range from the alteration of a few hundred base pairs to a few kilobases at the site of insertion. By contrast, marker systems based on SSRs are based on random small-scale changes (i.e., from one up to a few tens of nucleotides). Price et al. (2004) developed a multilocus IRAP marker system based on *copia*-like retrotransposons. The authors also reviewed marker methods based on retrotransposons and concluded that there was a scope for using these methods in oil palm breeding and diversity analysis as an alternative to AFLP.

4.2.2
Potential Applications of Markers – Simple Traits

Development of markers would shorten the process of selection, especially at the nursery stage. Currently, crossing and growth in the nursery stage take approximately 2 years. After 2 years, seedlings are field-planted (typically 143 palms per hectare). Although the seedlings start to produce fruit after 2.5 years, 5 years of recorded field data is necessary to properly assess the quality of a potential breeding palm. Marker-assisted selection (MAS) would allow the process to be speeded up. Some of the important areas of interest are shell thickness, *virescens*, and crown disease. A marker for shell thickness potentially could have a high commercial value in breeding programs. The value would be in determining whether selected palms are *pisifera* or sterile *tenera* before progeny testing. This marker could also assess the purity of commercial *tenera* seed lots (Mayes et al. 1995). Hartley (1977) states that *virescens* is possibly monofactorial and dominant. Exploitation of this gene, if a segregating population were available, could make spotting ripe bunches much easier and thus contribute to a higher yield, through decreasing the loss of loose fruit.

The development and establishment of technologies such as MAS (Mohan et al. 1997) which would allow selection of individuals to be based on the genetic marker information would represent a major step in oil palm breeding.

Markers would also assist breeding programs, in helping to maintain diversity within the populations used for breeding, in identifying outcrosses in breeding programs, and in allowing controlled introduction of foreign material into breeding programs. MAS would be used in conjunction with the existing selection based on general combining ability (GCA), specific combining ability (SCA), and other breeding values.

It has been noted by Corley and Tinker (2003) that the progress made in the breeding program depends both on the amount of variation present in the population before the selection starts, and on the heritability of the characteristics the breeder selects for. It is worth pointing out that the majority of characters measured by oil palm breeders are likely to be polygenic and these include bunch yield and its components, oil and kernel to bunch and their components, and carotene content. Although, the major effect on bunch composition is the shell thickness gene, the attempts made to find a marker for this gene were not entirely successful (the linkage was not sufficiently close) (Mayes et al. 1997; Moretzsohn et al. 2000; Billotte et al. 2001), until the recent development of the Billotte map (LINK2PALM EU Framework 6 program). Recently there has been interest in high carotene content (Rajanaidu 2000), for its high nutritional value. The author also showed that in the crosses derived from the Nigerian material the carotene content ranged from 180–2,500 ppm.

The systematic approach of extensive phenotypic, marker, and GCA analysis would allow the association of markers and phenotypic characters and permit quantitative trait locus (QTL) analysis.

4.2.3
Potential Applications of Markers – Complex Traits

Marker-based methods could provide a means of using QTL analysis to target regions of a genetic map and measuring their effects. Linkage mapping of QTLs depends on detecting the linkage disequilibrium between marker regions involved and the trait genes

themselves. Localization of a QTL depends more on the population size than on the density of markers (Kearsey and Pooni 1998) as well as upon the heritability of the trait studied. Firstly, the problem with the QTL mapping in oil palm is that population sizes are small and the basis of the genetic heritability of many quantitative traits is yet to be determined. The number of individuals in any one controlled cross is often limited (below 90). Secondly, QTLs are more easily identified for inbred lines but they are much more difficult to identify in outcrossing species where is much more background variation. Thirdly, for parents to provide linkage information, the hybrid F_1 must be heterozygous at both a marker and a linked QTL because only in this case can marker–trait associations be found in the progeny.

The availability of a large number of published SSR markers and dominant marker systems such as AFLP, REMAP, and IRAP and the relative ease with which those systems can be converted to automated marker typing should allow rapid identification of markers linked to agronomically important traits and subsequent QTL analysis.

The first attempts at QTL analysis in oil palm were reported by Rance et al. (2001). The authors investigated the underlying genetic basis of QTLs in oil palm and identified six marker regions associated with QTL effects; RFLP markers were identified linked to yield, oil/bunch and its components, and vegetative characters.

The recent development of the CIRAD genetic map will allow a detailed QTL analysis as the populations reach maturity. Already, an initial analysis has identified a QTL for palm height (LINK2PALM).

4.2.4
Potential Applications of Markers – Disease Resistance

The most serious diseases are *Fusarium* wilt (*Fusarium oxysporum* f. sp. *elaeidis*) in several parts of Africa and Latin America, *Ganoderma* in Asia, dry basal rot (*Ceratocystis*) in West Africa, and fatal yellowing and sudden withering on new plantations in Latin America (Gomez 2000). Unsuccessful attempts at finding RFLP markers linked to *Fusarium* wilt resistance have been made by Buchanan (1999). The present availability of highly polymorphic markers such as microsatellites makes this task much more feasible. *Ganoderma* trunk rot or basal stem rot has been a problem in some areas of Malaysia and Indonesia for the last 40 years, and in recent years it has also been the subject of much research in those countries. Most of this work has been summarized in Flood et al. (2000), which included also a general review of the current state of this disease in Asia by Ariffin (2000). De Franqueville et al. (2001) showed that there were significant differences between families in *Ganoderma* incidence and thus demonstrated the feasibility of breeding for resistance and of potential marker application. There has been extensive research on fatal yellowing (*Thielaviopsis paradoxa*, much of it reviewed in Gomez et al. (2000). To combat this disease, markers could be used to search for disease resistance factors within *E. guineensis* material and in interspecific hybrids with *E. oleifera*.

4.3
Genome Organization

Plant genomes are remarkably large and dynamic and contain up to 80% of repetitive DNA (San Miguel et al. 1996).

Bacterial artificial chromosome (BAC) libraries can be constructed to represent the majority of the genome possible (restriction endonucleases such as *Hind*III and *Eco*RI, which cut frequently and show no significant sensitivity to methylation of the genomic DNA) or to try to target the coding regions (rare cutting methylation-sensitive restriction endonucleases, such as *Mlu*I and *Not*I). Large insert clone libraries, such as BACs (Shizuya et al. 1992), are essential tools. A partial BAC *Hind*III library was reported by Singh et al. (2003) with the average size inserts of 40 kb, and a complete *Hind*III oil palm BAC library has been made recently by CIRAD (Pifanelli et al. 2002). Complete libraries have the advantage that they contain all of the clonable sequences in the genome, but with the disadvantage that this can require very large numbers of clones and significant infrastructure and resources to develop, maintain, and use. An alternative approach is to develop "targeted" BAC clones. These use methylation patterns within the genomic DNA to produce restriction enzyme cuts where there is no methylation present; lack of methylation is often indicative of coding and coding-associated regions (Martienssen 1998). The construction of an initial test BAC library for oil palm using the rare cutting methylation-sensitive restriction endonuclease *Mlu*I has allowed this possibility to be examined (Hafeez and Mayes, unpublished re-

sults). While the average insert size is relatively low (80 kbp), hybridization of specific sequence probes to colonies (see Price et al. 2002) and analysis of 600 *Mlu*I BAC end sequences, compared with *Hind*III-derived BAC clones and end sequence, confirm significant enrichment for low-copy sequences and the exclusion of high copy number classes of retrotransposons from *Mlu*I clones.

Understanding something about the structure of the oil palm genome could be a major advantage in the development of targeted markers for MAS.

4.4
Uses of Conserved Synteny

Perhaps the nearest relative of oil palm from within those major crops which have been studied in depth is cereals. It is estimated that the cereal group and oil palm diverged some 100 million years ago. Closer relatives of oil palm on which there is marker work are date palm (*Phoenix dactylus* and coconut palm (*Cocos nucifera*). Billotte et al. (2001) have demonstrated that SSR markers developed in one species could been used in another. Indeed, one of the targets of the recent EU Framework 6 INCO-DEV LINK2PALM proposal was to cross map markers in coconut and oil palm. While this work is under way and will provide useful information, oil palm is probably the most developed of these three species in terms of molecular tools and will only make limited gains from the other two species.

An initial study testing RFLPs which were used in cross mapping between a segment of chromosome 9 in rice with chromosome 5 in wheat (Foote et al. 1997) gave some evidence for conservation of gene order between cereals and oil palm (three out of five markers mapped in rice/wheat also showed linkage in oil palm (Hafeez, unpublished results); however, distances between markers were considerably greater in oil palm and with the lack of sequence data at the moment, this approach is unlikely to make a major impact in the next few years.

4.5
Vegetative Propagation of Oil Palm

The reasons for developing methods for vegetative production of oil palm are many, not least that it only has one vegetative meristem and cannot be propagated by taking cuttings. The ability to rapidly propagate elite genotypes has immense potential for a species with a selection cycle of 10–16 years. Yield advantages over seedling populations were predicted to be in the order of 30%. In spite of the fact that attempts at propagating oil palm by tissue culture started in the 1960s, the discovery of abnormal flowering and severe bunch failure caused a major setback, just as commercial exploitation had begun (Corley et al. 1986). At present, the application of clonal plant production in oil palm is still limited owing to the occurrence of these somaclonal variants. It has been shown that the frequency of abnormal flowering varies between the clones, with some clearly being more susceptible (Durand-Gasselin 1995). Furthermore, Eeuwens et al. (2002) showed that the tissue culture medium on which embryoids are grown has a large effect: short transfer intervals between media, a low level of auxin, and a high level of cytokinin will increase the risk of somaclonal variation. Better protocols are being developed alongside molecular work to try to understand the basis of the change in floral morphology and methods of exploitation have also been adapted. The use of tissue culture to produce "clonal seed" is seen as an intermediate step toward full commercial exploitation. Clonal seed has the advantage that one of the parents is an elite clone, while the other is a seed-derived palm (as *dura* palms are limiting for the production of commercial seed, the *dura* is often the clonal palm). This has the advantage that any recessive somaclonal mutations should be masked by the contribution of the seedling gamete.

It has been widely assumed that abnormal flowering is an "epigenetic" phenomenon. A definition of epigenetics as postulated by Russo et al. (1996) says "epigenetics is a study of mitotically and/or meiotically heritable changes in gene function that cannot be explained by the changes in DNA sequence." Cytosine methylation is one of the more prevalent and intriguing mechanisms for generating epigenetic change. Whereas symmetric CpG nucleotides are the major target for methylation in animals, methylcytosine in plants and fungi can occur at C residues at symmetric (CpG; CpNpG) and asymmetric CpXpX (where $\times$ is any base other than guanosine) sites (Gruenbaum et al. 1981; Finnegan et al. 1998). Methylation is required for the normal development of animals and plants (Kakutani et al. 1996) and functions as a global repressor of gene expression (Bird 2002). There is increasing evidence that reduced DNA methylation can result in abnormal plant development (Finnegan et al. 1996; Chen et al. 1998). Interestingly, the "mantled" somaclonal variation in oil palm has been shown to be correlated with DNA hypomethylation, thus showing that normal and abnormal plants differ in the degree of methylation of nuclear DNA (Jaligot et al. 2000; Rival et al. 2000; Matthes et al. 2001). Furthermore, Kaeppler et al. (2000), in the review of somaclonal variation in plants, suggested that variation in methylation levels as a result of tissue culture can be a possible cause for the abnormal flowering.

It has been suggested that methylation evolved as a genomic defense against invasive DNA, including transposable elements and viruses (Yoder et al. 1997). Although most plant TEs are not transcriptionally active (i.e., they are neutral), they can be reactivated under the conditions of abiotic or biotic stress, so-called genomic stress (Kumar and Bennetzen 1999). However, only *Tnt*1 and *Tto*1 have been observed to be actually transposing (Grandbastien et al. 1989; Hirochika, 1993; Pouteau et al. 1994). Further application of a marker system based on retrotransposons to the oil palm somaclonal variants of the same ploidy exhibiting differences in the genome size would perhaps help us to understand the potential role of retrotransposons in the generation of somaclonal floral variants. It is worth pointing out that so far molecular markers have been unsuccessful in finding differences between normal and mantled palms which would be sufficiently repeatable and efficient enough to be useful as a screening method (proteins – Marmey et al. 1991; cytokinins – Jones et al. 1995; DNA markers – Cheah et al. 1993; Paranjothy et al. 1995; Sharifah et al. 1999; messenger RNA – Rival et al. 1998, 2001; Toruan-Mathius et al. 1998; Rajinder et al. 2001; Jaligot et al. 2002; Tregear et al. 2002).

4.6
Transformation Technology

The first evidence for transient expression of a reporter gene (GUS) in oil palm tissues delivered by microprojectile bombardment was reported in 1993 (Mayes et al. 1993) and since then significant effort has gone into developing potential transformation systems for oil palm, with major focuses being oil composition, abscission of fruit (Henderson et al. 2001), and potentially disease resistance (Buchannan 1999). The potential of this approach to oil palm improvement was examined and reported by Corley and Stratford (1998) and initial transformation of oil palm with a marker gene has been reported (Praveez 2000). Corley and Stratford (1998) estimate that production of a transformant line in sufficient numbers for field planting could be 15 years, even after transformation has been achieved. Given the very mixed results using transformation in annual crops (and the sensitivity in the EU over transgenic products), it could be a couple of decades before we see major implementation of this technology for oil palm.

4.7
Conclusions – the Future of Genetic Improvement in Oil Palm

The impressive progress made over the last 90 years will be difficult to match in the future without a substantial contribution from MAS and transformation and tissue culture approaches.

These offer the potential to short-circuit the long breeding and selection cycles currently needed for the genetic improvement and multiplication of oil palm, as well as offering novel solutions to genetic and agronomic problems through transformation.

4.7.1
Molecular Genetics

Recent advances in the creation of generally accessible, codominant SSR markers by the EU LINK2PALM program (http://www.neiker.net/link2palm/OilP/DefOIL.htm) will provide resources which can be used to generate genetic maps that can be integrated and used to begin to dissect the genetic basis of some of the key traits in oil palm. Generating molecular markers for direct MAS will be one consequence of this, but possibly as important will be gaining an understanding of the genetic basis and inter-relations for a number of agronomic traits. This may allow breeding approaches to be modified to improve their efficiency without the intrinsic use of markers and their associated costs.

Also the creation and characterization of the first oil palm expressed sequence tag clone database should permit the development of the first slide-based microarrays and facilitate the first use of the potentially extremely powerful transcriptomics approach in oil palm (http://www.mpob.gov.my/).

A limited sequence currently exists for oil palm, but this is likely to expand rapidly and a complete genome sequencing program for this important oil crop can only be a matter of time.

4.7.2
Models and Comparative Genetics

Using model systems to investigate oil palm has great potential, despite the considerable genetic distance between oil palm and any of the information-rich model systems. Whether the direct use of conserved gene order between species such as rice and oil palm will be feasible is unclear, although the ability to characterize genes in model systems such as *Arabidopsis thaliana* and to use information from conserved biochemical pathways will be invaluable.

Important areas where comparative genetics may make an impact include control of cell abscission in fruit, to reduce the problem of loose fruit collection, engineering of oil quality and composition through transgenic expression of homologous or heterologous oil biosynthesis genes, or even approaches to reduce height increment without a reduction in yield as has occurred for many cereal crops with semi-dwarfing genes.

4.7.3
Transgenics and Tissue Culture

The intrinsic potential of clonal propagation should become realized in the coming years, once concerns over abnormal flowering subside and this should facilitate the generation of transgenic oil palm expressing specific traits to improve a number of characters and develop novel traits such as materials for bioplastics, PHB and polyhydroxyalkanoates (Masani Mat Usus et al. 2001; http://minihelix.mit.edu/malaysia/research/me1.htm).

Perhaps one of the most important applications of transgenic technology will be approaches to reduce pests and diseases. For diseases like *Ganoderma*, where there is variation for genetic susceptibility but the inheritance appears complex, transgenic approaches may be critical and would certainly justify the research investment, not least the 15–20-year timescale needed to produce and test a transgenic oil palm.

References

Ariffin D, Idris AS, Gurmit S (2000) Status of *Ganoderma* in Oil Palm. In: Flood J, Bridge PD, Holderness M (eds) Ganoderma Diseases of Perennial Crops. CABI Publ, UK, pp 49–68

Barcellos E, Second G, Kahn F, Amblard P, Lebrun P, Sequin M (1999) Molecular markers applied to the analysis of genetic diversity and to the biogeography of *Elaeis* (Palmae). In: Memoirs of The New York Botanical Garden 83:191–201

Beinaret A, Vanderweyen R (1941) Contribution à l'étude génétique et biométrique des variétés d'*Elaeis guineensis* Jacq. Publs INEAC Sér Sci 27:20–35

Billotte N, Lagoda PJL, Risterucci AM, Baurens FC (1999) Microsatellite enriched-libraries: applied methodology for the development of SSR markers in tropical crops. Fruits 54:277–288

Billotte N, Risterucci AM, Barcelos E, Noyer JL, Amblard P, Baurens FC (2001) Development, characterisation, and across-taxa utility of oil palm (*Elaeis guineensis* Jacq.) microsatellite markers. Genome 44: 413–425

Billotte N, Marseillac N, Brottier P, Noyer JL, Jacquemoud-Collet JP, Moreau C, Couvreur T, Chevallier MH, Pintaud JC, Risterucci AM (2004a) Nuclear microsatellite markers for the date palm (*Phoenix dactylifera* L.): characterization and utility across the genus *Phoenix* and in other palm genera. Mol Ecol Notes 4:256–258

Billotte N, Couvreur T, Marseillac N, Brottier P, Perthuis B, Vallejo M, Noyer JL, Jacquemoud-Collet, Risterucci AM, Pintaud JC (2004b) A new set of microsatellite markers for the peach palm (*Bactris gasipaes* Kunth): characterization and across-taxa utility within the tribe Cocoeae. Mol Ecol Notes 4(4):803–814

Billotte N, Marseillac N, Risterucci AM, Adon B, Brottier P, Baurens FC, Singh R, Herrán A, Asmady, Billot C, Amblard P, Durand-Gasselin T, Courtois B, Asmono D, Cheah SC, Rohde W, Ritter E, Charrier A (2005) Microsatellite-based High Density linkage map in oil palm (*Elaeis guineensis* Jacq.). Theor Appl Genet 110(4):754–765

Bird A (2002) DNA methylation patterns and epigenetic memory. Genes Dev 16:6–16

Buchanan AG (1999) Molecular genetic analysis of Fusarium wilt resistance in oil palm. PhD Thesis, Univ of Bath, Bath, UK

Casson A (2000) The hesitant boom: Indonesia's oil palm sub-sector in an era of economic crisis and political change. Centre for International Forestry Research, Jakarta, Indonesia

Chase M (2004) Monocot relationships: an overview. Am J Bot 91:1645–1655

Chase MW, Stevenson DW, Wilkin P, Rudall P J (1995) Monocot systematics: a combined analysis, In: Rudall PJ, Cribb DF, Cutler DF, Humphries CJ (eds) Monocotyledons: Systematics and Evolution. Royal Botanical Garden Kew, UK

Cheah SC, Siti Nor Akmar A, Ooi LCL, Rahimah AR, Madon M (1993) Detection of DNA variability in the oil palm using RFLP probes. In: Proc 1991 PORIM Intl Palm Oil Conf, Palm Oil Res Inst, Malaysia, pp 144–150

Chen RZ, Pettersson U, Beard C, Jackson-Grusby L, Jaenish R (1998) DNA hypomethylation leads to elevated mutation rates. Nature 395:89–93

Choo Y M, Cheah KY (2000) Biofuel. In: Basiron Y, Jalani BS, Chan KW (eds) Advances in Oil Palm Research. vol 2. Malaysian Oil Palm Board, Kuala Lumpur, pp 806–844

Chua K L, Singh R, Cheah S C (2001) Construction of oil palm (*Elaeis guineensis*) linkage maps using AFLP markers. In: Proc 2001 Intl Palm Oil Congr, Malaysian Palm Oil Board, Kuala Lumpur, pp 461–465

Corley RHV (1983) Potential productivity of tropical perennial crops. Exp Agric 19:217–237

Corley RHV, Lee C H, Law I H, Cundall E (1986) Abnormal flower development in oil palm clones. In: Proc 1987 Intl Oil Palm Conf, Palm Oil Res Inst Malaysia, Kuala Lumpur, pp 173–185

Corley RHV, Lee CH (1992) The physiological basis for genetic improvement of oil palm in Malaysia. Euphytica 60:179–184

Corley RHV, Donough CR (1995) Effects of defoliation on sex differentiation in oil palm clones. Exp Agric 31(17):177–189

Corley HRV, Stratford R (1998) Biotechnology and oil palm: opportunities and future impact. In: Proc 1998 Intl Oil palm Conf on 'Commodity of the Past and the Future' Indonesian Oil Palm Res Inst, Medan, Indonesia, pp 80–91

Corley HRV, Tinker PB (2003) The Oil Palm. 4th edn. Blackwell Science, Ames, Iowa, USA

Crone GR 1937) The voyages of Cadamosto and other documents on Western Africa in the second half of the fifteenth century. Hakluyt Society, Series II, p 80

de Franqueville H, Asmady H, Jacquemard J C, Hayun Z, Durand-Gasselin T (2001) Indications on sources of oil palm (*Elaeis guneensis* Jacq.) genetic resistance and susceptibility to *Ganoderma* sp., the cause of basal stem rot. In: Proc 2001 Intl Palm oil Congr, Malaysian Palm Oil Board, Kuala Lumpur, pp 420–431

Foote T, Roberts M, Kurata N, Sasaki T, Moore G (1997) Detailed comparative mapping of cereal chromosome regions corresponding to the *Ph1* locus in wheat. Genetics 147:801–807

Dahlgren RMT, Cliffort HT, Yeo PF (1985) The Families of Monocotyledons: Structure, Evolution and Taxonomy. Springer, Berlin Heidelberg New York

Dumortier F (2000) Utilisation of oil palm genetic resources at DAMI oil palm research station. Paper presented at Intl Symp on Oil Palm Genetic Resources and Utilization, 8–10 June, Malaysian Oil Palm Board, Kuala Lumpur

Durand-Gasselin T, Duval Y, Baudoin L, Maheran A, Konan K, Noiret M (1995) Description and degree of mantled flowering abnormality in oil palm (*Elaeis guineensis* Jacq.) clones produced using the Orston-CIRAD procedure. In: Rao V, Henson LE, Rajanaidu N (eds) Recent Developments in Oil Palm Tissue Culture and Biotechnology. Palm Oil Res Inst, Kuala Lumpur, Malaysia, pp 48–63

Duval M R, Clegg MT, Chase MW, Clark WD, Kress JW, Hills HG, Equiarte LE, Smith JF, Gaut BS, Zimmer EA, Learn GH (1993) Phylogenetic hypotheses for the monocotyledons constructed from *rbc*L sequence data. Ann Missouri Bot Gard 80:607–619

Elenga H, Schwartz D, Vincens A (1994) Pollen evidence of late quaternary vegetation and inferred climatic change in the Congo. Paleography Paleoclimat Paleoecol 109:345–346

Enig MG (1998) Palm oil and the anti-tropical campaign: good news towards counteracting a decades worth of damage. In: Proc 1998 Intl Oil palm Conf 'Commodity of the Past and the Future' Indonesian oil Palm Res Inst, Medan, Indonesia, pp115–126

Ergo A B (1997) New evidence for African origin of *Elaeis guineensis* Jacq. by the discovery of the fossil seeds in Uganda. Ann Gembloux 102:191–201

Eeuwens CJ, Lord S, Donough CR, Rao V, Vallejo G, Nelson S (2002) Effects of tissue culture conditions during embryoid multiplications on the incidence of 'mantled' flowering in clonally propagated oil palm. Plant Cell Tiss Org Cult 70:311–323

Eyre-Walker A, Gaut BS (1997) Correlated rates of synonymous site evolution across plant genomes. Mol Biol Evol 14:455–460

Falconer DS (1981) Introduction to Quantitative Genetics. 2nd edn. Longman, London, UK

Finnegan E, Peacock WJ, Dennis ES (1996) Reduced DNA methylation in *Arabidopsis thaliana* results in abnormal plant development. Proc Natl Acad Sci USA 93:8449–8454

Finnegan EJ, Genger RK, Peacock WJ, Dennis ES (1998) DNA methylation in plants. Annu Rev Plant Physiol Plant Mol Biol 49:223–247

Flood J, Cooper R-M, Lees PE (1989) An investigation into the pathogenicity of four isolates of *Fusarium oxysporum* from South America Africa and Malaysia to clonal oil palm. J Phytopathol 124:80–88

Flood J, Bridge PD, Holderness M (2000) *Ganoderma* Diseases of Perennial Crops. CABI Publ, Wallingford, UK

Gaut BS, Muse SV, Clark WD, Clegg MT (1992) Relative rates of nucleotide substitutions at the *rbc*L locus in monocoyledonous plants. J Mol Evol 35:292–303

Gaut BS, Muse SV, Clark WD, Clegg MT (1996) Substitution rates comparisons between grasses and palms: synonymous rate differences at the nuclear gene *Adh* and parallel rate differences at the plastid gene *rbc*L. Proc Natl Acad Sci USA 93:10274–10279

Ghesquiere M (1984) Enzyme polymorphism in oil palm (*Elaeis guineensis* Jacq). 1. Genetic–control of 9 enzyme–systems. Oleagineaux 39:561–574

Ghesquiere M (1985) Enzyme polymorphism in oil palm (*Elaeis guineensis* Jacq). 2. Variability and genetic–structure of 7 origins of oil palm. Oleagineaux 40:529–537

Grandbastien MA, Spielman A, Caboche M (1989) *Tnt*1, a mobile retroviral-like transposable element of tobacco isolated by plant cell genetics. Nature 337:376–380

Gruenbaum Y, Naveh-Many T, Cedar H, Razin A (1981) Sequence specificity of methylation in higher plant DNA. Nature 292:860–862

Gomez PL, Ayala L, Munevar F (2000) Characteristics and management of bud rot, a disease of oil palm. In: Pushparajah E (ed) Proc Intl Planters Conf 'Plantation Tree Crops in the New Millennium: the Way Ahead' Incorp Soc Planters, Kuala Lumpur, Malaysia, pp 545–553

Hahn WJ, Kress JW, Zimmer EA (1996) 18S nrDNA sequence phylogenetics of the monocots. Am J Bot 83:211–212

Hahn WJ (2002a) A phylogenetic analysis of the Arecoid Line of palms based on plastid DNA sequence data. Mol Phylogenet Evol 23:189–204

Hahn WJ (2002b) A molecular phylogenetic study of the Palmae (Arecaceae) based on *atp*B, *rbc*L, and 18S rDNA sequences. Syst Biol 51:92–112

Hardon JJ, Corley RHV, Lee CH (1987) Breeding and selecting the oil palm. In: Abbot AJ, Atkin RH (eds) Improving Vegetatively Propagated Crops. Academic Press, London, UK, pp 63–81

Hartley CWS (1977) The Oil Palm. 2nd edn. Longmans, London, UK

Hartley CWS (1988) The Oil Palm. 3rd edn. Longmans, London, UK

Henderson J, Davies HA, Heyes SJ, Osborne DJ (2001) The study of a monocotyledon abscission zone using microscopic, chemical, enzymatic and solid state ^{13}C CP/MAS NMR analyses. Phytochem 56:131–139

Heywood HV (1993) Flowering Plants of the World. BT Batsford, London, UK

Hirochika H (1993) Activation of tobacco retrotransposons during tissue-culture. EMBO J 12:2521–28

Houmiel KL, Slater S, Broyles D, Casagrande L, Colburn S, Gonzalez K, Mitsky TA, Reiser SE, Shah D, Taylor NB, Tran M, Valentin HE, Gruys KJ (1999) Poly(beta-hydroxybutyrate) production in oilseed leukoplasts of *Brassica napus*. Planta 209(4):5475–5485

Jack PL, Dimitrijevic TAF, Mayes S (1995) Assessment of nuclear mitochondrial and chloroplast RFLP markers in oil palm (*Elaeis guineensis* Jacq.). Theor Appl Genet 90:643–649

Jack PL, James C, Price Z, Groves L, Corley RHV, Nelson S, Rao V (1998) Application of DNA markers in oil palm breeding. In: Proc 1998 Intl Oil Palm Conf, Indonesian Oil Palm Res Inst, Medan, Indonesia, pp 315–324

Jacquin NJ (1793) *Selectarum stirpium*. Americanarum Historia

Jaligot E, Rival A, Beulé T, Dussert S, Verdeil J l (2000) Somaclonal variation in oil palm (*Elaeis guineensis* Jacq.): the DNA methylation hypothesis. Plant Cell Rep 19:684–690

Jaligot E, Beulé T, Rival A (2002) Methylation-sensitive RFLPs: characterisation of two oil palm markers showing somaclonal variation–associated polymorphism. Theo Appl Genet 104:1263–1269

Jones DL (1994) Palms Throughout the World. Reed Books, Australia

Jones H, Hanke DE, Euwens CJ (1995) An evaluation of the role of cytokinins in the development of abnormal inflorescences in oil palms (*Elaeis guineensis* Jacq.) regenerated from tissue culture. J Plant Growth Reg 14:135–142

Kaeppler SM, Kaeppler HF, Rhee Y (2000) Epigenetic aspects of somaclonal variation in plants. Plant Mol Biol 43:179–188

Kakutani T, Jeddeloh JA, Flowers SK, Munakata K, Richards EJ (1996) Developmental abnormalities associated and epimutations with DNA hypomethylation mutations. Proc Natl Acad Sci USA 93:12406–12411

Kamarudin N, Walker AK, Basri Wahid M, Mohd Isa Z, Maimon A (1999) Population studies of Oryctes rhinoceros in an oil palm replant using pheromone traps. In: Preprints, 1999 PORIM Intl Palm Oil Conf, Palm Oil Research Inst, Malaysia, Kuala Lumpur, pp 477–496

Kearsey MJ, Pooni HS (1996) The Genetical Analysis of Quantitative Traits. Chapman and Hall, London, UK

Koh MP, Rahim S, Mohd Nor MY, Kamarudin H, Jalani BS (1999) Manufacture of building materials from oil palm biomass. In: Singh Gurmit et al. (eds) Oil Palm and the Environment – a Malaysian Perspective. Malaysian oil Palm Growers' Council, Kulala Lumpur, pp 199–211

Kulartne RS, Shah FH, Rajanaidu N (2000) Investigation of genetic diversity in African natural oil palm populations and Deli dura using AFLP markers. Paper presented at Intl Symp 'Oil Palm Genetic Resources and Utilization', 8–10 June, Malaysian Oil Palm Board, Kuala Lumpur

Kumar A, Bennetzen JL (1999) Plant retrotransposons. Annu Rev Genet 33(1):479–532

Madon M, Clyde M (1995) Cytological analysis of *Elaeis guineensis*. Elaeis 17(2):124–134

Marmey P, Besse I, Verdeil JL (1991) A proteic marker found to differentiate 2 types of calli of the same clones in oil palm. CR Acad Sci, Paris, Ser III, 313:333–338

Martienssen R (1998) Transposons, DNA methylation and gene control. Trends Genet 14(7):263–264

Masani Mat Usus A, Ho CL, Parveez GKA (2001) Construction of PHB gene expression vectors for the production of biodegradable plastics in transgenic oil palm. In: Proc 2001 Intl Palm Oil Congr, Malaysian Oil Palm Board, Kuala Lumpur, pp 674–711

Matthes M, Singh R, Cheah SC, Karp A (2001) Variation in oil palm (*Elaeis guineensis* Jacq.) tissue culture-derived regenerants revealed by AFLPs with methylation-sensitive enzymes. Theor Appl Genet 102:971–979

Mayes S, Horner SF, Jack PL, Corley RHV (1993) The application of biotechnology to oil palm – prospects and progress. In: Proc. 1993 Int. Palm oil Congr. – Agriculture. Malaysian Palm Oil Board, Kuala Lumpur

Mayes S, James CM, Horner SF, Jack PL, Corley RHV (1995) The application of restriction fragment polymorphism to genetic fingerprinting in oil palm (*Elaeis guineensis* Jacq.). Mol Breed 2:175–180

Mayes S, Jack PL, Marshall DF, Corley HRV (1997) Construction of a RFLP genetic linkage map for oil palm (*Elaeis guineensis* Jacq.). Genome 40:116–122

Mayes S, Jack PL, Corley RH (2000) The use of molecular markers to investigate the genetic structure of an oil palm breeding programme. Heredity 85(3):288–293

Mohan M, Nair S, Bhagwat A, Krishna T, Yano M, Bhatia C, Sasaki T (1997) Genome mapping, molecular markers, and marker assisted selection in crop plants. Mol Breed 3:87–103

Moretzsohn MC, Nunes CDM, Ferriera ME, Grattapaglia D (2000) RAPD linkage mapping of the shell thickness locus in oil palm (*Elaeis guineensis* Jacq.). Theor Appl Genet 100:63–70

Morton B R, Gaut BS, Clegg MT (1996) Evolution of alcohol dehydrogenase genes in palm and grass families. Proc Natl Acad Sci USA 93: 11735–11739

Nadot S, Bittar G, Carter L, Lacroix R, Lejeune B (1995) A phylogenetic analysis of monocotyledons based on the chloroplast gene rps4, using parsimony and new numerical phenetic methods. Mol Phylogenet Evol 4:257–282

Paranjothy K, Ong LM, Sharifah S (1995) DNA and protein changes in relation to clonal abnormalities. In: Rao V, Henson IE, Rajanaidu N (eds) Recent Developments in Oil Palm Tissue Culture and Biotechnology. Palm Oil Res Inst, Kuala Lumpur, Malaysia, pp 86–97

Parveez GKA, Masri MM, Zainal A, Majid NA, Masani Mat Yunus A, Fadilah HH, Rasid O, Cheah SC (2000) Transgenic oil palm: production and projection. Biochem Soc Trans 28:969–972

Piffanelli P, Lagoda P, Clément D, Thibivilliers S, Vilarinhos AD, Sabau X, Billotte N, Séguin M, Chalhoub B, Glaszmann JC (2002) Bactrop: A BAC-based platform for physical mapping of tropical species. In: Plant, Animal and Microbe Genomes × Conf, San Diego, CA, USA

Price Z, Dumortier F, Mayes S (2003) The development and initial application of DNA-based genetic markers to the new Britain Palm Oil Ltd (PNG) Breeding Programme. Proc of the 2003 PIPOC International Palm Oil Congress, Kuala Lumpur, Malaysia, pp 885–898

Price Z, Schulman A, Mayes S (2004) Development of new marker systems: oil palm. Plant Genet Resour: Charact and Util 1(2/3):105–115

Pouteau S, Grandbastien MA, Boccara M (1994) Microbial elicitors of plant defence response activate transcription of a retrotransposon. Plant J 5:535–542

Purba AR, Noyer JL, Baudouin L, Perrier X, Hamon S, Lagoda PJ L (2000) A new aspect of genetic diversity of Indonesian oil palm (*Elaeis guineensis* Jacq.) revealed by isoenzyme and AFLP markers and its consequences to breeding. Theor Appl Genet 101:956–961

Rajanaidu N, Maizura I, Cheah SC (2000) Screening of oil palm natural populations using RAPD and RFLP molecular markers. Paper presented at Intl Symp 'Oil Palm Genetic Resources and Utilization,' 8–10 June, Malaysian Oil Palm Board, Kuala Lumpur

Rajinder S, Cheah SC, Madon M, Ooi LCL, Rahman O (2001) Genomic strategies for enhancing the value of oil palm. In: Proc 2001 Intl Palm oil Congr, Malaysian Palm Oil Board, Kuala Lumpur, pp 3–17

Rance KA, Mayes S, Price Z, Jack PL, Corley HRV (2001) Quantitative trait loci for yield components in oil palm (*Elaeis guineensis* Jacq.). Theor Appl Genet 103(8):1302–1310

Raynaud FI, Maley J, Wirrrmann D (1996) Vegetation and climate in the forest of S-W Cameroun since 4770 years BP. Pollen analysis of sediments from Lake Ossa. Acad Sci, ser IIA. Sciences de la terre et des plantes 332:479

Rival A, Beule T, Barre P, Hamon S, Duval Y, Noirot M (1997) Comparative flow cytometric estimation of nuclear DNA content in oil palm (*Elaeis guineensis* Jacq.) tissue cultures and seed derived plants. Plant Cell Rep 16:884–887

Rival A, Tregear J, Verdeil JL, Richaud F, Beule T, Duval Y, Hartman C, Rohde A (1998) Molecular search for mRNA and genomic markers of the oil palm 'mantled' somaclonal variants in oil palm (*Elaeis guineensis* Jacq.). Acta Hort 461:165–171

Rival A, Tregear J, Jaligot E, Morcillo F Aberlenc F, Billotte N, Richaud F, Beule T, Borgel A, Duval Y (2001) Oil palm biotechnology at CIRAD. In: Proc 2001 Intl Palm oil Congr, Malaysian Palm Oil Board, Kuala Lumpur, pp 51–82

Rosenquist EA (1985) The genetic base of oil palm breeding populations. Proc palm Oil Res Inst Malaysia 10:10–27

Rosenquist EA, Corley RHV, de Greef W (1991) Improvement of *tenera* populations using germplasm from breeding programmes in Cameroon and Zaire. *Proc. of the Workshop Progress of Oil Palm Breeding Populations.* PORIM Bangi p 37–69

Russo VEA, Martienssen RA, Riggs AD (1996) Epigenetic mechanisms of gene regulation. Cold Spring Harbor Lab Press, Cold Spring Harbor, NY, USA

Sambanthamurthi R, Sundram K, Tan Y (2000) Chemistry and biochemistry of palm oil. Prog Lipid Res 39:507–558

San Miguel P, Tikhonov A, Jin Y K, Motchoulskaia N, Zakharov D, Melake-Berlan A, Springer PS, Edwards KJ, Lee M, Avramova Z, Bennetzen JL (1996) Nested retrotransposons in the intergenic regions of the maize genome. Science 274:765–768

Shizuya H, Birren B, Kim U-J, Mancino V, Slepak T, Tachiiri Y, Simon M (1992) Cloning and stable maintenance of 300-kilobase-pair fragments of human DNA in *Escherichia coli* using an F-factor-based vector. Proc Natl Acad Sci USA 89:8794–8797

Shah FH, Rashid O, Simons AJ, Dunsdon A (1994) The utility of RAPD markers for determination of genetic variation in oil palm (*Elaeis guineensis* Jacq.). Theor Appl Genet 89:713–718

Sharifah SSA, Singh R, Cheah SC (1999) Molecular dissection of the floral clonal abnormality in oil palm. In: Preprints, 1999 PORIM Intl Palm Oil Conf, Palm Oil Research Inst, Malaysia, Kuala Lumpur, pp 477–496

Soltis DE, Soltis PS, Nickrent DL, Johnson LA, Hahn WJ, Hoot SB, Sweere JA, Kuzoff RK, Kron KA, Chase MW, Swensen SM, Zimmer EA, Chaw SM, Gillespie LJ, Kress WJ, Sytsma KJ, (1997) Angiosperm phylogeny inferred from 18S ribosomal DNA sequences. Ann Missouri Bot Gard 18:1–49

Sundram K, Sambanthamurthi R, Tan YA (2003) Palm fruit chemistry and nutrition. Asia Pac J Clin Nutr 12:355–362

Tautz D, Renz M (1984) Simple sequences are ubiquitous components of eucaryotic genomes. Nucl Acids Res 12:4127–4138

Tregear JW, Morcillo F, Richaud F, Berger A, Singh R, Cheah SC, Hartmann C, Rival A, Duval Y (2002) Characterization of a defensin gene expressed in oil palm inflorescences: induction during tissue culture and possible association with epigenetic somaclonal variation events. J Exp Bot 53:1387–1396

Toruan-Mathius N, Harris N, Ginting G (1998) Use of the biomolecular techniques in studies of abnormalities in oil palm clones. In: Proc 1998 Intl Oil palm Conf 'Commodity of the Past and the Future', Indonesian oil Palm Res Inst, Medan, Indonesia, pp 115–126

Uhl NW, Dransfield J (1987) Genera Palmarum. Intl Palm Soc and LH Bailey Hortorium, Ithaca, NY, USA

Uhl NW, Dransfield J, Davis JI, Luckow MA, Hansen KS, Doyle JJ (1995) Phylogenetic relationship among palms: Cladistic analyses of morphological and chloroplast DNA restriction site variation. Monocot systematics: a combined analysis, In: Rudall PJ, Cribb DF, Cutler DF, Humphries CJ (eds) Monocotyledons: Systematics and Evolution. Royal Botanical Garden Kew, UK

van Duijn G (2000) Technical aspects of trans-reduction in margarines. Oléagineux, Corps Gras, Lipides 7:95–98

Vos P, Hogers R, Bleeker M, Reijans M, van de Lee T, Hornes M, Frijters A, Pot J, Peleman J, Kuiper M, Zabeau M (1995) AFLP: a new technique for DNA fingerprinting. Nucl Acids Res 23 (21):4407–4414

Williams J, Kubelik A, Livak K, Rafalski J, Tingey S (1990) DNA polymorphisms amplified by arbitrary primers are useful as genetic markers. Nucl Acids Res 18:631–6535

Wilson MA, Gaut B, Clegg MT (1990) Chrloroplast DNA evolves slowly in the palm family (Arecaceae). Mol Biol Evol 7:303–314

Wolfe KH, Gouy M, Li WH, Sharp P M (1987) Rates of nucleotide substitution vary greatly among plant, mitochondrial, chloroplast and nuclear DNAs. Proc Natl Acad Sci USA 84:9054–9058

Zagre NM, Delpeuch F, Traissac P, Delisle H (2003) Red palm oil as a source of vitamin A for mothers and children: impact of a pilot project in Burkina Faso. Public Health Nutr 6:733–742

Zeven AC (1964) On the origins of the oil palm (*Elaeis guineensis* Jacq). Grana Palynol 5:121–123

Zietkiewicz E, Rafalski A, Labuda A (1994) Genome fingerprinting by simple sequence repeat (SSR)-anchored polymerase chain reaction amplification. Genomics 20:176–183

Yoder JA, Walsh CP, Bestor TH (1997) Cytosine methylation and the ecology of intragenic parasites. Trends Genet 13(8):335–340

5 Coffee

Philippe Lashermes and François Anthony

Institut de Recherche pour le Développement (IRD), GeneTrop, BP 64501, Montpellier cedex 5, 34394, France
e-mail: philippe.lashermes@mpl.ird.fr

5.1
Introduction

Coffee is one of the world's most valuable export commodities, ranking second on the world market after petroleum products. The total retail sales value exceeded US $70 billion in 2003 and about 125 million people depend on coffee for their livelihoods in Latin America, Africa, and Asia (Osorio 2002). Commercial production relies on two species, *Coffea arabica* L. and *C. canephora* Pierre. The cup quality (low caffeine content and fine aroma) of *C. arabica* makes it by far the most important species, representing 70% of the world production.

C. arabica has its primary center of genetic diversity in the highlands of southwestern Ethiopia and the Boma Plateau of Sudan. Wild populations of *C. arabica* have been also reported in Mount Imatong (Sudan) and Mount Marsabit (Kenya) (Thomas 1942; Anthony et al. 1987). Cultivation of *C. arabica* started in southwestern Ethiopia about 1,500 years ago (Wellman 1961). Modern coffee cultivars are derived from two base populations – known as Typica and Bourbon – that were spread worldwide in the eighteenth century (Krug et al. 1939). Historical data indicate that these populations were composed of progenies of very few plants, i.e., only one for the Typica population (Chevalier and Dagron 1928) and the few plants that were introduced to the Bourbon Island (now Reunion) in 1715 and 1718 for the Bourbon population (Haarer 1956). Breeders exploited these narrow genetic bases, resulting in Typica- and Bourbon-derived cultivars with homogeneous agronomic behavior characterized by high susceptibility to many pests and low adaptability (Bertrand et al. 1999).

Coffee species belong to the Rubiaceae family, one of the largest tropical angiosperm families. Variations in chloroplast DNA (cpDNA) classified the Coffeeae tribe into the Ixoroideae monophyletic subfamily, close to Gardenieae, Pavetteae, and Vanguerieae (Bremer and Jansen 1991). Two genera, *Coffea* L. and *Psilanthus* Hook. f., were distinguished on the basis of flowering and flower criteria (Bridson 1982). Each genus was divided into two subgenera on the basis of growth habit (monopodial vs. sympodial development) and type of inflorescence (axillary vs. terminal flowers). Approximately 100 coffee species has been identified so far and new taxa are still being discovered (Bridson and Verdcourt 1988; Stoffelen 1998; Stoffelen et al. 2006). All species are perennial woody bushes or trees in intertropical forests of Africa and Madagascar for the *Coffea* genus, and Africa, Southeast Asia, and Oceania for the *Psilanthus* genus. They differ greatly in morphology, size and ecological adaptations. Some species like *C. canephora* and *C. liberica* Hiern are widely distributed from Guinea to Uganda. Other species display specific adaptations, for example, *C. congensis* Froehner to seasonally flooded areas in the Zaire basin and *C. racemosa* Lour. to very dry areas in the coastal region of Mozambique. All species are diploid ($2n = 2x = 22$) and are generally self-incompatible, except for *C. arabica*, which is tetraploid ($2n = 4x = 44$) and self-fertile (Charrier and Berthaud 1985).

Molecular phylogeny of *Coffea* species has been established on the basis of DNA sequence data. The internal transcribed spacer 2 (ITS 2) region of the nuclear ribosomal DNA (rDNA; Lashermes et al. 1997) as well as the cpDNA variation (Lashermes et al. 1996a; Cros et al. 1998) were successfully used to infer phylogenetic relationships of *Coffea* species (Fig. 1). No major difference between coffee and other plants was observed in the arrangement of the chloroplast genome and in the structure of the ITS 2 region in nuclear rDNA. Furthermore, the results suggest a radial mode of speciation and a recent origin in Africa for the genus *Coffea*. Several major clades were identified, which present a strong geographical correspondence (i.e., Madagascar, East Africa, Central Africa, Central Africa, and West Africa). In addition, the *Psilanthus*

Fig. 1. Phylogenetic relationships of coffee species based on sequence variations of chloroplast and ribosomal genomes

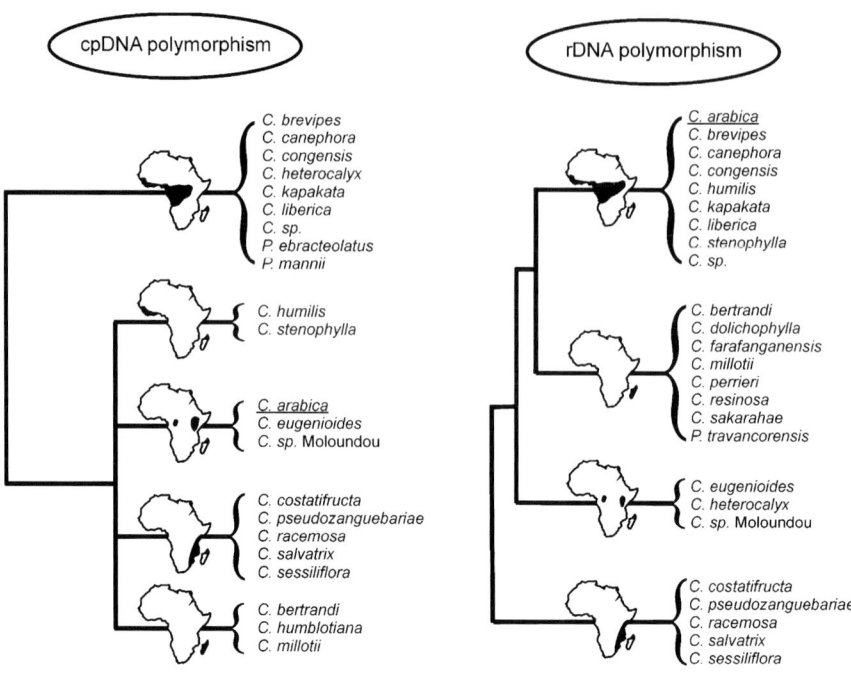

species did not differ from *Coffea* species, suggesting that the present division into two genera should be revised.

5.2
Genome Structure and Molecular Diversity

5.2.1
An Allotetraploid Species

The genome constitution and mode of speciation of *C. arabica* have been subjected to several investigations. On the basis of cytological observations and the fertility of interspecific hybrids, *C. canephora* and *C. congensis* were suggested to have a common ancestor with *C. arabica* (Carvalho 1952; Cramer 1957). Furthermore, *C. congensis* can be considered as an ecotype of *C. canephora* in light of the fertility of interspecific hybrids (Louarn 1993) and the genetic diversity detected by molecular markers (Prakash et al. 2005). These phylogenetic assumptions were consolidated by a high homology found in the ITS 2 region sequences of *C. arabica*, *C. canephora*, and *C. congensis* (Lashermes et al. 1997). Regarding cpDNA that exhibited a maternal inheritance in coffee (Lashermes et al. 1996a), *C. arabica* appeared to be similar

to two species, *C. eugenioides* Moore and *Coffea* sp. "Moloundou" (Cros et al. 1998).

Restriction fragment length polymorphism (RFLP) markers in combination with genomic in situ hybridization (GISH) were used to investigate the origin of *C. arabica*. Comparison of the RFLP patterns of potential diploid progenitor species with those of *C. arabica* allowed the source of the two sets of chromosomes, or genomes, combined in *C. arabica* to be specified. The genome organization of *C. arabica* was confirmed by GISH using simultaneously labeled total genomic DNA from the two putative genome donor species as probes (Raina et al. 1998; Lashermes et al. 1999). These results clearly suggested that *C. arabica* is an amphidiploid (i.e., $C^{a}E^{a}$ genomes) resulting from the hybridization between *C. eugenioides* (E genome) and *C. canephora* (C genome) or ecotypes related to those diploid species (Fig. 2). Results also indicated low divergence between the two constitutive genomes of *C. arabica* and those of its progenitor species, suggesting that the speciation of *C. arabica* took place very recently. Precise localization in Central Africa of the speciation process of *C. arabica* based on the present distribution of the coffee species appeared difficult since the constitution and extent of tropical forest varied considerably during the late quaternary period. Furthermore, investigations suggest that homoeologous chromosomes do not pair in *C. arabica*, not as a consequence of structural

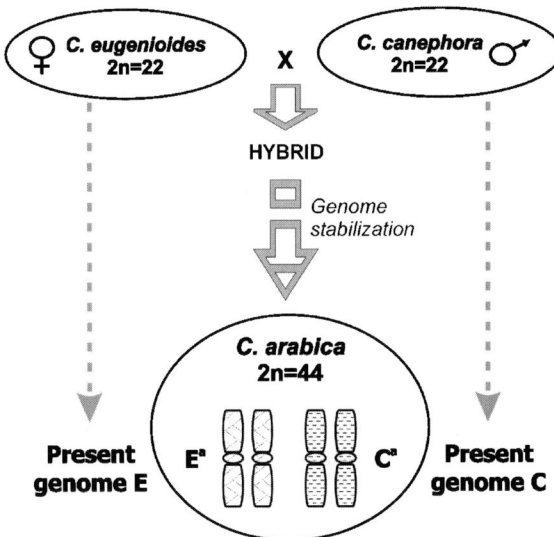

Fig. 2. Origin of the allotetraploid species *Coffea arabica*

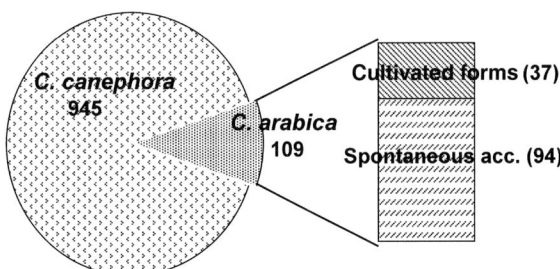

Fig. 3. The numbers of polymorphic markers (amplified fragment length polymorphism) observed among individuals within each group constituted by representative accessions of *C. arabica* and *C. canephora*, respectively. For *C. arabica*, cultivated forms (i.e., Bourbon- and Typica-derived cultivars) and wild accessions from Ethiopia were distinguished (Lashermes et al. 2000a)

differentiation, but because of the functioning of pairing regulating factors (Lashermes et al. 2000b).

5.2.2
An Extremely Low Genetic Diversity

In coffee, a whole range of different techniques have been used to detect polymorphism at the DNA level, including randomly amplified polymorphic DNA (Orozco-Castilho et al. 1994; Lashermes et al. 1996b; Anthony et al. 2001; Aga et al. 2003), cleaved amplified polymorphisms (Lashermes et al. 1996b; Orozco-Castilho et al. 1996), RFLPs (Paillard et al. 1996; Lashermes et al. 1999; Dussert et al. 2003), amplified fragment length polymorphism (AFLP; Lashermes et al. 2000a; Anthony et al. 2002), and simple sequence repeats or microsatellites (Mettulio et al. 1999; Combes et al. 2000; Anthony et al. 2002).

The use of molecular methods has opened up new possibilities for genetic analysis and provides new tools for the efficient conservation and use of coffee genetic resources. For instance, the genetic diversity in *C. arabica* appeared extremely reduced in comparison to the diversity observed in *C. canephora* (Fig. 3). This low genetic diversity has been attributed to the allotetraploid origin, reproductive biology, and evolution process of *C. arabica*. The base populations of cultivated coffee clearly originated from wild coffee collected in southwestern Ethiopia. The genetic differentiation was low between the Typica and Bourbon groups, but much higher between wild and cultivated coffee. Spontaneous accessions collected in the primary center of diversity appeared to constitute a valuable gene reservoir. These results should increase interest in wild coffee for the purpose of broadening the genetic base of cultivars. Moreover, a heterosis effect has been reported in F_1 hybrids resulting from crosses between subspontaneous Ethiopian accessions and improved cultivars (Bertrand et al. 1999).

5.3
Breeding Strategies

Transfer of desirable genes in particular for disease resistance from diploid species like *C. canephora* and *C. liberica* into tetraploid Arabica cultivars without affecting quality traits has been the main objective of Arabica breeding (Carvalho 1988; Van der Vossen 2001). To date, *C. canephora* provides the main source of disease and pest resistance traits not found in *C. arabica*, including coffee leaf rust (*Hemileia vastatrix*), coffee berry disease (*Colletotrichum kahawae*), and root-knot nematode (*Meloidogyne* spp.). Likewise, other diploid species present considerable interest in this respect. For instance, *C. liberica* has been used as a source of resistance to leaf rust (Srinivasan and Narasimhaswamy 1975), whereas *C. racemosa* constitutes a promising source of resistance to the coffee leaf miner (Guerreiro Filho et al. 1999). Exploitation of such genetic resources has so far relied on conventional procedures in which a hybrid is produced between an outstanding variety and a donor geno-

Fig. 4. Triploid and
tetraploid strategies of gene
introgression from diploid
coffee species into the
C. arabica genome

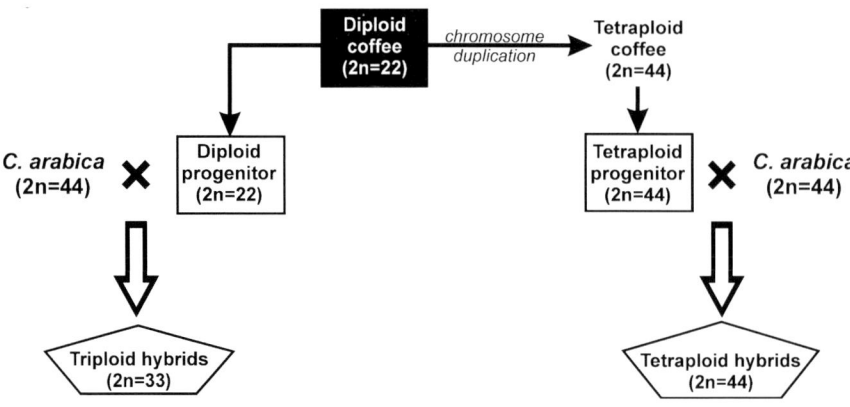

type carrying the trait of interest, and the progeny
is backcrossed to the recurrent parent. Undesirable
genes from the donor parent are gradually eliminated
by selection. In so doing, conventional coffee breeding
methodology faces considerable difficulties. In particular, strong limitations are due to the long generation
time of the coffee tree (5 years), the high cost of field
trials, and the lack of accuracy of current strategy.
A minimum of 25 years after hybridization is required
to restore the genetic background of the recipient cultivar and thereby ensure good quality of the improved
variety.

Genes from diploid coffee could be transferred
into the *C. arabica* genome through the production
of triploid ($2n = 33$) or tetraploid ($2n = 44$) interspecific hybrids (Fig. 4). Triploid hybrids are derived
from a direct cross between the diploid progenitor and
C. arabica, whereas tetraploid hybrids are obtained by
crossing both species after chromosome duplication
of the diploid coffee. The use of triploid hybrids is
more difficult than that of tetraploid hybrids because
of low fertility (Krug and Mendes 1940; Berthaud
1978a, b). However, selection of high producing lines
in early generations derived from triploid hybrids
(i.e., first backcross to *C. arabica* or second generation by selfing) has been reported (Orozco Castillo
1989).

5.3.1
Introgression Assessment

The presence of *C. canephora* DNA fragments was
sought in accessions derived from a tetraploid
interspecific hybrid known as Timor hybrid. The
Timor hybrid originated from a spontaneous cross
between *C. arabica* and *C. canephora* on the island
of Timor (Bettencourt 1973). Following a backcross

with a *C. arabica* cultivar (i.e., cultivar Caturra or
Villasarchi), progenies were selfed and selected over
three to five generations in several important coffee-
producing countries, such as Brazil, Colombia and
Kenya. Although varying among the Timor hybrid
derived inbreed lines, the amount of alien genetic material appeared substantial (Lashermes et al. 2000a).
Furthermore, the incidence of *C. canephora* gene
introgression on biochemical composition of beans
and cup quality in Arabica coffee breeding lines was
investigated (Bertrand et al. 2003). A high amount
of introgression appeared frequently associated with
lower quality factors. However, lines combining
resistance to leaf rust and root-knot nematode (*M.
exigua* and having good cup quality were successfully
developed.

A similar study was undertaken to analyze
the S.288 progeny of a putative natural hybrid
(*C. arabica* × *C. liberica*) and accessions (F_2 and F_4)
derived from the cross (S.288 × cultivar Kent)
(Prakash et al. 2002). The number of introgression
markers was found to be similar in the *C. liberica*
introgressed accessions and in the *C. canephora* introgressed accessions. Analysis of genetic relationships
in the introgressed lines suggested that introgression
was limited to a few fragments. Moreover, the alien
genetic material appeared to be fixed and there was
no elimination or counterselection over generations,
from introgressed parent to F_4.

5.3.2
Factors Controlling Gene Introgression

During the last few years, the behavior of interspecific
hybrids between *C. arabica* and the diploid species
C. canephora and *C. eugenioides* has been investigated. Numerous plant populations resulting from the

backcross (BC$_1$) of either triploid or tetraploid interspecific hybrids to *C. arabica* were analyzed (Herrera et al. 2002a, b, 2004). Flow cytometric analysis of the nuclear DNA content revealed that most of the BC$_1$ individuals were tetraploid or nearly tetraploid, suggesting that among the gametes produced by the interspecific hybrids, those presenting 22 chromosomes were strongly favored. Furthermore, molecular markers (i.e., RFLP, microsatellite, and AFLP) combined with evaluation of morphological characteristics and resistance to leaf rust were applied to verify the occurrence of gene transfer from the donor species into *C. arabica*, and to estimate the amount of introgression present in BC$_1$ individuals. While a high amount of introgression was observed in the progenies derived from the tetraploid interspecific hybrids, the BC$_1$ individuals generated from the triploid interspecific hybrids exhibited contrasting situations. The mean proportion of introgressed markers per plant was significantly lower in populations derived from *C. eugenioides* than from *C. canephora*. Moreover, the comparison of reciprocal progenies between *C. arabica* and triploid interspecific hybrids (*C. arabica* × *C. canephora*) used as a male or female parent revealed a very strong effect of the backcross direction. A severe reduction in frequency of *C. canephora* introgressed markers was observed when the triploid hybrids were used as the male parent. Breeding strategies based on gene introgression can now be designed according to the objectives of selection.

5.4
Genetic Mapping

The low polymorphism has been a major drawback for developing genetic maps of the *C. arabica* genome. Hence, the works reported so far (Pearl et al. 2004; Prakash et al. 2004) are restricted to alien DNA introgressed fragments. To overcome this difficulty, efforts were directed to the development of genetic maps in *C. canephora* or interspecific crosses. For the low differentiation between the C and C^a genomes (Lashermes et al. 1999; Herrera et al. 2002a), *C. canephora* maps are particularly relevant.

The earliest attempt was based on doubled haploid segregating populations (Paillard et al. 1996; Lashermes et al. 2001). *C. canephora* is a strictly allogamous species consisting of polymorphic populations and of strongly heterozygous individuals. Conventional segregating populations are therefore somehow difficult to generate and analyze. However, the ability to produce doubled haploid populations in *C. canephora* offers an attractive alternative approach. The method of doubled haploid production is based on the rescue of haploid embryos of maternal origin occurring spontaneously in association with polyembryony (Couturon 1982). Two complementary segregating plant populations of *C. canephora* were produced from the same genotype. One population (doubled haploid) comprised 92 doubled haploids derived from female gametes, whereas the other population was a testcross consisting of 44 individuals derived from male gametes.

A genetic linkage map of *C. canephora* was constructed with 160 DNA markers spanning 1,041 cM of the genome (Lashermes et al. 2001). Eleven linkage groups that putatively correspond to the 11 gametic chromosomes of *C. canephora* were identified (Fig. 5). Only two markers appeared unlinked, reflecting reasonably good genome coverage. The total genetic map length of the *C. canephora* genome was estimated at 1,400 cM. Although substantial variation over very short distance could be anticipated, one could estimate that 1 cM is equivalent to approximately 570 kb. In addition, this genetic linkage map comprised more than 40 specific sequence-tagged site markers, either single-copy RFLP probes or microsatellites, that are distributed on the 11 linkage groups. These markers constituted an initial set of standard landmarks of the coffee genome which has been used as anchor points for map comparison (Herrera et al. 2002a). Furthermore, the recombination frequencies in both populations were found to be almost indistinguishable. These results offer evidence in favor of the lack of significant sex differences in recombination in *C. canephora*. In connection with quantitative trait loci research, additional genetic maps of *C. canephora* have been recently reported (Crouzillat et al. 2004; T. Leroy, personal communication).

5.5
Gene Tagging

Gene tagging in *C. arabica* has been restricted so far to disease resistance genes. Special attention has been given to resistance traits introgressed from diploid species (Noir et al. 2003; Prakash et al. 2004).

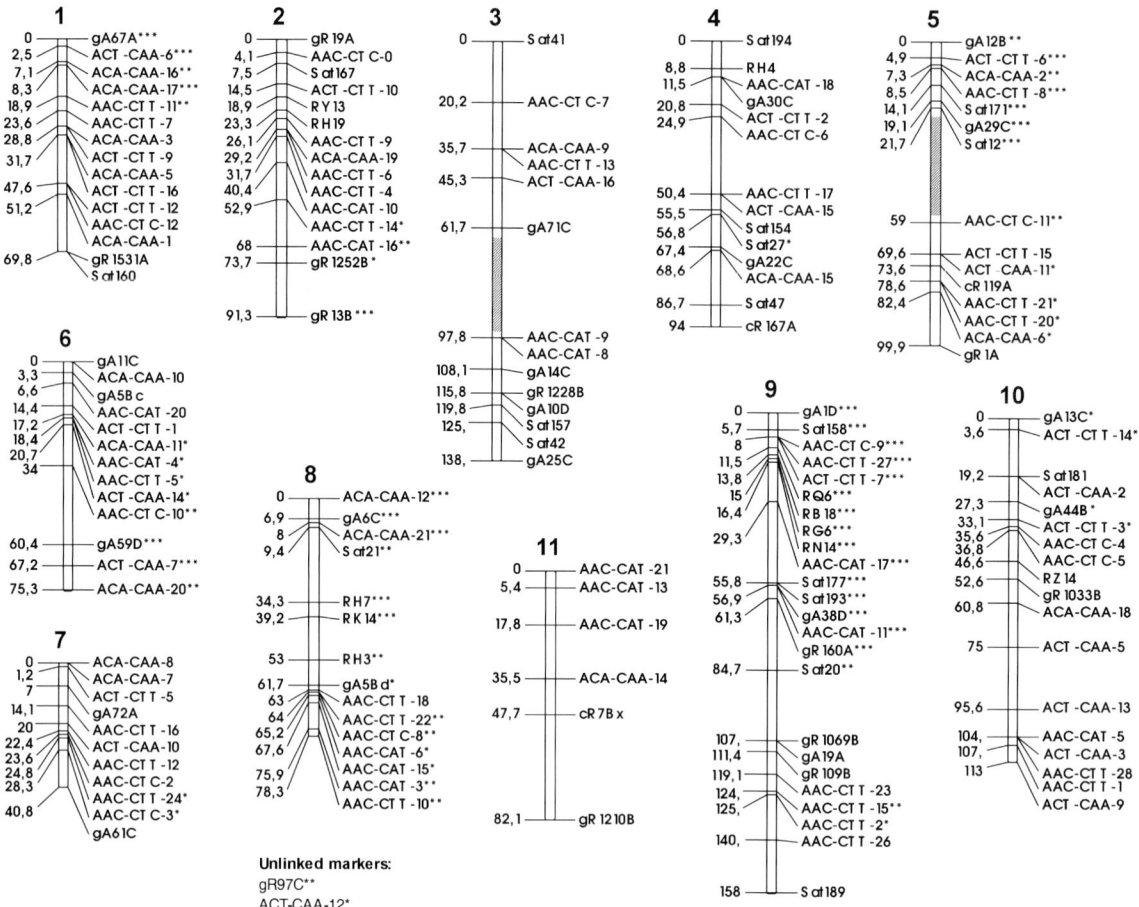

Fig. 5. Linkage map of *C. canephora* (Lashermes et al. 2001). Cumulated distances in centimorgans (Kosambi function) are indicated on the *left of the linkage groups*. All linkage groups were constituted with a logarithm of odds (*LOD*) greater than 5 except for two associations indicated by *hachures* (LOD=3.5)

Segregation data analysis of an F_2 progeny derived from a cross between the resistant introgression line T5296 and the susceptible accession Et6 showed that the resistance to *M. exigua* inherited from *C. canephora* is controlled by a simply inherited major gene, designated the *Mex-1* locus. The gall index distribution exhibited by the F_2 individuals suggested an incomplete dominant expression. AFLP markers were found associated with the resistance to *M. exigua* and a localized genetic map of the chromosome segment carrying *Mex-1* was constructed (Noir et al. 2003). Furthermore, the association of the identified AFLP markers with *Mex-1* was confirmed by the analysis of a set of genotypes involving 28 introgression Arabica lines either resistant or susceptible to *M. exigua* in field conditions. These results represent an important starting point to enhance backcross breeding pro-

grams and to perform an early selection of resistant seedlings.

Similar work was accomplished to gain insight into the mechanism of introgression of a leaf rust resistance gene from *C. liberica* (i.e., S_H3 resistance factor) into the *C. arabica* genome. An F_2 progeny (i.e., 101 individuals) derived from a cross between an Arabica accession (cultivar Matari) and a *C. liberica* introgressed line (S.288) was evaluated for resistance against three different races of *H. vastatrix* (Prakash et al. 2004). The data of segregation confirmed the hypothesis of a single dominant gene for the S_H3 resistance factor. Markers tightly linked to the S_H3 rust resistance gene were identified. All the markers linked to the S_H3 gene were distributed in a distance of 6.3 cM and recombination within the introgressed region was also evident as six recombinants were identified in the population. However, no recombinant individuals were detected between S_H3 and a marker named M8. In fact this marker was found to cosegregate perfectly with S_H3, which determines the high selection efficiency of this marker for resistant genotypes.

5.6
Genomics

Recent advances in coffee genomics consist of a huge collection of expressed sequence tags (ESTs) and bacterial artificial chromosome (BAC) libraries.

In the past few years, several projects of sequencing coffee ESTs have been initiated in the USA (Mueller et al. 2004), Brazil (http://www.cenargen.embrapa.br/genomacafe/), and Europe (Fernandez et al. 2004; Pallavicini et al. 2004). Libraries were constructed from various organs and tissues sampled at different developmental stages. Functional analysis has been initiated (Fernandez et al. 2004; Pallavicini et al. 2004). Furthermore, genetic transformation of coffee plants has been successfully achieved by several research groups (Hatanaka et al. 1999; Leroy et al. 2000) but still remains a tedious process. Recently, RNA interference technology was applied in coffee (Ogita et al. 2004).

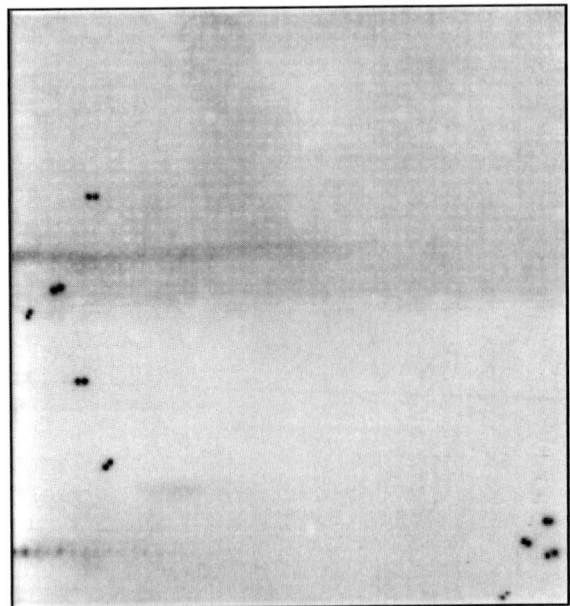

Fig. 6. Example of colony hybridization of bacterial artificial chromosome filters: 27,648 clones were double-spotted onto a nylon membrane and hybridized with nuclear single-copy probe gA6 (Noir et al. 2004)

The construction of an integrated genetic and physical map of the coffee genome has been undertaken. A BAC library of the allotetraploid species *C. arabica* was constructed (Noir et al. 2004). This large insert DNA library derived from a multidisease resistance line contains 88,813 clones with an average insert size of 130 kb, and represents approximately 8 times *C. arabica* haploid genome equivalents. The mapping approach combines hybridization with mapped markers and BAC fingerprinting (Fig. 6). For instance, hybridization with both low-copy RFLP markers distributed on the 11 chromosomes and probes corresponding to disease resistance gene analogs (Noir et al. 2001) was completed. Positive BAC clones from subgenomes E^a and C^a were assembled into separate contigs, showing the efficiency of the combined approach for mapping purposes.

5.7
Conclusions and Perspectives

In the last decade, overproduction has resulted in historically low coffee prices which have devastating effects on coffee producers. Farmers survive with great difficulties in coffee plantations or abandon coffee culture, switching to alternative crops. Consequently, lowering of bean quality and supply instability could affect the coffee market of consumer countries. In the twentieth century, coffee production benefited from the selection of spontaneous mutants (e.g., the compact cultivar Caturra), natural hybrids between species (e.g., the Timor hybrid), and new cultural practices (e.g., open sun monoculture). Coffee germplasm was collected but little has been done at the molecular level to exploit biodiversity in coffee species as well as their genomic resources.

The recent development of high-capacity methods for analyzing the structure and function of genes, which may be collectively termed "genomics", represents a new paradigm which broad implications. Although currently available for only a few model plants, it seems likely that such information will rapidly become available for most widely studied plant species such as *C. arabica*. The advent of large-scale molecular genomics will provide access to previously inaccessible sources of genetic variation which could be exploited in breeding programs. Anticipated outcomes in coffee breeding include (1) rapid characterization and managing of germplasm resources, (2) enhanced understanding of the genetic control of priority traits, (3) identification of candidate genes or tightly linked genomic regions underlying important traits, and (4) identification of accessions in genetic collections with variants of genomic regions or alleles of candidate genes having a favorable impact on priority traits. In this way, the recent efforts to set up an international commitment (ICGN, http://www.coffeegenome.org) to work jointly for the development of common sets of genomic tools, plant populations, and concepts would be extremely useful.

References

Aga E, Bryngelsson T, Bekele E, Salomon B (2003) Genetic diversity of forest arabica coffee (*Coffea arabica* L.) in Ethiopia revealed by random amplified polymorphic DNA (RAPD) analysis. Hereditas 138:36–46

Anthony F, Berthaud J, Guillaumet JL, Lourd M (1987) Collecting wild *Coffea* species in Kenya and Tanzania. Plant Genet Resour Newsl 69:23–29

Anthony F, Bertrand B, Quiros O, Wilches A, Lashermes P, Berthaud J, Charrier A (2001) Genetic diversity of wild coffee (*Coffea arabica* L.) using molecular markers. Euphytica 118:53–65

Anthony F, Combes MC, Astorga C, Bertrand B, Graziosi G, Lashermes P (2002) The origin of cultivated *Coffea arabica* L. varieties revealed by AFLP and SSR markers. Theor Appl Genet 104:894–900

Berthaud J (1978a) L'hybridation interspécifique entre *Coffea arabica* L et *Coffea canephora* Pierre. Obtention et comparaison des hybrides triploïdes, Arabusta et hexaploïdes. Première partie. Café Cacao Thé 22:3–12

Berthaud J (1978b) L'hybridation interspécifique entre *Coffea arabica* L et *Coffea canephora* Pierre. Obtention et comparaison des hybrides triploïdes, Arabusta et hexaploïdes. Deuxième partie. Café Cacao Thé 22:87–112

Bertrand B, Aguilar G, Santacreo R, Anzueto F (1999) El mejoramiento genético en América Central. In: Bertrand B, Rapidel B (eds). Desafíos de la caficultura centroamericana. IICA, San José (Costa Rica), pp 407–456

Bertrand B, Guyot B, Anthony F, Lashermes P (2003) Impact of *Coffea canephora* gene introgression on beverage quality of *C. arabica*. Theor Appl Genet 107:387–394

Bettencourt A (1973) Consideraçoes gerais sobre o 'Hibrido de Timor'. Circular no 31. Instituto Agronômico de Campinas

Bremer B, Jansen RK (1991) Comparative restriction site mapping of chloroplast DNA implies new phylogenetic relationships within *Rubiaceae*. Am J Bot 78:198–213

Bridson D (1982) Studies in *Coffea* and *Psilanthus* (*Rubiaceae* subfam. *Cinchonoideae*) for Part 2 of 'Flora of Tropical East Africa': *Rubiaceae*. Kew Bull 36:817–859

Bridson D, Verdcourt B (1988) *Coffea*. In: Polhill RM (ed) Flora of Tropical East Africa. Rubiaceae (Part 2). Balkema, Rotterdam, The Netherlands, pp 703–727

Carvalho A (1952) Taxonomia de *Coffea arabica* L. Caracteres morfologicos dos haploides. Bragantia 12:201–212

Carvalho A (1988) Principles and practice of coffee plant breeding for productivity and quality factors: *Coffea arabica*. In: Clarke RJ, Macrae R (eds) Coffee, vol 4: Agronomy. Elsevier Applied Science, London, UK, pp 129–165

Charrier A, Berthaud J (1985) Botanical classification of coffee. In: Clifford MN, Willson KC (eds) Coffee: Botany, Biochemistry and Production of Beans and Beverage. Croom Helm, London, UK, pp 13–47

Chevalier A, Dagron M (1928) Recherches historiques sur les débuts de la culture du caféier en Amérique. Communications et Actes de l'Académie des Sciences Coloniales (Paris) 5:1–38

Combes MC, Andrzejewski S, Anthony F, Bertrand B, Rovelli P, Graziosi G, Lashermes P (2000) Characterisation of microsatellite loci in *Coffea arabica* and related coffee species. Mol Ecol 9:1178–1180

Couturon E (1982) Obtention d'haploïde spontanés de *Coffea canephora* Pierre par l'utilisation du greffage d'embryons. Café Cacao Thé 26 (3):155–160

Cramer PJS (1957) Review of literature of coffee research in Indonesia. Miscellaneous publication no 15. Interamerican Institute of Agricultural Sciences, Turrialba

Cros J, Combes MC, Trouslot P, Anthony F, Hamon S, Charrier A, Lashermes P (1998) Phylogenetic relationships of *Coffea* species: new evidence based on the chloroplast DNA variation analysis. Mol Phylogenet Evol 9:109–117

Crouzillat D, Rigoreau M, Bellanger L, et al. (2004) A Robusta consensus map using RFLP and microsatellites markers for the detection of QTL. In: Proc 20th International Scientific Colloquium on coffee, Banagalore (India). ASIC ed, Paris, France

Dussert S, Lashermes P, Anthony F, Montagnon C, Trouslot P, Combes MC, Noirot M, Hamon S (2003) Coffee (*Coffea canephora*). In: Hamon P, Seguin M, Perrier X, Glaszmann C (eds) Genetic Diversity of Cultivated Tropical Plants. Science Publ, Plymouth, UK, pp 239–258

Fernandez D, Santos P, Agostini C, Bon MC, Petitot AS, Silva MC, Guerra-Guimaraes L, Ribeiro A, Argout X, Nicole M (2004) Coffee (*Coffea arabica* L.) genes early expressed during infection by the rust fungus (*Hemileia vastatrix*). Mol Plant Pathol 5(6):527–536

Guerrero Filho O, Silvarolla MB, Eskes AB (1999) Expression and mode of inheritance of resistance in coffee to leaf miner *Perileucoptera coffeella*. Euphytica 105:7–15

Haarer AE (1956) Modern Coffee Production. Leonard Hill (books) Ltd, London, UK

Hatanaka T, Choi YE, Kusano T, Sano H (1999) Transgenic plants of coffee *Coffea canephora* from embryogenic callus via *Agrobacterium Tumefaciens*-mediated transformation. Plant Cell Rep 19:106–110

Herrera JC, Combes MC, Anthony F, Charrier A, Lashermes P (2002a) Introgression into the allotetraploid coffee (*Coffea arabica* L.): segregation and recombination of the *C. canephora* genome in the tetraploid interspecific hybrid (*C. arabica* × *C. canephora*). Theor Appl Genet 104:661–668

Herrera JC, Combes MC, Cortina H, Alvarado G, Lashermes P (2002b) Gene introgression into *Coffea arabica* by way of triploid hybrids (*C. arabica* × *C. canephora*). Heredity 89:488–494

Herrera JC, Combes MC, Cortina H, Lashermes P (2004) Factors regulating gene introgression into the allotetraploid *Coffea arabica* L. from its diploid relatives. Genome 47:1053–1060

Krug CA, Mendes JET (1940) Cytological observations in *Coffea*. J Genet 39:189–203

Krug CA, Mendes JET, Carvalho A (1939) Taxonomia de *Coffea arabica* L. Technical bulletin no 62. Instituto Agronômico do Estado, Campinas

Lashermes P, Cros J, Combes MC, Trouslot P, Anthony F, Hamon S, Charrier A (1996a) Inheritance and restriction fragment length polymorphism of chloroplast DNA in the genus *Coffea* L. Theor Appl Genet 93:626–632

Lashermes P, Trouslot P, Anthony F, Combes MC, Charrier A (1996b) Genetic diversity for RAPD markers between cultivated and wild accessions of *Coffea arabica*. Euphytica 87:59–64

Lashermes P, Combes MC, Trouslot P, Charrier A (1997) Phylogenetic relationships of coffee tree species (*Coffea* L.) as inferred from ITS sequences of nuclear ribosomal DNA. Theor Appl Genet 94:947–955

Lashermes P, Combes MC, Robert J, Trouslot P, D'Hont A, Anthony F, Charrier A (1999) Molecular characterisation and origin of the *Coffea arabica* L. genome. Mol Gen Genet 261:259–266

Lashermes P, Andrzejewski S, Bertrand B, Combes MC, Dussert S, Graziosi G, Trouslot P, Anthony F (2000a) Molecular analysis of introgressive breeding in coffee (*Coffea arabica* L.). Theor Appl Genet 100:139–146

Lashermes P, Paczek V, Trouslot P, Combes MC, Couturon E, Charrier A (2000b) Single-locus inheritance in the allotetraploid *Coffea arabica* L. and interspecific hybrid *C. arabica* × *C. canephora*. J Hered 91:81–85

Lashermes P, Combes MC, Prakash NS, Trouslot P, Lorieux M, Charrier A (2001) Genetic linkage map of *Coffea canephora*: effect of segregation distortion and analysis of recombination rate in male and female meioses. Genome 44:589–595

Leroy T, Henry AM, Royer M, Altosaar I, Frutos R, Duris D, Philippe R (2000) Genetically modified coffee plants expressing the *Bacillus thuringiensis cry1Ac* gene for resistance to leaf miner. Plant Cell Rep 19:382–389

Louarn J (1993) Structure génétique des caféiers africains diploïdes basée sur la fertilité des hybrides interspécifiques. In: Proc 15th International Scientific Colloquium on Coffee, Montpellier, France, ASIC ed, Paris, France, pp 243–252

Mettulio R, Rovelli P, Anthony F, Anzueto F, Lashermes P, Graziosi G (1999) Polymorphic microsatellites in *Coffea arabica*. In: Proc 18th International Scientific Colloquium on Coffee, Helsinki, Finland, ASICed, Paris, France, pp 344–347

Mueller L, Lin C, Mc Carthy J, Pétiard V, Crouzillat D, Ilut D, Tanksley S (2004) Generation and analysis of a coffee EST database: deductions about genome content and comparison with tomato/potato. In: Proc 20th International Scientific Colloquium on Coffee, Banagalore (India), ASIC ed, Paris, France

Noir S, Combes MC, Anthony F, Lashermes P (2001) Origin, diversity and evolution of NBS-type disease-resistance gene homologues in coffee trees (*Coffea* L.). Mol Genet Genom 265:654–662

Noir S, Anthony F, Bertrand B, Combes MC, Lashermes P (2003) Identification of a major gene (*Mex*-1) from *Coffea canephora* conferring resistance to *Meloidogyne exigua* in *Coffea arabica*. Plant Pathol 52:97–103

Noir S, Patheyron S, Combes MC, Lashermes P, Chalhoub B (2004) Construction and characterisation of a BAC library for genome analysis of the allotetraploid coffee species (*Coffea arabica* L.). Theor Appl Genet 109:225–230

Ogita S, Uefuji H, Morimoto M, Sano H (2004) Application of RNAi to confirm theobromine as the major intermediate for caffeine biosynthesis in coffee plants with potential for construction of decaffeinated varieties. Plant Mol Biol 54 (6):931–941

Orozco Castillo FJ (1989) Utilización de los híbridos triploides en el mejoramiento genético del café. In: Proc 13th International Scientific Colloquium on Coffee, Paipa (Colombia), ASIC ed, Paris, France, pp 485–495

Orozco-Castillo C, Chalmers KJ, Waugh R, Powell W (1994) Detection of genetic diversity and selective gene introgression in coffee using RAPD markers. Theor Appl Genet 87:934–940

Orozco-Castillo C, Chalmers KJ, Powell W, Waugh R (1996) RAPD and organelle specific PCR re-affirms taxonomic relationships within the genus *Coffea*. Plant Cell Rep 15:337–341

Osorio N (2002) The global coffee crisis: a threat to sustainable development. ICO, London, UK

Paillard M, Lashermes P, Pétiard V (1996) Construction of a molecular linkage map in coffee. Theor Appl Genet 93:41–47

Pallavicini A, de Nardi B, Dreos R, del Terra L, Sondhal MR, Guerreiro-Filho O, Asquini E, Martellossi C, Rajkumar R, Graziosi G (2004) Transcriptomics of resistance response in coffee (*C. arabica* L.). In: Proc 20th International Scientific Colloquium on Coffee, Banagalore (India), ASIC ed, Paris, France

Pearl HM, Nagai C, Moore PH, Steiger DL, Osgood RV, Ming R (2004) Construction of a genetic map for arabica coffee. Theor Appl Genet 108:829–835

Prakash NS, Combes MC, Dussert S, Naveen S, Lashermes P (2005) Analysis of genetic diversity in Indian robusta coffee genepool (*Coffea canephora*) in comparison with a representative core collection using SSRs and AFLPs. Genet Resour Crop Evol 52:333–343

Prakash NS, Combes MC, Somanna N, Lashermes P (2002) AFLP analysis of introgression in coffee cultivars (*Coffea arabica* L.) derived from a natural interspecific hybrid. Euphytica 124:265–271

Prakash NS, Marques DV, Varzea VMP, Silva MC, Combes MC, Lashermes P (2004) Introgression molecular analysis of a leaf rust resistance gene from Coffea liberica into Coffea arabica L. Theor Appl Genet 109:1311–1317

Raina SN, Mukai Y, Yamamoto M (1998) *In situ* hybridisation identifies the diploid progenotor of *Coffea arabica* (Rubiaceae). Theor Appl Genet 97:1204–1209

Srinivasan KH, Narasimhaswamy RL (1975) A review of coffee breeding work done at the Government coffee experiment station, Balehonnur. Indian Coffee 34:311–321

Stoffelen P (1998) *Coffea* and *Psilanthus* (Rubiaceae) in tropical Africa: a systematic and palynological study, including a revision of the West and Central African species. Doctoral thesis, Katholieke Universiteit Leuven, Faculteit Wetenschappen, Belgium

Stoffelen P, Noirot M, Couturon E, Anthony F (2006) A new caffeine-free coffee species in the deep rain forest of Cameroon. Taxon (in press)

Thomas AS (1942) The wild arabica coffee on the Boma Plateau, Anglo-Egyptian Sudan. Emp J Exp Agric 10:207–212

Van der Vossen HAM (2001) Agronomy I: Coffee Breeding Practices. In: Clarke RJ, Vitzthum OG (eds) Coffee: Recent Developments. Blackwell Science, UK, pp 184–201

Wellman FL (1961) Coffee: Botany, Cultivation and Utilization. Leonard Hill Books, London:, UK

6 Tea

Junichi Tanaka[1] and Fumiya Taniguchi[1]

Genetic Resources and Breeding Team & Tea Genome Research Team, National Institute of Vegetable and Tea Science, National Agriculture and Bio-Oriented Research Organization, Seto-cho 87, Makurazaki, Kagoshima 898-0087, Japan
e-mail: tanajun@affrc.go.jp

6.1
Introduction

6.1.1
Tea in a Botanical Context

Botanically speaking, *Camellia sinensis* (L.) O. Kuntze is a woody evergreen plant of the genus *Camellia* in the Theaceae family, native to southern China. It has a genome of 4,000 Mb (Tanaka et al. 2004) with a base number of $n = 15$. Known varieties are *sinensis* and *assamica*. The former grows in the wild, mainly in southern China, and the latter is found over a wide area of southeastern Asia, from southern China to India. There is no known natural growth of *C. sinensis* in Sri Lanka, a prominent tea producer. Variety *sinensis* is a low shrub characterized by small rounded leaves and cold hardiness. Conversely, variety *assamica* is an arboreous plant characterized by large, pointed leaves more serrated than those of variety *sinensis* and low tolerance to cold. Although many intermediate morphologies exist in southern China, presumably resulting from interbreeding of the two varieties, a comparison of typical variety *sinensis* and typical variety *assamica* finds significant morphological differences as if the two varieties are unrelated. Fig. 1 compares the leaves of the small-leaved variety *sinensis* and the large-leaved variety *assamica*. Although their morphological differences are remarkable, no clear reproductive isolation can be found between the two varieties; they can freely cross-breed, and their offspring are just as fertile. The Taiwanese mountain-type tea plant native to mountainous parts of Taiwan also has large, well-serrated leaves. Although this is classified as variety *assamica*, it is much more tolerant to cold than variety *assamica*. It is also known to lack trichomes and to have a characteristic odor in the plant itself. Results of DNA analyses using randomly amplified polymorphic DNA (RAPD) and am-plified fragment length polymorphism (AFLP) point to significantly less affinity with variety *sinensis* or typical variety *assamica* (Wachira et al. 2001). Although *C. sinensis* shows remarkably wide morphological variation, there is no clear reproductive isolation between variety *sinensis* and variety *assamica*, or between Taiwanese mountain-type tea and either of the two varieties. Cross-breeding successfully produces offspring.

6.1.2
Camellia sinensis as a Beverage Crop

The custom of brewing leaves of the tea plant for a beverage has its origin in China, with numerous records dating back more than 2,000 years. It is believed that, by the period of the Three Kingdoms (the third century) at the latest, tea had been established as a beverage for the ordinary citizenry. In the sixteenth century, British colonists began cultivating tea in India with seeds (variety *sinensis*) brought from China. They produced Chinese-style green tea at the beginning. The 1823 discovery of a tall tea tree (variety *assamica*) growing in the wild in Assam province was followed by many more discoveries of wild tea trees growing in the area. Tea producers began to use the seeds from this newly discovered variety in developing tea plantations. Furthermore, as black tea, which was fermented longer than the Chinese-style green tea, gained popularity in Europe, tea production in India gradually converted to black tea. Sri Lanka has a relatively short history in tea production, having started in the middle of the nineteenth century.

Today, tea is widely cultivated and drunk as green or black tea not only in China but all around the world. Planted area totals about 2.3 million hectares, with China, India, Sri Lanka, Kenya, and Indonesia being the major producers. These five countries account for

Fig. 1. Comparison of size of the leaves in tea (*Camellia sinensis*). The small one on the *left* is the leaf of the smallest type of variety *sinensis* (from Nigata, Japan), the one in the *middle* is the leaf of large type of variety *sinensis* (Japanese cultivar Yabukita) and one on the *right* is the typical variety *assamica*

75% of the total planted area. Tea is a major export product for these major producers, except China, and an important crop as a source of foreign currency. Major importers include the UK, Russia, Pakistan, the USA, Egypt, and Japan. Black tea accounts for about 70% of the world production, and green tea for most of the remainder.

6.1.3
Breeding Objectives for Tea

We discuss the main breeding objectives and the possibility of utilizing marker-assisted selection (MAS) for tea.

Quality
As tea is a recreational beverage, quality is the most important breeding objective. Taste and aroma are important measures of the quality of tea. Characters that denote high quality, however, are significantly different among the types of tea, such as black, green, or oolong tea. For example, tea rich in catechins will be considered high quality for black tea, but the opposite for green tea. Because samples processed into various types of tea are evaluated by sensory tests, it takes a long time before a stable evaluation of quality can be established.

If dominant alleles are detected, MAS may be available for tea breeding.

Yield
As is the case with any crop, yield is a very important breeding objective in tea. Harvesting tea, however, is done by picking the shoots of new leaves. Since new leaves grow over time, an early harvest results in a small yield, and a late harvest in a reduced yield. This makes an evaluation of yield as a genetic capability very difficult. Green tea breeders in Japan actually harvest shoots at a specific stage and weigh them to evaluate yield. As the age of the tea plant affects yield significantly, it takes a long time before a stable evaluation of characteristics can be established.

Resistance to Pests and Diseases
As is the case with any crop, resistance to diseases and pests is a very important breeding objective. Tea diseases include webbed blister blight, anthracnose, and gray blight. Pests include mites, aphids, plant hoppers, and scale. In Japan, mulberry scale has been especially rampant in recent years. With clear varietal differences in resistance to many of these diseases and pests, they are possible candidates for MAS. MAS for mulberry scale resistance has been prac-

ticed in Japan (Tanaka and Taniguchi 2003), and will be discussed later.

Harvest Time

Harvested tea shoots cannot be stored. They need to be processed into products immediately after harvest. In the temperate zone, new shoots grow out in the spring as the temperature rises. With a single cultivar, the entire crop reaches the harvesting stage all at once. To avoid this, a producer will need to plant several cultivars with different harvest times. If the locus of a gene which determines early or late harvest can be identified, it may be possible to use MAS for genetic control of harvest time.

Other Objectives

One of the breeding objectives is adaptability to inclement environments, such as water stress, high or low temperatures, salinity, and excess moisture. Nitrogen fertilizers used on tea plantations are incriminated in groundwater pollution in Japan. Consequently, it is important to produce characteristics which will improve the ability of tea plants to absorb nutrients while maintaining quality and yielding ability with reduced nitrogen.

6.1.4
Challenges for Tea Breeding and Significance of MAS

Tea is an arboreous crop which requires a long time to establish itself so as to provide reliable yields. This also means that it will be a long time before its characteristics can be assessed. In the case of green tea breeding in Japan, it takes an average of 30 years before the characteristics of a cultivar can be assessed, the varietal competence is recognized, and it is actually registered as a new cultivar. Such cultivars are often used for cross-breeding only after they have been recognized, requiring generational turnover taking several decades. As a result, tea breeding has not reached the stage of applying selection pressure through repeated genetic recombination processes. Many genetic resources, however, have been collected and preserved in China, India, and Japan. Such resources represent very wide genetic diversity. With this diversity as the foundation, if an appropriate selection pressure is applied through a series of genetic recombination processes, tea breeding should improve exponentially.

Under the present conditions for green tea breeding in Japan, it is technically feasible to turn one generation over to the next in 3–5 years at the shortest by keeping the tea plant in a pot to restrict the root zone or treating it with a dwarfing agent. It is, however, often unclear as to which individual plant should be used in cross-breeding for such a generational turnover. Providing an early selection index, especially a direct genotype assessment, has to wait for the development of practical MAS.

6.2
Use of Markers and Genetic Analysis with DNA Markers

6.2.1
Use of DNA Markers for Nongenetic Analyses

Objectives of DNA-marker related work on tea often focus on assessment of genetic diversity, varietal classification, and identification of parentage, rather than genetic analysis. Matsumoto et al. (1994) assessed genetic diversity with RFLP using an phenylalanine ammonia-lyase sequence, while Wachira et al. (1997) used RAPD for their assessment. Kaundun and Matsumoto (2003) have demonstrated that it is possible to classify many tea cultivars using cleaved amplified polymorphism sequence markers. DNA markers have also been used in some assessments. For example, one such assessment resulted in discovery of an erroneous identification of registered parentage in some cultivars (Tanaka and Yamaguchi 1996). DNA markers were also used to identify the parentage of the mulberry scale resistant cultivar Sayamakaori (Tanaka et al. 2001), as well as interspecific crossing between tea (*C. sinensis*) and *C. japonica* (Tanaka et al. 2003).

6.2.2
Genetic Analysis with DNA Markers

Because tea is a self-incompatible allogamous plant, the existence of homozygous individuals is doubtful. Even if a homozygous individual can be produced, it will suffer extreme inbreeding depression. Because of this, linkage maps have been constructed based on pseudo-testcross strategy (Grattapaglia and Sederoff 1994). This method generates two maps for both parents in a group if a dominant maker technology is used. Tanaka et al. (1995) constructed a preliminary

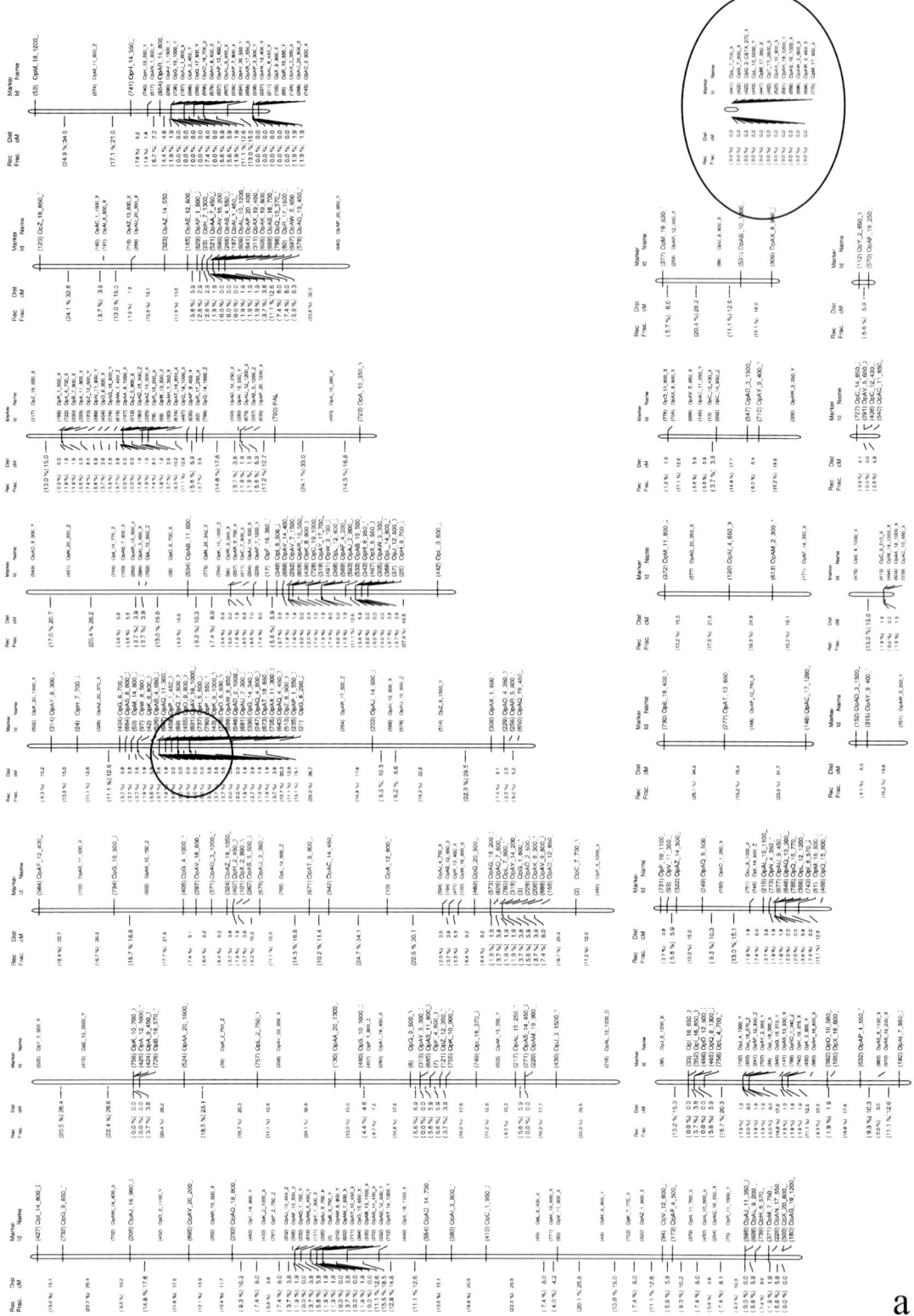

Fig. 2. Linkage maps of tea clones (**a**) Sayamakaori and (**b**) Kana-Ck17

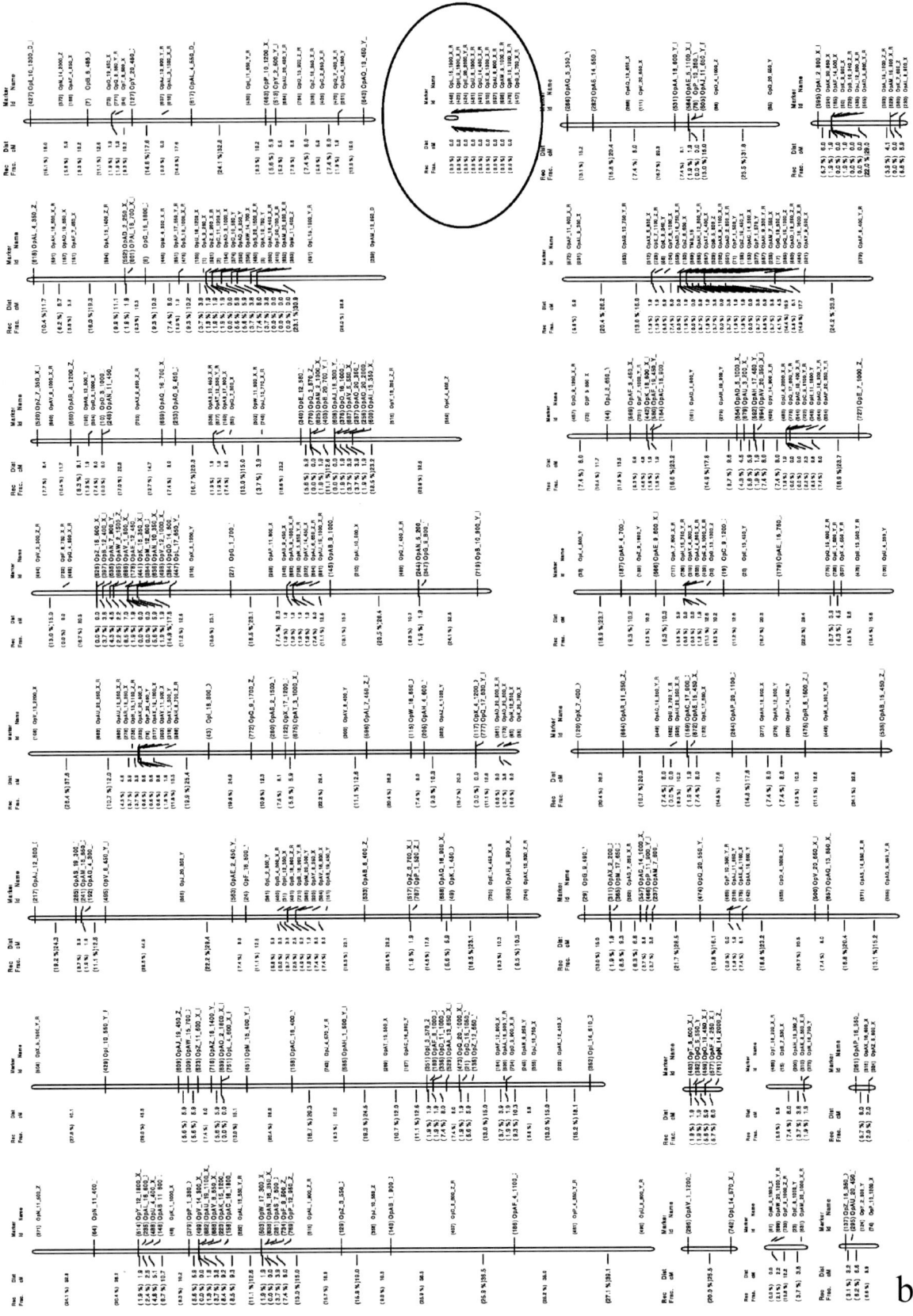

Fig. 2. (continued)

b

linkage map with RAPDs. In addition, Hackett et al. (1997) constructed the maps with AFLPs, and Ota and Tanaka (1999) with RAPDs. Fig. 2 shows linkage maps with the largest number of markers to date. We produced the maps by building on the maps generated by Ota and Tanaka (1999). As seen in these maps, the markers in the linkage groups enclosed by ovals, where genetic recombination does not occur, represent cytoplastic inheritance, presumably derived from organellar DNA. The cytogenetic use of RAPD is explained in detail in the report by Tanaka and Ota (2002).

In addition to the quantitative trait loci (QTLs) associated with anthracnose, leaf angle, and leaf color on the maps in Fig. 2, a mulberry scale resistant QTL was detected in the linkage group of Sayamakaori at the location indicated by the circle. The dominance of this QTL was as high as 56%. It was also detectible as a single dominant gene (MSR-1).

6.3
MAS for Mulberry Scale Resistance

Mulberry scale lives inside the tree crown where pesticides do not penetrate. The window for control is limited to the larval period, making it a typical tea pest highly difficult to control. Damage from mulberry scale has been steadily increasing over the last decade. Verification of mulberry scale resistance, however, is extremely complex, and MAS using MSR-1 was expected to be a very valuable means for control.

The use of MAS with MSR-1, however, needed to overcome two major hurdles. The first was the range of applicable cross-breeding combinations. There were many segregated populations in offspring of scale-resistant Sayamakaori, but a marker which could be used for both the parent and the offspring was needed. Very rare allele markers linked to coupling were critical for solving this problem.

Fortunately, MSR-1 was found at a locus densely populated by many markers, resulting in many linked DNA markers, from which markers closely linked to coupling but rarely detected in most other parents were selected. These established markers which were not affected by crossing parents and could be used for selection of offspring as long as the linkage remained intact.

The second hurdle was labor, time, and cost for detecting the DNA markers, most of the labor and cost being needed for DNA extraction. In particular, the significant amounts of polysaccharides and polyphenols contained in tea required a complex technique to extract high-quality DNA. Tanaka and Ikeda (2002) successfully developed a protocol that could extract high-quality DNA from tea leaves quickly with the diatomaceous earth used in plasmid extraction kits for coliform bacteria.

These technological solutions have made MAS available for breeding mulberry scale resistant tea in Japan.

6.4
Advanced Works

We are investigating polymorphism using a collection of 2,000 simple sequence repeat sequences so as to establish landmark markers on linkage maps. It is possible to make mass detection of single nucleotide polymorphisms (SNPs) available at low cost. Currently, many studies on expressed sequence tags (EST) in *C. sinensis* are being carried out (Park et al. 2004; Chen et al. 2005; Sharma and Kumar 2005). We have developed EST and bacterial artificial chromosome libraries and are now analyzing them. We need to pick out SNPs from these sequences and place them on linkage maps.

References

Chen L, Zhao L, Gao Q (2005) Generation and analysis of expressed sequence tags from the tender shoots cDNA library of tea plant (*Camellia sinensis*). Plant Sci 168:359–363

Grattapaglia D, Sederoff R (1994) Genetic linkage maps of *Eucalyptus grandis* and *Eucalyptus urophylla* using a pseudo-testcross: Mapping strategy and RAPD markers. Genetics 137:1121–1137

Hackett CA, Wachira FN, Paul S, Powell W, Waugh R (2000) Construction of a genetic linkage map for tea (*Camellia sinensis*). Heredity 85:346–355

Kaundun SS, Matsumoto S (2003) Development of CAPS markers based on three key genes of the phenylpropanoid pathway in tea, *Camellia sinensis* (L.) O. Kuntze, and differentiation between *assamica* and *sinensis* varieties. Theor Appl Genet 106 375–383

Matsumoto S, Takeuchi A, Hayatsu M, Kondo S (1994) Molecular cloning of phenylalanine ammonia-lyase cDNA and classification of varieties and cultivars of tea plants (*Camellia sinensis*) using the tea PAL cDNA probe. Theor Appl Genet 86:671–675

Ota S, Tanaka J (1999) RAPD-based linkage mapping using F₁ segregating populations derived from crossings between tea cultivar "Sayamakaori" and strain "Kana-Ck17". Breed Res 1(Suppl 1):16

Park J, Kim J, Hahn B, Kim K, Ha S, Kim J, Kim Y (2004) EST analysis of genes involved in secondary metabolism in *Camellia sinensis* (tea), using suppression subtractive hybridization. Plant Sci 166:953–961

Sharma P, Kumar S (2005) Differential display-mediated identification of three drought-responsive expressed sequence tags in tea [*Camellia sinensis* (L.) O. Kuntze]. J Biosci 30(2):231–235

Tanaka J, Ikeda S (2002) Rapid and efficient DNA extraction method from various plant species using diatomaceous earth and a spin filter. Breed Sci 52:151–155

Tanaka J, Ota SM (2002) Detection of maternally inheritable RAPDs using the F_1 populations derived from reciprocal crossing on tea. Breed Res 4:215–222

Tanaka J, Taniguchi F (2003) Allele-specific e-RAPD MSRS8E can screen efficiently the mulberry scale resistant individuals by resistant gene MSR-1. http://www.naro.affrc.go.jp/top/seika/2003/vegetea/ve03030.html

Tanaka J, Yamaguchi S (1996) Use of RAPD markers for the identification of parentage of tea cultivars. Bull Natl Res Inst Veg Ornam Plants Tea B9:31–36

Tanaka J, Sawai Y, Yamaguchi S (1995) Genetic analysis of RAPD markers in tea. Jpn J Breed 45(Suppl 2):198

Tanaka J, Yamaguchi N, Nakamura Y (2001) Pollen parent of tea cultivar Sayamakaori with insect and cold resistance may not exist. Breed Res 3:43–48

Tanaka J, Ota SM, Takeda Yoshiyuki (2003) The garden-variety Camellia 'Robiraki' derived from crossing between *Camellia japonica* as a seed parent and *C. sinensis* as a pollen parent. Application of RAPD and an SSR marker analysis to tea breeding by interspecific hybridization. Breed Res 5:149–154

Tanaka J, Hirai N, Taniguchi F, Yamaguchi S (2004) Estimations of the genome-sizes of *Camellia sinensis, C. japonica* and their interspecific hybrids by flow cytometry. Tea Res J 98(Suppl):88–89

Wachira FN, Powel W, Waugh R (1997) An assessment of genetic diversity among *Camellia sinensis* L. (cultivated tea) and its wild relatives based on randomly amplified polymorphic DNA and organelle-specific STS. Heredity 78:603–611

Wachira FN, Tanaka J, Takeda Y (2001) Genetic variation and differentiation in tea germplasm revealed by RAPD and AFLP variation. J Hort Sci Biotechnol 76:557–563

7 Cacao

Ranjana Bhattacharjee[1,2] and P. Lava Kumar[2]

[1] International Institute of Tropical Agriculture (IITA), Ibadan, Oyo State, Nigeria
 e-mail: r.bhattacharjee@cgiar.org
[2] International Crops Research Institute for the Semi-Arid Tropics (ICRISAT), Patancheru 502 324, Andhra Pradesh, India

7.1
Introduction

Cacao, *Theobroma cacao*, a diploid ($2n = 2x = 20$) tropical tree, belongs to the family Sterculiaceae (alternatively Malvaceae *sensu lato*) and order Malvales. It is native to humid topics of the central and northern parts of South America. It is commonly grown in hot, rainy climates between 20°N and 20°S of the equator, with maximum cultivation between 10°N and 10°S (Fig. 1). The plant is the source of chocolate and four intermediary products: cocoa cake, cocoa butter, cocoa powder, and cocoa liquor. In this chapter an overview of *T. cacao*, the methods used for crop improvement, and recent developments in *T. cacao* molecular breeding is presented.

7.1.1
History and Distribution

Cacao has its origin in ancient Central America where the Maya and the Aztecs cultivated it for its seeds (beans), which are used for extracting a drink called *chocolatl*, a precursor to the modern chocolate (Young 1994). Olmec and Mayan civilizations believed that cacao had a divine origin and regarded it as "food of the gods" and thus its scientific name *Theobroma*, (*Theo* meaning "food" and *Broma* meaning "gods") (Coe and Coe 1996). Archaeological evidence suggests that the cacao plantations in the northern and central parts of South America date back to 2,000 years before Spanish contact (Bergmann 1969; Whitlock et al. 2001). The cacao plantations extend from Mexico to Costa Rica, and other locations in South America and the Caribbean (Wood and Lass 1985). The primary center of origin was believed to be the region extending from the forests of the Amazon to the Orinoco and Tabasco in southern Mexico (Whitlock et al. 2001). However, two theories were proposed to explain the origin and domestication of cacao: some studies hypothesized a Central American origin and others a South American origin (Van Hall 1914; Cheesman 1944; Schultes 1984). The hypothesis of Cuatrecasas (1964) of separate simultaneous origins in South and Central America was widely supported (Miranda 1962; Cope 1976; Wood and Lass 1985; Laurent et al, 1994; Whitkus et al. 1998). However, a recent study based on restriction fragment length polymorphism (RFLP) and microsatellite analysis strongly suggests that cacao originated in the Upper Amazon of South America and was later introduced by humans to Central America (Motamayor et al. 2002).

The spread of cacao to other continents was triggered during the post-Colombian era. Hernando Cortez in the sixteenth century was reported to discover the bitter drink used by the Aztecs, and sent the beans and recipes back to Europe (Opeke 1969), where the recipes were refined, and during the nineteenth century technologies were developed that facilitated the roasting and pressing of cacao beans, leading to the creation of eating chocolate, and gained global popularity (Coe and Coe 1996). Cultivation of cacao in other continents was initiated during the colonial era of the eighteenth and nineteenth centuries. It was believed that William Pratt in the 1840s had introduced cacao into a Spanish colony, Fernando Po (presently Bioko in Equatorial Guinea) in Africa, to sustain an interest in cacao drinks, and to produce cacao beans at lower prices from Spanish colonies. From there, cacao was spread to Nigeria, Ghana, Côte d'Ivoire, and other West African countries by trading companies, missionaries, chiefs, and soldiers. Countries such as Sierra Leone, Togo, and Benin embraced cacao cultivation in the early part of the twentieth century (Johns and Gibberd 1951). The early development of the cacao industry in West Africa was entirely due to the initiative and entrepreneurship of the West African peasant farmers (Johns and Gibberd 1951). Owing to good

Fig. 1. *Theobroma cacao* producing regions of the world (Source: International Cacao Organization)

adoption of the tree and improved trade, West Africa has become a hub for cacao production, contributing 70% of the total world production of cacao beans (Gray 2001).

Currently cacao is cultivated in 6.9 million hectares worldwide with an annual productivity of 3.9 Mt (FAOSTAT data 2006). Over half of the world supply of commercial cacao comes from Côte d'Ivorie (39%), Ghana (19%), and Indonesia (13%) (UNCTAD 2005; Table 1). Brazil, Cameroon, Ecuador, Madagascar, Nigeria, Sri Lanka, and Venezuela export significant amounts. It is also cultivated for export in Columbia, the People's Republic of the Congo, the Democratic Republic of the Congo, Costa Rica, Cuba, the Dominican Republic, Fiji, Gabon, Grenada, Haiti, Jamaica, Malaysia, south central Mexico, Panama, Papua New Guinea, Peru, the Philippines, Sau Tome, Sierra Leone, Togo, Trinidad, and Western Samoa (Soberanis et al. 1999; Duguma et al. 2001; Kraus and Soberenis 2001; Ramirez et al. 2001; Fig. 1). Almost 80% of cacao is grown by smallholder farmers and is often cultivated as a component of complex agro-ecosystems, providing both economic and ecological benefits, like habitat and water conservation, soil stabilization, and carbon sequestration (Wood and Lass 1985; Alves 1990).

Table 1. World production of cacao

Country	Production (metric tons)		
	2002/2003	2003/2004	2004/2005
Côte d'Ivoire	1,367	1,547	1,331
Ghana	498	605	560
Indonesia	413	460	470
Nigeria	178	175	190
Cameroon	152	160	178
Brazil	163	163	171
Ecuador	87	119	110
Malaysia	21	25	26
Other America	172	170	162
Other Africa	39	44	45
Other Asia	64	62	71
Total	4,154	3,530	3,314

Source: LMC International, International Cacao Organization

7.1.2
General Characteristics of Cacao

Cacao is a wide-branching evergreen tree with "cauliflorous" flowers (and later fruits) protruding directly from the woody branches and trunk (Fig. 2). The plant grows to a height of 4–8 m, rarely up to 20 m. The plant produces branches from 1 to 1.5 m height and bears large leaves and produce inflorescence on the trunk and branches. The fruits (commonly referred to as pods) are 10–32 cm long and spherical to cylindrical in shape. The pods are indehiscent and contain 20–60 seeds (referred as beans) arranged in five rows. Unripe pods are white, green, or red, but turn green, yellow, red, or purple when fully ripened (Reed 1976). Beans from mature pods are removed and fermented to stimulate biochemical activities required for flavor development, but this process destroys seed viability. Usually, cacao seeds are viable for a short time (10–13 weeks) and require up to 50% moisture for germination. In young trees, flowers are produced mainly on the trunk; in adult trees, they emerge all over the plant. An adult tree was reported to produce more than 50,000 flowers per year, less than 5% of which pollinate and only 0.5–2% bear fruits (Alvim 1984). Flowers that were not pollinated within the first 8–10 h after emergence were reported to drop in 24–48 h (Alvim 1984). The potential yields of cacao were reported as 3,375 kg of dry beans per hectare on good plantations, but on-farm productivity of cacao ranges between 29 and 2,000 kg of dry beans per hectare. Trade from dried cacao beans has an annual estimated value exceeding US $3 billion (Gray 2001).

Cacao is commonly cultivated in subtropical dry to wet regions in tropical forest zones (Fig. 1). Commercial cacao plantation is through stem cuttings, buds, or grafts, but direct seed sowings are also done for cultivation purposes. Cacao is often intercropped with other trees of economic value, such as bananas, rubber, oil palm, or coconut. Plants require uniformly high temperatures and thrive well in climates with high humidity and rainfall, but are sensitive to drought (Reed 1976). Plants are shade-tolerant, and thrive in rich, organic, well-drained, moist, deep soils (Reed 1976). Although cacao plants can flower from 2–3 years old, plants are maintained in the vegetative state for 3–5 years by removing flower buds. Subsequently, the plants are allowed to bear flowers and pods (Alvim 1984). Fertilized flowers would take about 4–6 months to mature for harvesting. Although fruits mature throughout the year, usually only two harvests are made. In West Africa, the main harvest season is between September and February, and a second harvest period is between May and June. Matured pods are cut from trees and allowed to mellow on the ground. Then, the pods are cracked, beans are removed, and husks are burned. Beans are fermented in leaf-lined kegs for 2–8 days before drying in the sun. Beans are then bagged and shipped for trade. Further processing includes roasting, crushing, and separating out the kernel, grinding the nibs, and extraction of fat (Fulton 1989).

7.1.3
Botany

T. cacao was the name given to the cacao tree by Linnaeus in the first edition of his *Species Plantarum* published in 1753. The genus *Theobroma*, together with the genera *Herrania*, *Guazuma*, and *Cola*, has been placed in the family Sterculiaceae. However, recent classification based on molecular phylogeny by Angiosperm Phylogeny Group II System suggests *T. cacao* as a species in the subfamily Sterculioideae, of the family Malvaceae *sensu lato* (order Malvales) (Alverson et al. 1999; Angiosperm Phylogeny Group 2003; Baum et al. 2004; Tate et al. 2005). This nomenclature has since been adopted in a number of recent references (Arnold et al. 2003; Sounigo et al. 2005; Sereno et al. 2006).

Phylogeny, Types, and Populations

The natural habitat of the genus *Theobroma* is in the evergreen rain forests of the western hemisphere between 18°N and 15°S, spreading from Mexico to the southern edge of the Amazon forest, and this region is generally considered as the primary center of diversity for cacao (Motamayor et al. 2002). A considerable and useful variation has been recorded in this region through direct observation of trees and also from research on material introduced from this area. Cacao-growing regions around the world are largely centered in important biodiversity hotspots, impacting 13 of the world's most biologically diverse regions (Piasentin and Klare-Repnik 2004).

Cuatrecasas (1964) divided the genus *Theobroma* into six sections consisting of 22 species. *T. cacao* is the only species, which is cultivated widely. The other well-known species in the genus are *T. bicolor* and *T. grandiflorum*. Cacao consists of a large number of morphologically variable populations that are intercrossable with each other. On the basis of the genetic origin, these populations were grouped as autogamous or allogamous (Lanaud et al. 2001). A system of gametophytic–sporophytic self-incompatibility was reported to increase the allogamy of certain cacao populations (Knight and Rogers 1955; Bouharmont 1960; Cope 1962; Glendinning 1962).

Classification of cacao populations into two groups was first proposed by Morris (1882) and named as Criollo and Forastero. From its probable center of origin in the high Amazon region cacao was reported to spread in two main directions,

Fig. 2. *T. cacao*: **a** Mature plant, **b** pod, **c** Transverse section of the ripe pod, showing mucilage covered beans (Photos: R. Bhattacharjee)

which resulted in two principal races: (1) the Criollo, grown in Venezuela, Colombia, Ecuador, northern Central America, and Mexico; and (2) the Forastero, cultivated in northern Brazil and Guyana (Cheesman 1944). Pittier (1935) designated Forastero and Criollo as different species and named them as *T. lelocarpum* and *T. cacao*, respectively. However, both these populations being interfertile were designed as a single species, *T. cacao* (Cuatrecasas 1964). Cuatrecasas (1964) proposed two subspecies, *T. cacao* subsp. *cacao* for Criollo and *T. cacao* subsp. *sphaerocarpum* for Forastero, which are similar to the groupings proposed on the basis of the morphogeographic features by Cheesman (1944). A third group was later recognized, namely, Trinitario, consisting of the hybrids between these two subspecies (Dias 2001). The Forastero group covers the majority of world cacao production. The characters of Criollo, Forestero, and Trinitario are presented in Table 2 (Wood and Lass 1985).

Diversity in Cacao Populations

Owing to the outbreading nature of the species, most cacao populations were reported to show a degree of variation for all the morphological characters and even variation at the genetic level (Cuatrecasas 1964). Certain morphological characters of pods and beans

were used as the basis to classify cacao into different categories such as varieties, cultivars, types, or populations (Dias 2001). The appearance of the pod, or its morphology, was considered an important character in defining the types and populations in cacao. On the basis this cacao was grouped as follows: (1) Angoleta (deeply ridged, warty, square at the stalk end); (2) Cundeamo (similar to Angoleta but characterized by a bottleneck); (3) Amelonado (smooth, shallow furrows, melon-shaped with a blunt end and slight bottleneck); and (4) Calabacillo (small and nearly spherical). However, taxonomists could not establish a correlation between pod shape and other traits and thus this system was not adopted. The attempts of categorizing cacao genotypes into horticultural races by morphological descriptors (Engels 1986) or isoenzymes (Lanaud 1987) failed owing to large overlaps between the groups.

Characterization of genotypes using molecular markers was believed to provide unambiguous classification and various studies in this direction have generally demonstrated the genetic difference between Upper and Lower Amazon Forasteros, Trinitarios, and Criollos (reviewed in Dias 2001). However, some of these studies have reported difficulties in distinguishing the genotypes/races clearly owing to the occurrence of overlap among the genotype clusters due to

Table 2. Characters and types of populations of *Theobroma cacao*

Criollo

Slender trees, green pods with or without anthocyanin pigmentation. It is regarded as the most anciently cultivated type. The beans ferment quickly and possess excellent flavor, but trees have poor vigor and are extremely susceptible to diseases like bark canker (*Phyophthora* spp.) and Ceratocystis wilt (*Ceratocystis fimbriata*) and are highly susceptible to mirid bug leaf damage. Criollos are subdivided into two geographical groups: Central American Criollos and South American Criollos. Until the middle of the eighteenth century, Criollos dominated the market and accounted for most of the exports to Europe, but at present, only a few pure Criollo types exist. Different types of Criollos have been described by Soria (1970) and Pound (1938) and are as follows:

- *Mexican Criollo*: Occurs only as scattered stands in a few plantings in the state of Chiapas, Mexico. The shape and the size of pods and beans are highly variable, but the beans are white. The color of the pods is between green and red, and the pods have a pointed tip.
- *Pentagona or Lagarto*: In Mexico and Guatemala, within the plantings of Criollo and Trinitario, there are trees that bear pods with very thin, leathery husks; warty with only five angular ridges, red or rarely green in color, and contain seeds with varying shades of purple.
- *Nicaraguan Criollo or Cacao Real*: These populations exist in certain areas of Nicaragua as small populations or isolated groups and are characterized by intense red colored pods with very pronounced bottled neck.
- *Columbian Criollo*: Population based on the color of the pods, which are green and deep purple types, both with a smooth surface and are uniform in their pod characteristics.

Forastero

This is a large group consisting of cultivated populations as well as semiwild and wild types, of which the Amelonado populations are most extensively cultivated.

- *Amelonado*: This population covers a vast cacao area in the state of Bahia, Brazil, and in West Africa (except Cameroon). It is predominantly uniform and homozygous in all its characters. The pod is light green in color with a smooth surface and ridges bottlenecked with a pointed tip that is not very pronounced. The average number of beans per pod is around 40 and the beans are dark purple in color. It is a hardy and productive type, but low in production.
- *Comum*: This variety with typical Amelonado pods was introduced in the late 1700s from the Lower Amazon region as planting material to initiate the cacao industry of Bahia and until recently it covered roughly 90% of the mature plantings.
- *West African Amelonado*: In 1824, a few pods of cacao were successfully transferred by the Portuguese from Brazil across the Atlantic to the island of Sao Tome. By the end of the century, Sao Tome became a major exporter of cacao. The main cacao variety though looks very similar to Comum, but is commonly referred to as West African Amelonado. Toward the end of the 1850s, cacao plants were introduced from Sao Tome to the island of Bioko, Equatorial Guinea. The variety resembled typically as the West African Amelonado.
- *Cacao Nacional*: This is an old variety of Ecuador, but following the incidence of witches'-broom in the 1920s this type was reported to have been virtually wiped out. This variety produces large pale purple beans with distinctive Arriba flavor. The pods are large and green with a rough surface and fairly deep ridges. It is generally considered to be an Amelonado type of cacao.
- *Matina or Ceylan*: This is considered as the Amelonado variety of Central America and is grown commonly in Costa Rica and Mexico. Both probably have a common origin, which may well have been Brazil or Surinam (Soria 1970).
- *Guiana wild Amelonado*: This variety was first discovered and reported by Stahel (1920) in the forests of Surinam toward the western border. Later Myers (1930, 1934) confirmed the general occurrence of this wild cacao in the forests of western Guyana. The pod shape is uniform but typically resembles Amelonados, the seeds are also of the same size and shape as those of Amelonado but their color is bright violet and taste bitter with aromatic pulp. All the trees of this type are heavily infected with witches'-broom.
- *Amazon populations*: These encompass all the populations described and collected by expeditions in the vast Amazon river basin. The pod color is pale to dark green except for some trees with pods having splashes of red and reported from the western extremity of the area where the Napo river extends into Ecuador. Pound (1938) referred to this population as *Criollo de la mantagne* (the native cacao in the forest). The pod shape and husk texture are highly variable, having mostly dark purple cotyledons, but occasionally pale purple beans are reported.

Table 2. (continued)

Trinitario

These have features intermediate to those of Criollo, and have descended from an initial cross between Criollo and Forastero (Cheesman 1944, Cuatrecasas 1964). The pod and bean characters are variable. The initial cross resulted in highly vigorous, prolific, hardy trees and these characters continued for a few generations; however in advanced generations the vigor declined but remained higher than that of old Criollo trees. Evidence for this phenomenon has been found in Bioko (Swarbrick et al. 1964), Central America (Soria 1970), and Indonesia. Trinitarios were identified in Trinidad (Pound 1932), Cameroon (Preuss 1901), Samoa, Sri Lanka, Java, Papua New Guinea, and Fiji.

Catongo

A population selected for white beans occurs in Brazil, and is propagated mainly through seeds. Catongo types possess seed resembling Criollo, the characters of the pod husk resemble the Amelonado type, and the seed number resembles the Forestero character. The original tree was found in an Amelonado population and was considered to be a mutant. Therefore, the Catongo population was thought to be Forestero type and was described as white-beaned Amelonado.

Djati Roenggo

In Indonesia, about 10,000 ha of budded, clonal cacao named as Djati Roenggo has been reported, which has all the pod characters of Criollo types, an exception being that the number of beans per pod is 35, a Forastero or Trinitario character.

---

the great heterogeneity within the races and the large number of natural hybrids (Lerceteau et al. 1993, 1997; Figueira et al. 1994). However, recent studies utilizing cacao genomics and simple sequence repeat (SSR) markers have paved a way to capture important levels of diversity to understand the origin and for unambiguous classification (Sereno et al. 2006). A set of 15 microsatellites (SSRs) that can detect a high degree of polymorphism have been identified and are proposed as international molecular standards for DNA fingerprinting of *T. cacao* and the development of a unified database of cacao germplasm (Saunders et al. 2001, 2004).

7.1.4
Self-Incompatibility

Self-incompatibility is an efficient common phenomenon in several tropical trees, mostly to avoid inbreeding among them (Bullock 1985), and species in *Theobroma* are no exception. The self-incompatibility system in cacao has been widely described (Knight and Rogers 1955; Cope 1962; Yamada et al. 1982). The incompatibility phenomenon in cacao was reported to be due to the failure of gamete nuclei fusion in the embryo sac. Incompatible mating therefore leads to flower abscission in 72 h after pollination and is characterized by the presence of nonfused nuclei in the ovary. In contrast, Aneja et al. (1994) worked with a self-compatible cacao genotype (IMC 30) and proposed that the self-incompatibility system

occurs at pollen germination and gametic fusion stages.

7.1.5
Importance and Uses of Cacao

Cacao originated in lowland rainforests of the Amazon basin but is currently grown throughout the world in the humid tropics and is the main source for the world chocolate industry, which is estimated to exceed US $55 billion per year. A number of products have been derived from the cacao pods and beans. The pod consists of about 42% beans, 2% mucilage, and 56% husk. The bean is composed of the nibs (cotyledons and embryo) and the testa. Cacao contains an alkaloid, theobromine. It is a close structural relative of caffeine. Some references suggest cacao to contain 0.2% caffeine, but this was unfounded. About 3.5% of theobromine is present in fat-free dry cacao beans. It is a stimulant for muscular activity; however, high consumption of theobromine was shown to cause harmful symptoms such as excessive stimulation of kidneys, heart, and smooth muscles. Cacao beans contain about 50% fat that is used in confections and in manufacture of tobacco, soap, and cosmetics. Cacao sweatens (mucilage) is a viscous liquid surrounding the seeds within the pod and provides several by-products (Fig. 3). The mucilage is free of alkaloids and toxic substances, and consists of pectin (5%) and glucose (11%), which are used in the making of jams, jellies, and weight-reducing dietary formula-

Fig. 3. Industrial derivatives with *T. cacao* mucilage (Source: Wood and Lass, 1985)

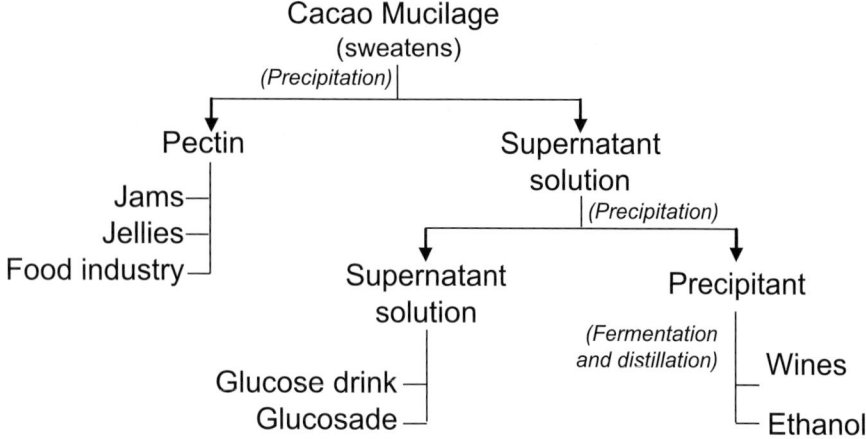

tions, and glucose is fermented to produce wines and other alcoholic drinks (Coe and Coe 1996). Cacao is also a folk remedy for burns, cough, dry lips, fever, malaria, rheumatism, snakebite, and wounds. It is reported to be an antiseptic and a diuretic. Emerging studies have shown that cocoa and chocolate are rich in plant antioxidant flavonoids with beneficial cardiovascular properties favoring antioxidant activity, vasodilation and blood pressure reduction, inhibition of platelet activity, and decreased inflammation, and contribute to heart and vascular protection (Engler and Engler 2006; Schroeter et al. 2006). The cacao bean testa (shell) is an important by-product in the chocolate industry and is used as ruminant feed and an organic fertilizer.

7.1.6
Production Constraints

Several diseases caused by fungi, viruses, and nematodes were reported in cacao (Wood and Lass 1985; Bowers 2001), but only a few of them are recognized to be of global importance and most of them are only regionally or locally important. The diseases of global importance are black pod, caused by *Phytophthora* spp. (*P. palmivora*, *P. capsici*, *P. citrophthora*, *P. heveae*, and *P. megakaria*); witches'-broom, caused by *Crinipellis perniciosa*; swollen shoot caused by *Cocoa swollen shoot virus* (CSSV); and monilia pod rot, caused by *Moniliophthora roreri* (Dias 2001). These pathogens contribute to reduction of the potential annual yields by as much as 40%, but the reduction is reported to reach up to 90–100% in

cases of diseases such as witches'-broom and monilia pod rot in certain locations (Wood and Lass 1985). CSSV transmitted by mealy bugs is reported to be destructive in West African countries (Muller and Sackey 2005). Diseases such as cacao wilt (Ceratocystis blight), caused by *Ceratocystis fimbriata*, Verticillium wilt, caused by *Verticillium dahliae*, and pink disease, caused by *Corticium salmonicolor*, are of local occurrence, although they are sometimes of relative economic importance. Wood and Lass (1985) have given detailed information on symptoms, taxonomy, origin, dissemination, transmission, and disease control.

Chemical and nonchemical management strategies to control important cacao diseases were found to be ineffective owing to high costs and difficulties in application. The combined practice of phytosanitary pruning with adequate fungicide application is widely used for the control of most diseases (Wood and Lass 1985; Dias 2001; Krauss and Soberanis 2001). But such practices are expensive and not ecofriendly. Extensive efforts are being made to replace susceptible cultivars with durable resistant cultivars. However, progress in this direction is slow owing to limited sources of genetic resistance in cacao against most of these pests and pathogens (Eskes and Lanaud 1997), but recent advances in cacao genetics and molecular techniques are enhancing this process (Eskes and Lanaud 1997; Bennett 2003; Risterucci et al. 2003; Clement et al. 2004). Furthermore, protocols for transformation and regeneration of cacao plants have been established and these can contribute to the development of transgenic resistance against major pathogens (Maximova et al. 2005; Sec. 7.6).

7.2
Cacao Breeding

In the case of cacao, it has been estimated that about 30% of the total cultivated area is planted with selected varieties (Paulin and Eskes 1995; Eskes 2001), consisting mostly of mixtures of biparental crosses (hybrids) between local and introduced clones. The remaining 70% of cultivation has been with traditional populations such as Trinitario, Amelonado, F3 Amazon, and open-pollinated populations derived from selected hybrid varieties. Studies have shown that farmers tend to increasingly use seeds from their own preferred trees (Opoku et al. 2006), which is expected to result in partial inbreeding and a narrow genetic base of cultivated cacao, making it vulnerable to various diseases (Bennet 2003), and loss of vigor and yielding capacity (INGENIC 1995, 1999). The past and on-going cacao improvement programs involve germplasm management, characterization, and evaluation; development of breeding tools (early screening methods, application of molecular tools); genetic studies; creation and selection of new varieties (on-farm varietal trials); and multiplication and distribution of new planting materials (INGENIC 1999; Eskes 2001). The major breeding methods employed in cacao are described next.

7.2.1
Clone Selection

Cacao breeding started in the 1920s when clones were selected in commercial plantations. Since then, clone selection has been the major breeding method carried out in most of the cacao-producing countries. These clones were further used to establish collections of local material for use in making crosses with introduced genotypes and to obtain new hybrid cultivars. In some cases, clones were selected on the basis of higher yield and quality and were used as commercial varieties, such as Trinitario clones (ICS, DR). These clones are still being used as cultivars in Trinidad and Indonesia (Thévenin et al. 2004). Since 1970s, new clones have been selected mainly to obtain rapid progress for resistance to devastating diseases such as vascular streak dieback in Southeast Asia and to witches'-broom (Trinidad selected hybrid clones) in Trinidad, and more recently in Brazil. Recently, large-scale selections of commercial clones with high resistance to black pod have also been started (Eskes et al. 1998; Efron 2000).

7.2.2
Hybrid Selection

Heterosis observed in individuals obtained from crosses between genetically distinct genotypes is the basis for hybrid selection. In worldwide cacao breeding programs between the 1950s and the 1990s, selection of new hybrid cultivars was the main activity. Crosses were generally made using local and introduced clones available in the germplasm collections as parents. Looking into the progress made with the introduction of these hybrids for agronomic traits such as early production (precocity), yield capacity, and vigor, large-scale utilization of hybrids started in cacao breeding. However, this method did not obtain good disease resistance.

Significant improvement in cacao productivity with the development of superior hybrids was reported from Brazil, and there is a consensus occurs among experts that hybridization is and will remain the main breeding method for cacao (Gotsch 1997). It was demonstrated by Dias et al. (1998) that in a large field trial there is a need for using cacao tree hybrids along with traditional local cultivars. Generally cacao tree hybrids show wide adaptability, low interaction with time, and better performance for yield and its components along with some resistance to diseases, in comparison to unimproved traditional local cultivars (Dias and Kageyama 1995). However, the major drawback in cacao hybridization is its empirical use of geographic divergence as an indicator for genetic divergence and this is still a common practice worldwide, though in many instances the positive correlation between geographic and genetic divergence does not exist (Dias et al. 1997).

The other frequently adopted criterion in cacao breeding is complementation of traits of interest, mainly to get rid of the deficiencies in each parent. Crosses are conducted randomly, primarily when information about the parental clones is not available. Because the yield ability of the clones per se is not associated with their performance as progenitors, combining ability tests are used to overcome part of the randomness of the hybridization process (Dias et al. 2003). However, according to Bos and Sparnaaij (1993), the use of a combining ability test in a diallel crossing scheme to study a complex trait like yield has lost its appeal because of the restricted resources in terms of land and labor which limit the number of parents that can be tested, and the almost consistent conclusion that variance due to general combining ability effects exceeds the variance due to specific combining ability. Since the long juvenile period of cacao lasts from 3 to 5 years and a single selection cycle might last one decade, new predictive tools provided by quantitative and molecular genetics must be used more intensively in cacao breeding programs.

Most of the breeding efforts in cacao have been concentrated toward selection based on seed yield and disease resistance. Very little is known about the available diversity for fat content within *T. cacao* for this trait (Kennedy et al. 1987; Lockwood and Yin 1993). There is also little knowledge available about the mode of inheritance of this trait and the effect of environmental factors on fat content.

7.2.3
Recurrent Selection and Prebreeding

Generally the traditional procedure of selecting hybrids does not lead to continuous genetic progress, therefore, breeders use successive breeding cycles to increase the frequency of favorable alleles in parental populations (Toxopeus 1972; Kennedy et al. 1987). Recurrent selection procedure is a robust method that exploits general and specific combining ability effectively, predominantly for most traits in cacao, mainly when populations are based on genetically related individuals, such as Lower Amazon types (Baudouin et al. 1997). In Côte d'Ivoire, a recurrent selection program was initiated using two base populations (Clement et al. 1994). Two cycles of selection were proposed in order to increase the frequency of favorable alleles for traits with relatively high heritability (e.g., disease resistance, self-compatibility, and pod index) in the base population. Countries such as Brazil, Ghana, and Malaysia also initiated this program to increase the number of favorable alleles in the locally adapted germplasm.

Prebreeding is a specific form of recurrent selection of improving genetically distinct base populations for specific traits in large germplasm collections before distributing the materials to users. The Cocoa Research Unit of the University of the West Indies, managing the International Cocoa Genebank in Trinidad, has started a prebreeding program with emphasis on disease resistance (Iwaro and Butler 2002).

Most of the ongoing breeding programs are concentrating on developing disease and pest resistance, especially in those cases where suitable screening methods are available. In addition to breeding for yield, breeders are selecting for more efficient and smaller trees that can be easily managed by farmers. Quality, such as flavor traits, is becoming a major selection criterion for breeding (INGENIC 1999)

7.3
Genetic Linkage and Gene Mapping

Cacao is a diploid with a chromosome number, $2n = 20$, and a small genome (0.4 pg/C) (Lannaud et al. 1992), about 2.8 times the size of that of *Arabidopsis thaliana* (L.) Heynh. (Couch et al. 1993). Lannaud et al. (1995) published the first linkage map of cacao developed from an F_1 cross of an Amazonian Forastero clone UPA402 with a Trinitario selection, UF676, from Costa Rica. UPA402 in turn was produced from sib-mating of two Ecuadorian Forastero clones, IMC60 and Na34 (Lannaud et al. 1995). The initial map from 100 individual trees was developed on the basis of five isozyme loci, four functional genes, 55 RFLPs of genomic DNA, and 28 randomly amplified polymorphic DNAs, resulting in ten linkage groups, which correspond to the number of chromosome pairs in cacao. The recombination-based genome length map thus obtained was reported to cover 759 cM (Lanaud et al. 1995).

Risterucci et al. (2000) saturated the map with additional markers (424 in total) and produced the first high-density linkage map of cacao using 81 additional trees. The high-density map of Risterucci et al. (2000) covered 885.4 cM. Both the genetic maps resulted in ten linkage groups having good agreement for marker alignment (Lannaud et al. 1995; Risterucci et al. 2000). Pugh et al. (2004) reported a genetic map using the same parental populations used by Lanaud et al. (1995) and 135 progenies. This map included the codominant markers from previous maps and 201 new SSR (microsatellite) markers and 16 resistance gene analog based markers (Lanaud et al. 2004). A total of 465 markers were used to develop a map that has the genome length of 782.8 cM (Pugh et al. 2004). A slight reduction in distance compared with the high-density map of Risterucci et al. (2000) was attributed to amplified fragment length polymorphism marker bias toward AT-rich, heterochromatin sites located near the centromeres and telomeres (Pugh et al. 2004). However, the colinearity was reported to be very high between the two maps, and genetic distances between loci common to both maps were generally of the same magnitude (Risterucci et al. 2000; Pugh et al. 2004).

Brown et al. (2005) used 146 progenies and 204 codominant markers (174 SSR markers identified by Lanaud et al. 1999b; Risterucci et al. 2000; Pugh et al. 2004; and 18 SSR markers and eight resistance gene homologs from Kuhn et al. 2003; and four WRKY genes from Borrone et al. 2004) to developed a map using the Kosambi mapping function and JoinMap 3.0 (van Ooijen and Voorrips 2001). Ten linkage groups were reported from this study, corresponding to the ten chromosomes of cacao with a total genome length of 671.9 cM. This map was 213.5 cM shorter than the high-density linkage map (885.4 cM) produced by Risterucci et al. (2000) and 110.9 cM shorter than the F_1 codominant map of Pugh et al. (2004).

The first genetic map was used for the identification of the quantitative trait loci (QTL) for resistance to black pod disease (Lanaud et al. 1999a). Subsequently, QTLs for yield and yield-related agronomic traits and resistance to black pod disease caused by *Phytophthora pulmivora* were reported by Crouzillat et al. (2000a, b) and Flament et al. (2001), respectively. The QTLs identified were reported to account for moderate (17%) to relatively high (48%) levels of resistance to black pod disease. However, Flament et al. (2001) observed that none of the QTLs were common across the three different measurements for resistance, and found results from artificial inoculation data to be poorly related to results based on field resistance, as did Lanaud et al. (1999a). A three-way test cross was used by Clement et al. (2003a, b) with two Trinitario clones (DR1 and S52) and one Upper Amazon clone (IMC78) crossed onto one homozygous Lower Amazon clone (Catongo) for map construction and QTL colocalization for yield and yield-related traits, and for black pod resistance. This effort was successful in mapping several yield and yield-related traits with good reliability and commonality among correlated traits. Resistance QTLs for black pod were found in DR1 and IMC78 in similar regions of chromosome 4 to those reported earlier by Flament et al. 2001.

Brown et al. (2005) used an F_2 population (derived by selfing TSH516, a clone from a cross of Sca6 with ICS1, selected in Trinidad) segregating for resistance to witches'-broom to construct a map of *T. cacao*. Quieroz et al. (2003) used earlier a subset of an F_2 population comprising 82 progenies and mapped mostly dominant markers, and identified a putative QTL conferring resistance to witches'-broom on the map containing 25 linkage groups.

MapQTL 4.0 (van Ooijen and Maliepaard 1996) was used to locate the putative QTLs for phenotypic traits, using simple interval mapping, followed by the multiple QTL mapping procedure of Jansen (1993). Two major QTLs controlling resistance to witches'-broom were detected, one located on linkage group 9 with a very high logarithm of odds score (10.55) explaining nearly 51% of the variance for the trait, and the other located on linkage group 1 with a logarithm of odds score of 3.38 explaining 6.7% of the variance.

The availability of the F_2 map corresponding to the specific chromosome number will allow better comparisons of the cacao genome with that of many other annual crops, in which maps are more often made from F_2 or recombinant inbred populations. Further, development of anchor markers with other important members of the genus *Theobroma* L. such as *T. grandiflorum* (Willd. ex Spreng.) Schum (de Sousa Silva et al. 2001), and with members of the Malvaceae family such as *Gossypium hirsutum* L. and other dicotyledons would be useful for genomic comparisons and for gene expression studies.

The established molecular maps of the cacao genome are now providing a rational direction for systematic breeding programs, most importantly to broaden the genetic base of cultivated trees as well as to characterize the existing *T. cacao* germplasm collections (Bennett 2003; Clement et al. 2004; Sereno et al. 2006).

7.4
Marker-Assisted Selection

QTL analysis has a greater potential for identification of markers for marker-assisted breeding, especially for incorporation of disease resistance. There have been several studies that showed QTLs associated with traits of interest in *Theobroma* (Crouzillat et al. 1996, 2000a, b, 2001). Major QTLs were identified for resistance to *C. perniciosa* (agent of witches'-broom) (Quieroz et al. 2003) and for yield traits such as pod weight (Clement et al. 2003a, b). The prospect of progress from identification of QTLs to the isolation of gene(s) by map-based cloning has not yet been applied in cacao. Such an approach requires a very large population to accumulate a suitable number of recombinant individuals (around 1,000), which is not only expensive in terms of time and resources but is also extremely difficult in the case of cacao (Wilkinson 2000). The bacterial artificial chromosome (BAC) library developed recently by Clement et al. (2004) can be a valuable resource for cloning genes corresponding to some important QTLs.

7.5
Application of Genomic Tools and Gene Discovery

Relatively less attention has been directed toward gene discovery and expression studies in cacao. Recently, a variety of genetic and genomic resources were applied to identify genes involved in disease resistance using a high-density linkage map, and the identification of disease-associated QTLs (Bennett 2003), expressed sequence tags (ESTs)/array technologies (Jones et al. 2002; Verica et al. 2004), and BAC libraries (Clement et al. 2004). Resistance gene analogs and defense gene analogs were isolated from cacao genomic DNA using degenerate primers designed from conserved domains of several plant resistance defense genes (Kuhn et al. 2003; Lanaud et al. 2004). Most of the resistance and defense gene analogs were mapped in segregating populations and QTLs for resistance to *Phytophthora* spp. were identified (Crouzillat et al. 2000b; Clement et al. 2003a; Resterucci et al. 2003). The cacao BAC library was constructed with 36,864 clones from the genotype Scavina-6 (Clement et al. 2004), and was used to identify QTLs for resistance to *Phytophthora*, witches'-broom, and black pod disease. More than 5,500 ESTs have been sequenced, from which a set of 1,380 unique gene sequences have been derived from beans and leaf complementary DNA (cDNA) libraries (Verica et al. 2004). These could be used to identify differentially expressed sequences for both bean and leaf traits. Microarrays have been constructed and used in demonstration experiments to evaluate genotype and tissue-specific gene expression using the publicly available sequences from the *T. cacao* unique gene set, and the expression was validated by real-time polymerase chain reaction (Jones et al. 2002). Verica et al. (2004) used subtractive hybridization of cDNA libraries, macroarray hybridization analysis, and high-throughput DNA sequencing to identify cacao genes in the plants treated with inducers of defense response and identified a unigene set of 1,256 members, including 330 members representing genes induced during the defense response. Unigene sets of Jones et al. (2002) and Verica et al. (2004) were shown to have 10% of the sequences overlapping, giving rise to a combined unigene set of about 2,500 sequences (Verica et al. 2004). These sequences are a potentially valuable resource for identifying novel defense genes. In addition, candidate defense genes can be used to search for colocalization with QTLs associated with resistance to cacao diseases and also are valuable when used in combination with the germplasm resources to evaluate the underlying molecular basis of disease response as well as bean quality traits that contribute to bean yield, composition, and flavor (Bennett 2003).

7.6
Genetic Transformation

The application of tissue culture and genetic transformation complemented by traditional plant breeding programs plays an important role in enhancing productivity and bean supply (Eskes and Lanaud 1997; Wilkinson 2000). This approach opens up an opportunity to incorporate novel sources of resistance genes or any other valuable traits into the genome and then can be used in breeding programs (Hansen and Wright 1999; Persley 2000). While these approaches have been utilized for many plant species, they have been limited in cacao because of several technical difficulties posed by the species. The first cacao transformation attempt resulting in transformed callus cells was by Sain et al. (1994). Subsequently, Maximova et al. (2003) reported the development of an efficient genetic transformation system for cacao using somatic embryogenesis as a regeneration system, coupled with *Agrobacterium tumefaciens* cocultivation that resulted in transgenic plants. The transformed plants were grown to maturity and segregation and expression of the transgenes in the subsequent T_1 generation was demonstrated (Maximova et al. 2003). This technology was used to develop the world's first transgenic cacao overexpressing class I chitinase gene (*TcChi1*) isolated from cacao to enhance resistance against *Colletotrichum gloeosporioides* (Maximova et al. 2005). It was reported that the transgenic line has up to a sixfold increase of endochitinase activity compared with nontransgenic and transgenic control plants (Maximova et al. 2005). This study demonstrated for the first time the utility of the transformation system as a tool for gene functional analysis and the potential utility of the cacao chitinase gene for increasing fungal pathogen resistance in cacao and also the scope for incorporating resistance to other fungal and viral pathogens, and insect pests. However, genetically modified cacao material has not yet been released and the scope of transgenic cacao will depend upon economic, social, environmental, and political factors of the introducing country. Nonetheless, the efficient tissue culture system on its own contributes to clonal propagation of elite germplasm.

7.7
Future Prospects

The cacao plant originated in the Amazon basin of South America has emerged as one of the most successful commercial tree crops. Despite its continuing importance as a chief crop supplying raw material for the multibillion-dollar confectionary industry, research on genetic improvement of *T. cacao* has been slow. However, efforts for systematic genetic improvement of *T. cacao* have gained pace during the past decade, leading to development of a high-density linkage map, SSR markers, and even efficient tissue culture and transformation protocols.

Several new technologies of recent emergence have dramatically increased the genetic knowledge on cacao and provided tools to accelerate the breeding/selection programs to develop elite cacao varieties. The established molecular maps of the cacao genome are contributing to a rational direction for systematic breeding programs, most importantly to broaden the genetic base of cultivated trees as well as to characterize the existing cacao germplasm collections, leading toward unambiguous classification of germplasm. There is a need to improve the resolution and accuracy of QTL analysis in cacao. This requires large mapping populations across several geographically dispersed sites and repetition of phenotypic scoring over several years to ensure stability of the identified QTLs. It is necessary to understand the components of cacao crop management to grow sustainable cacao by reducing the impact of diseases such as black pod, witches'-broom, frosty pod rot, and swollen shoot. Private and public partnerships for research for genetic improvement of cacao can boost the development of much needed technologies to enhance cacao production and quality. As stated by

Bennett (2003), in the coming years, cacao is likely to become a model for other tree crops for being transformed through genome-based breeding.

References

Alverson WS, Whitlock BA, Nyffler R, Bayer C, Baum DA (1999) Phylogeny of the core Malvales: evidence from *ndh*F sequence data. Am J Bot 86:1474–1486

Alves MC (1990) The role of cacao plantations in the conservation of the Atlantic forest of southern Bahia, Brazil. Masters Thesis, Univ of Florida, USA

Alvim PT (1984) Flowering of cocoa. Cocoa Growers Bull 35:22–31

Angiosperm Phylogeny Group (2003) An update of the Angiosperm Phylogeny Group classification for the orders and families of flowering plants: APG II. Bot J Linn Soc 141:399–436

Aneja M, Gianfagna T, Ng E, Badilla I (1994) Carbon-dioxide treatment partially overcomes self-incompatibility in a cacao genotype. HortScience 29:15–17

Arnold EA, Mejía LC, Kyllo D, Rojas EI, Maynard Z, Robbins N, Herre EA (2003) Fungal endophytes limit pathogen damage in a tropical tree. Proc Natl Acad Sci USA 100:15649–15654

Baudouin L, Baril C, Clément-Demange A, Leroy T, Paulin D (1997) Recurrent selection of tropical tree crops. Euphytica 96:101–114

Baum DA, Smith SD, Yen A, Alverson WS, Nyffeler R, Whitlock BA, Oldham RL (2004) Phylogenetic relationships of Malvatheca (Bombacoideae and Malvoideae; Malvaceae sensu lato) as inferred from plastid DNA sequences. Am J Bot 91:1863–1871

Bennett AB (2003) Out of the Amazon: *Theobroma cacao* enters the genomic era. Trends Plant Sci 8:561–563

Bergmann JF (1969) The distribution of cacao cultivation in Pre-Columbian America. Ann Assoc Am Geogr 59:85–96

Borrone JW, Kuhn DN, Schnell RJ (2004) Isolation, characterization, and development of WRKY genes as useful genetic markers in *Theobroma cacao* L. Theor Appl Genet 109 (3):495–507

Bos I, Sparnaaij LD (1993) Component analysis of complex characters in plant breeding. II. The pursuit of heterosis. Euphytica 70:237–245

Bouharmont J (1960) Recherches cytologiques sur a fructification et l'incompatibilitê chez *Theobroma cacao* L. PubI de l'lnearc, Serie Scientifique 59:117

Brown JS, Schnell RJ, Motamayor JC, Lopes U, Kuhn DN, Borrone JW (2005) Resistance gene mapping for Witche's broom disease in *Theobroma cacao* L. in an F_2 population using SSR markers and candidate genes. J Am Soc Hort Sci 130(3):366–373

Bowers JH, Bailey BA, Hebbar PK, Sanogo S, Lumsden RD (2001) The impact of plant diseases on world chocolate production. Online. Plant Health Progress doi:10.1094/PHP-2001-0709-01-RV

Bullock SH (1985) Breeding systems in the flora of tropical deciduous forest in Mexico. Biotropica 17:287–301

Cheesman EE (1944) Notes on the nomenclature, classification and possible relationships of cocoa populations. Trop Agri Trinidad 27:144–159. Reprinted 1982 in Arch Cocoa Res 1:98–116

Clement D, Eskes AB, Sounigo O, N'Goran J (1994) Amélioration génétique du cacaoyer en Côte d'Ivoire: presentation d'un nouveau schema de selection. In: 11th Intl Cocoa Res Conf, Yamoussoukro, Ivory Coast, July 1993, Lagos, Nigeria, Cocoa Producer's Alliance, pp 451–455

Clement D, Risterucci AM, Grivet L, Motamayor JC, N'Goran J, Lanaud C (2003a) Mapping QTL for yield components, vigor, and resistance to Phytophthora palmivora in Theobroma cacao L. Genome 46:204–212

Clement D, Risterucci AM, Grivet L, Motamayor JC, N'Goran J, Lanaud C (2003b) Mapping quantitative trait loci for bean traits and ovule number in Theobroma cacao L. Genome 46:103–111

Clement D, Lanaud C, Sabau X, Fouet O, Le Cunff L, Risterucci AM, Glaszmann JC, Piffanelli (2004) Creation of BAC genomic resources for cocoa (Theobroma cacao L.) for physical mapping of RGA and containing BAC clones. Theor Appl Genet 108:1727–1634

Coe SD, Coe MD (1996) The true history of chocolate. Thames and Hudson, New York, USA

Cope FW (1962) The mechanism of pollen incompatibility in Theobroma cacao L. Heredity 17:157–182

Cope FW (1976) Cacao. Theobroma cacao L. (Sterculiaceae). In: Simmonds NW (ed) Evolution of Crop Plants. Longman, London, UK, pp 207–213

Couch J, Zintel HA, Fritz P (1993) The genome of the tropical tree, Theobroma cacao L. Mol Gen Genet 237:123–128

Crouzillat D, Lerceteau E, Pétiard V, Morera J, Rodriguez H, Walker D, Phillips W, Ronning C, Schnell R, Osei J, Fritz P (1996) Theobroma cacao L.: a genetic linkage map and quantitative trait loci analysis. Theor Appl Genet 93:205–214

Crouzillat D, Menard B, Mora A, Phillips W, Petiard V (2000a) Quantitative trait analysis in Theobroma cacao using molecular markers. Yield QTL detection and stability over 15 years. Euphytica 114:13–23

Crouzillat D, Phillips W, Fritz P, Petiard V (2000b) Quantitative trait analysis in Theobroma cacao using molecular markers. Inheritance of polygenic resistance to Phytophthora palmivora in two related cacao populations. Euphytica 114:23–36

Crouzillat D, Rigoreau M, Cabigliera MA, Bucheli P, Pétiard V (2001) QTL studies carried out for agronomic, technological and quality traits of cocoa in Ecuador. In: Proc Intl Workshop on New Technologies and Cocoa Breeding, IN-GENIC, 16-17 Oct, 2000, Kota Kinabalu Sabah, Malaysia, pp 120–126

Cuatrecasas J (1964) Cacao and its allies. A taxonomic revision of the genus Theobroma. Contributions to the United States National Herbarium 35(6):379–614

de Sousa Silva CR, de Oliveira Figueira AV, Spaggiari Souza CA (2001) Diversity in the genus Theobroma (in Portuguese). In: Dias LAS (ed) Genetic Improvement of Cacao. FUNAPE, UFG, Viçosa, MG, Brazil, pp 40–80

Dias LAS, Kageyama PY (1995) Combining ability for cacao (Theobroma cacao L.) yield components under southern Bahia conditions. Theor Appl Genet 90:534–541

Dias LAS, Kageyama PY, Castro GCT (1997) Divergência fenética multivariada na preservação de germoplasma de cacau (Theobroma cacao L.). Agrotrópica 9:29–40

Dias LAS, Souza CAS, Augusto SG, Siqueira PR, Müller MW (1998) Performance and temporal stability analyses of cacao cultivars in Linhares, Brazil. Plant Rec Dev 5:343–355

Dias LAS (ed) (2001) Genetic Improvement of Cacao. Food and Agriculture Organization, Rome, Italy

Dias LAS, Marita J, Cruz CD, Barros EG, Salomão TMF (2003) Genetic distance and its association with heterosis in cacao. Brazilian Arch Biol Technol 46(3):339–348

Duguma B, Gochowski J, Bakala J (2001) Smallholder cacao (Theobroma cacao Linn.) cultivation in agroforestry systems of West and Central Africa: Challenges and opportunities. Agrofor Syst 51:177–188

Engels JMM (1986) The identification of cacao cultivars. Acta Hort 182:195–202

Efron Y (2000) Provisional release of three varieties of hybrid clones. Bull PNG Cocoa and Coconut Res Inst, Rabaul, Papua New Guinea, p 8

Engler MB, Engler MM (2006) The emerging role of flavonoid-rich cocoa and chocolate in cardiovascular health and disease. Nutr Rev 64(3):109–18

Eskes AB, Lanaud C (1997) Cocoa. In: Charrier A, Jacquot M, Hamon S, Nicolas D (eds) L' Amelioration des plantes tropicales. CIRAD/ORSTROM, Paris, p 623

Eskes AB, Engels J, Lass T (1998) The CFC/ICCO/IPGRI project: a new initiative on cocoa germplasm utilization and conservation. Plantations Recherche Développement 5(6):412–422

Eskes AB (2001) Introductory notes. Proc Intl Workshop on New Technologies for Cocoa Breeding, Kota Kinabalu, Malaysia, INGENIC, London, UK, pp 8–11

FAOSTAT (2006) http://apps.fao.org

Flament MH, Kebe I, Clement D, Pieretti I, Risterucci AM, N'-Goran JAK, Cilas C, Despreaux D, Lanaud C (2001) Genetic mapping of resistant factors in Phytophthora palmivori in cacao. Genome 44:79–85

Figueira A, Janick J, Levy M, Goldsbrough PB (1994) Re-examining the classification of Theobroma cacao L. using molecular markers. J Am Soc Hort Sci 119:1073–1082

Fulton RH (1989) The cacao disease trilogy: black pod, Monilia pod rot and witches' broom. Plant Dis 73(7):601–603

Glendinning DR (1962) Natural pollination of cocoa. Nature 193(4822):1305

Gotsch N (1997) Cocoa biotechnology: status, constraints and future prospects. Biotechnol Adv 15:333–352

Gray A (2001) The world cocoa market outlook. Ghana conf, LMC Intl, London, UK

Hansen G, Wright MS (1999) Recent advances in the transformation of plants. Trends Plant Sci 4:226–231

INGENIC (1995) The present status of cocoa breeding at ICCRI: results and future programmes. In: Proc Intl Workshop on Cocoa Breeding Strategies. 18–19 Oct 1994, Kuala Lumpur, Malaysia INGENIC, Reading, UK, p 195

INGENIC (1999) Breeding for resistance to diseases in Malaysia with special reference to Vascular Streak Dieback. In: Proc Intl Workshop on the Contribution of Disease Resistance to Cocoa Variety Improvement. 24–26 Nov 1996, Salvador, Bahia, Brazil, INGENIC, Reading, UK, p 219

Iwaro AD, Butler DR (2002) Germplasm enhancement for resistance to black pod and witches' broom diseases. In: 13th Intl Cocoa Res Conf, Kota Kinabalu, Malaysia, 9–14 Oct 2000, Lagos, Nigeria, Cocoa Producer's Alliance, pp 3–10

Jansen R (1993) Interval mapping of multiple quantitative trait loci. Genetics 135:205–211

Johns R, Gibberd AV (1951) Review of cocoa industry in Nigeria. Report presented during 6th Cocoa Conf, London, UK, Cocoa Chocolate Confectionary Alliance Ltd, p 137

Jones PG, Allaway D, Gilmour MD, Harris C, Rankin D, Retzel ER, Jones CA (2002) Gene discovery and microarray analysis of cacao (Theobroma cacao L.) varieties. Planta 216:255–264

Kennedy AJ, Lockwood G, Mossu G, Simmonds NW, Tan GY (1987) Cocoa breeding: past, present and future. Cocoa Grower's Bull 38:5–22

Knight R, Rogers HH (1955) Incompatibility in Theobroma cacao. Heredity 9:69–77

Krauss U, Soberanis W (2001) Rehabilitation of diseased cacao field in Perú through shade regulation and timing of biocontrol. Agrofor Syst 53:179–184

Kuhn DN, Heath M, Wisser RJ, Meerow A, Brown JS, Lopes U, Schnell RJ (2003) Resistance gene homologues in Theobroma cacao as useful genetic markers. Theor Appl Genet 107:191–202

Lanaud C (1987) Nouvelles données sur la biologie du cacaoyer (Theobroma cacao L.): diversité des populations, systeme d'incompatibilité, haploides spontanes, leurs consequences pour l'amelioration de cette espece. Université de Paris-Sud, Centre d'Orsay (PhD Thesis) (Cited in Dias 2001)

Lanaud C, Hamon P, Duperray C (1992) Estimation of the nuclear DNA content of Theobroma cacao L. by flow cytometry. Café, Cacao, Thé 36:3–8

Lanaud C, Risterucci AM, N'Goran JAK, Clement D, Flament MH, Lauent V, Flaque M (1995) A genetic linkage map of Theobroma cacao L. Theor Appl Genet 91:987–993

Lanaud C, Kebe I, Risterucci AM, Clement D, N'Goran JAK, Grivet L, Tahi M, Cilas C, Pieretti I, Eskes AB, Despreaux D (1999a) Mapping quantitative trait loci (QTL) for resistance

to Phytophthora palmivora in T. cacao L. Proc Intl Cocoa Res Conf 12:99–105

Lanaud C, Risterucci AM, Pieretti I, Falque M, Bouet A, Lagoda PJL (1999b) Isolation and characterization of microsatellites in Theobroma cacao L. Mol Ecol 8:2141–2152

Lanaud C, Motamayor JC, Risterucci AM (2001) Implications of new insight into the genetic structure of Theobroma cacao L. for breeding strategies. In: Proc Intl Workshop on New Technologies for Cocoa Breeding, Kota Kinabalu, Malaysia, London: Ingenic Press, pp 89–107

Lanaud C, Risterucci AM, N'Goran JAK, Fargeas D (2004) Characterization and mapping of resistance and defense gene analogs in cocoa (Theobroma cacao L.). Mol Breed 13:211–227

Lockwood G, Yin JPT (1993) Utilization of cocoa germplasm in breeding for yield. In: Proc Intl Workshop on Conservation, Characterization and Utilization of Cocoa Genetic Resources in the 21st century. Cocoa Res Unit, Sept 13–17, 1992, Port-of-Spain, Trinidad and Tobago

Laurent V, Risterucci AM, Lanaud C (1994) Genetic diversity in cocoa revealed by cDNA probes. Theor Appl Genet 88:193–198

Lerceteau E, Crouzillat D, Petiard V (1993) Use of random amplified polymorphic DNA (RAPD) and restriction length polymorphism (RFLP) to evaluate genetic variability within the Theobroma genus. In: Intl Workshop on Conservation, Characterization and Utilization of Cocoa Genetic Resources in the 21st Century, Proceedings. CRU/The Univ of the West Indies, St Augustine, pp 332–344

Lerceteau E, Robert T, Pétiard V, Crouzillat D (1997) Evaluation of the extent of genetic variability among Theobroma cacao accessions using RAPD and RFLP markers. Theor Appl Genet 95:10–19

Maximova SN, Miller C, Antúnez de Mayolo G, Pishak S, Young A, Guiltinan MJ (2003) Stable transformation of Theobroma cacao L. and influence of matrix attachment regions on GFP expression. Plant Cell Rep 21:872–883

Maximova SN, Marelli JP, Young A, Pishak S, Verica JA, Guiltinan MJ (2005) Over-expression of a cacao class I chitinase gene in Theobroma cacao L. enhances resistance against the pathogen, Colletotrichum gloeosporioides. Planta 16:1–10

Miranda F (1962) Wild Cacao in the Lacandona forest, Chiapas, Mexico. Cacao (Turrialba) 7:7. CATIE: Costa Rica

Morris D (1882) Cocoa: how to grow and how to cure it. Jamaica, pp 1–45

Motamayor JC, Risterucci AM, Lopez PA, Ortiz CF, Moreno A, Lanaud C (2002) Cacao domestication I: the origin of the cacao cultivated by the Mayas. Heredity 89:380–386

Muller E, Sackey S (2005) Molecular variability analysis of five new complete cacao swollen shoot virus genomic sequences. Arch Virol 150:53–66

Myers JG (1930) Notes on wild cacao in Surinam and in British Guiana. Kew Bull 1:1–10

Myers JG (1934) Observations on wild cacao and wild bananas in British Guiana. Trop Agri Trinidad 11:263–267

Opeke LK (1969) Annual Report of the Cocoa Research Institute of Nigeria 1968–69. Ibadania, Nigeria, 163 p

Opoku SY, Bhattacharjee R, Kolesnikova-Allen M, Enu-Kwesi L, Asante EG, Adu-Ampomah Y (2006) Impact of breeders collections on cocoa plantings of Ghana: Assessment by molecular marker analysis and farmers field survey. J Ghana Sci Assoc (in press)

Paulin O, Eskes AB (1995) Le cacaoyer: strategies de selection. Plantations Recherche Développement 2:5–11

Persley GJ (2000) Agricultural biotechnology and the poor promethean science. In: Persley GJ, Lantin MM (eds) Agricultural Biotechnology and the Poor. CGIAR, Washington DC, USA, pp 3–21

Piasentin F, Klare-Repnik L (2004) Biodiversity conservation and cocoa agroforests. Gro Cocoa 5:7–8. http://www.cabicommodities.org/Acc/ACCrc/

Pittier H (1935) Degeneration of cocoa through natural hybridization. Heredity 36:385–390

Pound FJ (1932) The genetic constitution of the cacao crop. First Annual Report on Cacao Research, 1931, Trinidad, pp 10–24

Pound FJ (1938) Cacao and witches broom disease of South America. Port-of-Spain, Trinidad. Reprinted 1982 in Arch Cocoa Res 1:20–72

Preuss P (1901) Expedition nach Central and Sudamerika. Verlag des Kolonial-Wirtschaftlichen Komitees, Berlin, Germany

Pugh T, Fouet AM, Brottier P, Abouladze M, Deletrez C, Courtois B, Clement D, Larmande P, N'Goran JAK, Lanaud C (2004) A new cacao linkage map based on codominant markers: Development and integration of 201 new microsatellite markers. Theor Appl Genet 108:1151–1161

Quieroz VT, Guimarães CT, Anhert D, Schuster I, Daher RT, Pereira MG, Miranda VRM, Loguercio LL, Barros EG, Moreira MA (2003) Identification of a major QTL in cocoa (*Theobroma cacao* L.) associated with resistance to witches' broom disease. Plant Breed 122:268–272

Ramirez OA, Somarriba E, Ludewigs T (2001) Financial returns, stability and risk of cacao-plantain-timber agroforestry systems in central America. Agrofor Syst 51:141–154

Reed CF (1976) Information summaries on 1000 economic plants. Typescripts submitted to the USDA

Risterucci AM, Grivet L, N'Goran JAK, Pieretti I, Flament MH, Lanaud C (2000) A high-density linkage map of *Theobroma cacao* L. Theor Appl Genet 101:948–955

Risterucci AM, Paulin D, Ducamp M, N'Goran JAK, Lanaud C (2003) Identification of QTL related to cocoa resistance to three species of *Phytophthora*. Theor Appl Genet 108:168–174

Sain SL, Oduro KK, Furtek DB (1994) Genetic transformation of cocoa leaf cells using *Agrobacterium tumefaciens*. Plant Cell Tiss Org Cult 37:342–351

Schultes RE (1984) Amazonian cultigens and their northward and westward migrations in pre-Columbian times. In: Stone D (ed) Pre-Columbian Plant Migration, Papers of the Peabody Museum of Archaeology and Ethnology. Vol 76. Massachusetts:Harvard Univ Press: Cambridge, pp 69–83

Saunders JA, Hemeida AA, Mischke S (2001) USDA DNA fingerprinting program for identification of *Theobroma cacao* accessions. In: Proc Intl Workshop on New Technologies for Cocoa Breeding. Kota Kinabalu, Malaysia, INGENIC, London, UK, pp 108–114

Saunders JA, Mischke S, Leamy EA, Hemeida AA (2004) Selection of international molecular standards for DNA fingerprinting of *Theobroma cacao*. Theor Appl Genet 110:41–47

Schroeter H, Heiss C, Balzer J, Kleinbongard P, Keen CL, Hollenberg NK, Sies H, Kwik-Uribe C, Schmitz HH, Kelm M (2006) (-)-Epicatechin mediates beneficial effects of flavanol-rich cocoa on vascular function in humans. Proc Natl Acad Sci USA 103(4):1024–1029

Sereno ML, Albuquerque PSB, Vencovsky R, Figueira A (2006) Genetic diversity and natural population structure of cacao (*Theobroma cacao* L.) from the Brazilian Amazon evaluated by microsatellite markers. Conserv Genet 7:13–24

Soberanis W, Ríos R, Arévalo E, Zúñiga L, Cabezas O, Krauss U (1999) Increased frequency of phytosanitary pod removal in cacao (*Theobroma cacao*) increases yield economically in eastern Peru. Crop Protec 18:677–685

Soria J de V (1970) Principal varieties of cocoa cultivated in Tropical America. Cocoa Grower's Bull 15:12–21

Sounigo O, Umaharan R, Christopher Y, Sankar A, Ramdahin S (2005) Assessing the genetic diversity in the International Cocoa Genebank, Trinidad (ICG,T) using isozyme electrophoresis and RAPD. Genetic Res Crop Evol 52:1111–1120

Stahel G (1920) Een wild cacaobosch aan de Mamaboen Kreek. De Indische Mercuur 43e Jaarg, no 39:681–682

Swarbrick JT, Toxopeus H, Hislop EC (1964) Estate cocoa in Fernando Po. World Crops 16 (2):35–40

Tate JA, Aguilar, JF, Wagstaff, SJ, La Duke JC, Bodo Slotta TA, Simpson BB (2005) Phylogenetic relationships within the tribe Malveae (Malvaceae, subfamily Malvoideae) as inferred from ITS sequence data. Am J Bot 92:584–602

Thévenin JM, Ducamp M, Kebe I, Tahi M, Nyassé S, Eskes A (2004) Planting material screening by controlled inoculation. In: Cilas C, Despréaux D (eds) Improvement of Cocoa Tree Resistance to *Phytophthora* Diseases. CIRAD, Montpellier, France, pp 103–146

Toxopeus H (1972) Cocoa Breeding: a consequence of mating system, heterosis and population structure. In: Cocoa and Coconuts Conf in Malaysia, Kuala Lumpur, Malaysia, Incorporated Society of Planters, pp 3–12

UNCTAD (2005) Based on the data from International Cocoa Organization, quarterly bulletin of cocoa statistics 2004–05. http://www.unctad.org

Van Hall CJJ (1914) Cocoa. Macmillan, London, UK

van Ooijen JW, Maliepaard C (1996) MapQTL version 3.0: Software for the calculation of QTL position on genetic maps. Plant Research International, Wageningen, The Netherlands

van Ooijen JW, Voorrips RW (2001) JoinMap version 3.0, software for the calculation of genetic linkage maps. Plant Research International, Wageningen, The Netherlands

Verica JA, Maximova SN, Strem MD, Carlson JE, Bailey BA, Guiltinan MJ (2004) Isolation of ESTs from cacao (*Theobroma cacao* L.) leaves treated with inducers of the defense response. Plant Cell Rep 23(6):404–13

Whitlock BA, Bayer C, Baum DA (2001) Phylogenetic relationships and floral evolution of the Byttnerioideae ("Sterculiaceae" or Malvaceae s.l.) based on sequences of the chloroplast gene *ndh*F. Syst Bot 26:420–437

Whitkus R, De la Cruz M, Mota-Bravo L (1998) Genetic diversity and relationships of cocoa (*Theobroma cacao* L.) in southern Mexico. Theor Appl Genet 96:621–627

Wilkinson MJ (2000) The applications and constraints of new technologies in plant breeding. In: Proc Intl Workshop on New Technologies and Cocoa Breeding, INGENIC, 16–17 Oct 2000, Kota Kinabalu Sabah, Malaysia, pp 12–24

Wood GAR, Lass RA (1985) Cocoa 4th Edition. Longman Scientific & Technical, Essex, UK, 620 p

Yamada MM, Bartley BGD, Melo GRP (1982) Herança do fator compatibilidade em *Theobroma cacao* L. I. Relações fenotípicas na família PA (Parímarí). Revísta Theobroma 12:163–167

Young AM (1994) The Chocolate Tree: A Natural History of Cacao. Smithsonian Institution Press, Washington and London

8 Rubber

P. M. Priyadarshan

Rubber Research Institute of India, Regional Station, Agartala, 799006, India
e-mail: pmpriyadarshan@gmail.com

8.1
Introduction

Natural rubber, a polyisoprene elastomer (*cis*-1, 4-polyisoprene), has wide utility from erasers to aviation tires, making it an undeniable, durable commercial commodity with greater resilience. The prime source of commercial rubber, *Hevea brasiliensis* (Willd. ex A. de. Juss. Müell-Arg.), is a deciduous perennial tree of the family Euphorbiaceae (Fig. 1). The predominant constituent of rubber derived from *Hevea* is *cis*-1,4 polyisoprene ($C_5H_8)_n$, where n may range from 150 to 2,000,000 (Pushparajah 2001). The scientific advancement through the discovery of vulcanization by Goodyear in 1839 adjudged rubber as a prime industrial commodity which had otherwise been unknown to mankind for over 450 years, since Christopher Columbus gave the first description of rubber in the fifteenth century (Priyadarshan and Clément-Demange 2004). It provided almost 40% of the export revenue of Brazil till 1940 (Dean 1987). Since then, Brazil and the adjoining areas of Latin America have shared only 2%, of the total production, mainly owing to the occurrence of South American leaf blight (SALB) caused by *Microcyclus ulei* (P. Henn.) Von. Arx. The Southeast Asian countries continue to enjoy dominance in rubber production and trade through maneuvering more than 90% of the 7.97 milliontons of rubber produced worldwide in 2003 (Sekhar 2004). Thailand with 2.3 million tons is at the helm of rubber producers, followed by Indonesia, India, Malaysia, China, Vietnam, Côte d'Ivoire, Liberia, Sri Lanka, Brazil, the Philippines, Cameroon, Nigeria, Cambodia, Guatemala, Myanmar, Ghana, the Democratic Republic of Congo, Gabon, and Papua New Guinea in descending order of production.

The aim of rubber breeding is to provide a genetically superior planting material in the form of a clone that can be multiplied through bud grafting. Although breeding is a long-term activity, introducing superior clones will be the best way to maintain the profitability of rubber plantations. Nevertheless, biotechnologies hold new promises for deriving improved clones and increasing the efficiency of rubber breeding.

8.1.1
Commercial Importance

Rubber is one of the most important polymers naturally produced by plants because it is a strategic raw material used in more than 40,000 products, including more than 400 medical devices (Mooibroek and Cornish 2000). Primarily owing to its molecular structure and high molecular weight (more than 1×10^6 Da), rubber has resilience, elasticity, abrasion resistance, efficient heat dispersion (minimizing heat buildup under friction), and impact resistance that cannot easily be mimicked by artificially produced polymers. The search for alternative sources of natural rubber production has already resulted in a large number of interesting plants and prospects for immediate industrial exploitation of guayule (*Parthenium argentatum*) as a source of high-quality latex. Metabolic engineering will permit the production of new crops designed to accumulate new types of valued isoprenoid metabolites, such as rubber and carotenoids, and new combinations extractable from the same crop. Currently, experiments are under way to genetically improve guayule rubber production strains in both quantitative and qualitative respects. Although economic considerations may prevent commercial exploitation of new rubber-producing microorganisms, transgenic yeasts and bacteria may yield intermediate or alternative (poly)isoprenes suitable for specific applications.

Fig. 1. **a** Immature plantation. **b** Tapping by a tribal woman from Tripura (northeast India). **c** The tapping panel

8.1.2
Botanical Aspects

Rubber is synthesized in over 7,500 plant species, confined to 300 genera of seven families, viz., Euphorbiaceae, Apocynaceae, Asclepiadaceae, Asteraceae, Moraceae, Papaveraceae, and Sapotaceae (Cornish et al. 1993). The genus *Hevea* includes ten species (Webster et al. 1989; Wycherley 1992; Table 1). A few species are intercrossable (Clément-Demange et al. 2000). Consequently, the *Hevea* species can be considered as a species complex. Since an elaborate description of taxonomical and botanical aspects of *Hevea* is out of the scope of this chapter, readers may refer to other sources (Schultes 1977a, Wycherley 1992; Priyadarshan 2003a; Priyadarshan and Gonçalves 2003; Priyadarshan and Clément-Demange 2004) for narrations on the subject. The natural habitats of *Hevea* species are Bolivia, Brazil, Colombia, Ecuador, French Guiana, Guyana, Peru, Surinam, and Venezuela (Webster et al. 1989). All species are diploids having $2n = 36$ chromosomes ($x = 9$), with the exception of one triploid clone of *H. guianensis* ($2n = 54$) and the existence of one genotype of *H. pauciflora* with 18 chromosomes (Baldwin 1947). However, *H. brasiliensis* behaves as an amphidiploid (Ramaer 1935; Ong 1975; Wycherley 1976; Priyadarshan and Gonçalves 2003). All species probably evolved in Amazonian forests over 100,000 years (Clément-Demange et al. 2000).

8.1.3
Historical Aspects

Fusée Aublet was the first to give a botanical description of the genus *Hevea* in 1775. Five distinguished men played pivotal roles for rubber domestication, viz., Clement Markham (of the British India Office), Joseph Hooker (Director of Royal Botanic Gardens, Kew), Henry Wickham (naturalist), Henry Ridley (Director of Singapore Botanic Gardens), and R.M. Cross (Kew gardner). Kew Botanic Gardens played the crucial role for rubber procurements and distribution. As per directions of Markham, Wickham collected 70,000 seeds from the Rio Tapajoz region of the Upper Amazon (Boim district) and transported them to Kew Botanic Gardens during June 1876 (Schultes 1977b; Baulkwill 1989). Of the 2,700 seeds germinated, 1911 were sent to Botanical Gardens, Sri Lanka during 1876, where 90% of them survived. Later, during September 1877, 100 *Hevea* plants specified as "Cross material" were sent to Sri Lanka. However, in June 1877, "22 seedlings," not specified either as Wickham or Cross, were sent from Kew to Singapore, and were distributed in Malaya and formed the prime source of 1,000 tappable trees found by Ridley during 1888. An admixture of Cross and Wickham materials might have occurred, as the "22 seedlings" were unspecified (Baulkwill 1989). Seedlings from the Wickham collection of Sri Lanka were also distributed worldwide. Rubber trees covering millions of hectares in Southeast Asia are believed to be somehow derived from a very small number of few plants of Wickham's original stock from the banks of the Tapajoz (Imle 1978). After reviewing the history of rubber tree domestication in East Asia, Thomas (2001) drew the conclusion that the modern clones invariably originated from the 1911 seedlings sent to Sri Lanka during 1876. Hence, the contention that the modern clones were derived from "22 seedlings" is debatable. Moreover, if the modern clones are derived from 1911 seedlings, then the

Table 1. Occurrence and features of *Hevea* species (modified from Priyadarshan and Goncalves 2003)

Species	Occurrence	Notable features[a]
H. benthamiana Muell.-Arg.	North and west of Amazon forest basin, upper Orinoco basin (Brazil)	Complete seasonal defoliation Medium-sized tree Habitat: swamp forests
H. brasiliensis (Willd. ex A. L. Juss.) Muell.-Arg.	South of Amazon river (Brazil, Bolivia, Ecuador, Peru)	Complete defoliation Medium to large tree size Habitat: well-drained soils
H. camargoana Pires	Restricted to Marajo island of Amazon river delta (Brazil)	Possibility of natural hybridization with *H. brasiliensis*. 2–25-m tree height Habitat: seasonally flooded swamps
H. camporum Ducke	South of Amazon between Marmelos and Manicoré rivers tributaries of Madeira river	Retains old leaves until new leaves appear Maximum 2-m tall Habitat: dry savannahs
H. guianensis Aublet	Throughout the geographic range of the genus (Brazil, Venezuela, Bolivia, French Guyana, Peru, Colombia, Surinam, Ecuador)	Retains old leaves until new leaves and inflorescences appear. Grows at higher altitudes (1,100 m above sea level) Medium-sized tree Habitat: well-drained soils
H. microphylla Ule	Upper reaches of Negro river in Venezuela. It is not found in other region of geographic range of the genus	Complete defoliation Small trees. They live on flooded areas (*igapós*) Habitat: sandy soils
H. nitida Mart. ex Muell.-Arg.	Between the rivers Uaupes and Icana, tributaries of the upper Negro river (Brazil, Peru, Columbia)	Inflorescences appear when leaves are mature Small to medium-sized trees (2 m)
H. pauciflora (Spr. ex Benth.) Muell.-Arg.	North and west of Amazon river (Brazil, Guyana, Peru). Distribution discontinuous owing to habitat preferences	Retains old leaves until new leaves and inflorescences appear. No wintering Small to big trees Habitat: well-drained soils, rocky hillsides
H. rigidifolia (Spr. ex Benth.) Muell-Arg.	Among Negro river and its effluents. Uaupes and Içana rivers (Brazil, Colombia and Venezuela)	Retains old leaves even after inflorescences appear Small tree from savannahs. Sometime tall, with small crown on the top Habitat: well-drained soils
H. spruceana (Benth.) Muell.-Arg.	Banks of Amazon, Negro and lower Madeira (Brazil)	Retain old leaves until new leaves and inflorescences appear. Flowers reddish purple Medium-size tree Habitat: muddy soils of islands
H. paludosa Ule[b]	Marshy areas of Iquitos, Peru	Small leaflets, narrow and thin in the fertile branches Habitat: marshy areas

After Wicherley (1992), Schultes (1977a), Goncalves et al. (1990), Pires (1973) and Brazil (1971)

[a]Deciduous characteristics mentioned here have a bearing on the incidence of fungal diseases especially through secondary leaf fall (*Oidium*) since retention of older leaves may make the tree "escape oidium." Dwarf types are desirable for the possible wind tolerance. All species are diploid ($2n = 36$) (Majumder 1964), and are intercrossable (Clement-Demange et al. 2000).

[b]Pires (1973) considered 11 species, including *H. paludosa*; Brazil (1971) considered 11 species.

argument that they originated from a "narrow genetic base," as believed even now, needs to be reviewed.

P.J.S. Cramer conducted experiments on variations observed among 33 seedlings introduced from Malaysia in 1883 from which the first clones of the East Indies were derived (Dijkman 1951). Along with van Helten, a horticulturist, he standardized vegetative propagation by 1915. The first commercial planting with bud-grafted plants was undertaken during 1918 on Sumatra's east coast. Ct3, Ct9, and Ct38 were the first clones identified by Cramer (Dijkman 1951; Tan et al. 1996). Commercial ventures gradually spread to China, Thailand, India, Sri Lanka, and Vietnam and rubber became an integral part of the economy of Southeast Asia toward the latter half of the twentieth century. Around 1950, bud-grafted clones proved to be overwhelmingly popular because of higher productivity.

Progress in yield improvement in *Hevea* resulted in a gradual increment, from 650 kg/ha in unselected seedlings during the 1920s to 1,600 kg/ha in the best clones during the 1950s. The yield potential was further enhanced to 2,500 kg/ha in Prang Besar Rubber Estate (PB; Malaysia), Rubber Research Institute of Malaysia (RRIM), Rubber Research Institute of India (RRII), Rubber Research Institute of Ceylon (RRIC), Institut de Recherches sur le Caoutchouc en Afrique (IRCA; Côte d'Ivoire), Balai Penelitian Medan (BPM; Brazil), and Rubber Research Institute of Vietnam (RRIV) clones during the 1990s. During these 70 years of rigorous breeding and selection, notable clones like RRIM 501, RRIM 600, RRIM 712, PB 217, PB 235, PB 260, RRII 105, RRIC 100, IRCA 18, IRCA 230, IRCA 331, and BPM 24 were derived (Tan 1987; Simmonds 1989; Clément-Demange et al. 2000; Priyadarshan 2003a). Primary clones selected during the aforesaid period (PB 56, Tjir 1, Pil B84, Pil D65, Gl 1, PB 6/9, and PB 86) became parents of improved clones. It must also be acknowledged that primary clones like GT 1 and PR 107 are still widely used although their identification traces back to the 1920s.

8.1.4
Propagation Systems

Rubber is currently planted in the form of grafted trees, at a density of about 450 trees per hectare. It experiences an immature phase varying from 5 to 9 years, depending on climate, soil conditions, and management. Multiplication through grafting onto seedlings enables the production of elite genotypes as clones as the almost exclusive planting material. The high level of homogeneity in bud-grafted trees should exhibit intraclonal variation in yield to a minimum, barring factors like (1) soil heterogeneity, (2) difference in juvenility of buds, and (3) variable seedling rootstocks. In contrast, such clonal populations exhibit significant variations. In an experiment with RRII 105, the total volume of latex and dry rubber yield ranged between 5.0 to 325.0 ml and from 1.8 to 144.0 g, respectively (Chandrashekar et al. 1997). The differences exhibited are significant and refutable for a homogeneous population. Owing to lack of an efficient cloning technique, the root system directly affects soil–plant relationships, like water and mineral uptake, water stress resistance, and resistance to wind uprooting (Ahmad 2001). Moreover, efficient breeding for growth of budded clones and the increasing use of fast-growing clones may have generated an imbalance between stock and scion, so emphasizing uprooting hazard (Clément-Demange et al. 1995). Consequently, cloning the root system is a major challenge for rubber tree breeding, as it would greatly facilitate growth, yield improvement, and adaptation to various environments.

8.1.5
Breeding Objectives

Improving dry rubber yield is the exclusive objective of *Hevea* breeding. Growth of the trunk during the immature phase, yield per tree over a specific period, stability of the stand per unit area, and resistance to stresses (tapping panel dryness, wind damage, varied diseases, low temperature, higher altitude, and moisture deficit) are a few of the factors that govern productivity levels. Latex yield and growth are not correlated, obviously owing to differential partitioning of assimilates. Breeding and selection are exclusively applied to scion and the choice of rootstock is very limited. The possibility of cloning a whole plant in vitro would allow breeding to be applied to the root system for resistance to root diseases, for better adaptation to specific soils, and for anchorage. This led to the concept of "compound tree" with three different genetic components, viz., roots, trunk, and canopy, each selected for its own requirements (Simmonds 1985, 1989). Though high-yielding trunks with canopies resistant to SALB had been experimented on by the way of crown budding, it failed commercially.

Adaptation and the yield potential of clones to specific environments are optimized through multi-location trials and localized experimentation. Characterization of the architecture of the trees in connection with wind risk and phenology assessed in relation to susceptibility to leaf diseases (*Colletotrichum gloeosporioides* Penz. Sacc., *M. ulei*) is vital (Priyadarshan et al. 2001). Studies on adaptation of clones to new environments, especially to suboptimal or marginal areas, are gaining momentum (Priyadarshan 2003a, b). In all these aspects, diversification of clones allows large plantations to mitigate risks. Among those clones, the more stable ones are identified for recommendation to smallholders, since smallholders represent a predominant share. A selection focused on fast-growing trees with effective competence toward weed growth, canopy adapted to multicropping, clones adapted to uneven and intensive tapping systems, and climatic variations needs to be made.

Derivation of clones for timber has emerged as a recent objective. An estimation from RRIM shows that a hectare of rubber plantation can yield around 190 m^3 of rubber wood, and 2.7 × 10^6 m^3 of *Hevea* wood would be available from Malaysia (Arshad et al. 1995). Also, there has been some interest generated among scientists to evolve rubber as a factory producing useful chemicals, especially life-saving drugs (Yeang et al. 2002). Possibilities of using rubber trees for reforestation or carbon sequestration may come up in future, which breeders may have to take up with required priority.

8.1.6
Genetic Resources and Variability

H. brasiliensis is believed to be an amphidiploid ($2n = 4x = 36$) that became stabilized during the course of evolution. This contention is amply supported by the observance of tetravalents during meiosis (Ong 1975). However, for practical purposes, *Hevea* is considered as a diploid genus ($2n = 2x = 36$). In situ hybridization studies revealed two distinct 18S-25S ribosomal DNA (rDNA) loci and one 5S rDNA locus, suggesting a possible allotetraploid origin with the loss of 5S rDNA during the course of evolution (Leitch et al. 1998). Hence, as long as a potential ancestor with $2n = 18$ is not known, the rubber tree would be considered an amphidiploid (Priyadarshan and Gonçalves 2003). Locus duplications are infrequent in the *Hevea* genome, and they could have

occurred owing to chromosomal modifications after the polyploidization event (Seguin et al. 2003). Consequently, the two ancestral genomes of *Hevea* would have strongly diverged. Only a comprehensive molecular analysis will reveal the details of the origin.

Allied species of *Hevea* make up a gene pool for breeding purposes, especially for the identification and introduction of genes of resistance to leaf diseases (Priyadarshan and Gonçalves 2003). Within *H. brasiliensis*, a clear distinction needs to be made between the "Wickham" population and the "Amazonian" population. Whereas the Wickham population was domesticated and bred for more than a century, Amazonian populations are still under evaluation, and despite poor yield they display a fairly high resistance to leaf diseases such as *Microcyclus* or *Corynespora cassiicola* Berk et Curt. Wei. (Clément-Demange et al. 2000).

During 1951–1952, 1,614 seedlings of five *Hevea* species (*H. brasiliensis*, *H. guianensis*, *H. benthamiana*, *H. spruceana*, and *H. pauciflora*) were introduced to Malaysia (Tan 1987). Brookson (1956) has given an account of these introductions. In Sri Lanka, 11 clones of *H. brasiliensis* and *H. benthamiana*, and 105 hybrid materials were imported during 1957–1959, through triangular collaboration of USDA, Instituto Agronomico do Norte (IAN; Brazil), and Liberia. Many of these clones were later given to Malaysia (Tan 1987). During 1981, owing to initiative taken by the International Rubber Research and Development Board (IRRDB), 63,768 seeds, 1,413 m of bud wood from 194 high-yielding trees, and 1,160 seedlings were collected from Brazilian Amazonia (Tan 1987; Simmonds 1989). This collection was carried out over three states, viz., Acre, Rondonia, and Mato Grosso, from 60 different locations spread over16 districts. Of the seeds collected, 37.5% were sent to Malaysia and 12.5% to Côte d'Ivoire. Half of the collections were maintained in Brazil. The accessions from budwood collection were sent to Malaysia and Côte d'Ivoire after quarantine against SALB. After the establishment of two IRRDB Germplasm Centers in Malaysia and in Côte d'Ivoire, other IRRDB member countries were supplied with material according to their request.

Attempts to improve yield of wild accessions through Wickham × Amazonian crosses resulted in recombinants with low yield, ranging between 30 and 50% of the level of GT 1, probably owing to the important genetic gap lying between the two populations. Conversely, a wide variability was found within these crosses for growth, with probable heterotic effects enabling the selection of very vigorous Wickham × Amazonian clones. It is quite evident that the Wickham population, though originally meager in number, was subjected to natural prebreeding. This must have occurred in two ways: one through indirect selection of ortets exhibiting adaptation to specific hydrothermal environment and the other by evaluation of useful recombinants. A clear difference in branching habit could be observed between accessions from Acre and Rondonia, which more often have tall trunks with poor branching located at high height, and those from Mato Grosso, which display trees with abundant branching at low height (Clément-Demange et al. 1998). Further during 1995 an expedition was launched by RRIM to collect rubber seeds from Brazil. From this collection, about 50,231 seedlings were planted in Malaysia, including allied species (RRIM Annual Report 1997; MRB Annual report 1999).

8.1.7
Breeding Methodologies and Achievements

Breeding methodologies utilized for maximizing genetic gain are based on breeding objectives with the specific aim of providing farmers with high-yielding clones. Such methodologies are backed by the theory of quantitative genetics, which derives clones well adapted to a given environment. Elements of breeding methodologies are available with major contributions of Dijkman (1951), Shepherd (1969), Wycherley (1969), Tan (1987), Simmonds (1989), Clément-Demange et al. (2000), Priyadarshan (2003a), and Priyadarshan and Clément-Demange (2004). These ideas are discussed here with a separate section on biotechnology.

Primary Clones

The first clones released out of seedlings were those of Cramer's *Cultuurtuin* (Ct3, Ct9, Ct88) selected from 33 seedlings planted in Penang through Java in Indonesia (Dijkman 1951). Mixed planting of these clones gave a yield over 1,700 kg/ha, against unselected seedlings (496 kg/ha) (Tan et al. 1996). During 1924, Major Gough selected 618 seedlings from a population of about one million in the Kajang district of Malaysia that yielded prominent primary clones like Pil A44, Pil B84, Pil B16, PB 23, PB 25, PB 86, PB 186, and Gl 1 (Tan et al. 1996). By 1930, it was understood that the primary clones had reached a plateau of yield (Tan 1987); hence, the emphasis shifted from primary clones to recombinants issued from controlled pollination. Simultaneously, polyclonal seed gardens were organized with improved clones for raising polyclonal seedlings for ensuring supplementary planting materials. Thus, the best seedlings came from Prang Besar Isolated Gardens (PBIG), Gough Gardens (GG), and Prang Besar further proof trials (Tan et al. 1996). By 1970, polyclonal seedling areas extended to 7,700 ha. Both yield and secondary attributes were given deserving importance while selecting clones based on 65 and 35% scores for yield and secondary attributes, respectively (Ho et al. 1979; Tan et al. 1996). The procedure involved field selection in the estates, nursery selection applied to seedlings, small-scale selection with 16 plants per genotype, and large-scale testing with 128 plants per genotype.

Polyclonal seed gardens involving clones with high general combining ability (GCA) ensure panmictic conditions for deriving seedlings with high genetic divergence. Selection for both vigor and high yield can be exercised in such seedlings (Simmonds 1986). After popularization of clones in 1950s, the potentiality of extending rubber to marginal areas was realized. This seems to be an appreciable option since results on the yield of polyclonal seedlings from nontraditional areas like Tripura (northeast India) and Konkan (western India) are encouraging (Sasikumar et al. 2001; Chandrashekar et al. 2002). There is contention that yield and girth variation can be largely accounted for by additive genetic variance (Gilbert et al. 1973; Nga and Subramaniam 1974; Tan 1981). As per general genetic principles, selection based on genotypic values as reflected by GCA would be more reliable and desirable. GCA could be estimated by the evaluation of seedling progenies, in order to select the best parent clones. DNA fingerprinting can contribute significantly to assess molecular diversity of parents and their progenies. The optimum number of parents is crucial while constituting seed gardens and Simmonds (1986) suggested a layout involving nine clones with all heteroneighbors as the best (Fig. 2).

The extent of selfing due to lack of self-incompatibility may reduce the vigor of the first-generation population (SYN$_1$), since there is no evidence of self-incompatibility. Since inbreeding reduces the ability of zygotes to germinate, the presumption is that only cross-pollinated seeds will survive (Simmonds 1986). Till recently, such SYN$_1$ progenies were considered as class I planting material in Malaysia and must be of better use in nontraditional/marginal areas. However, factors like

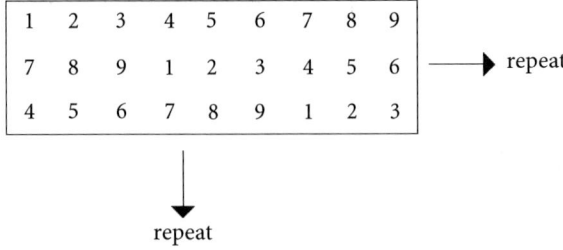

Fig. 2. Two-dimensional design for the production of polycross progenies (after Simmonds 1986)

agronomic performance of such synthetic seedlings, the long time taken to attain seed production, and supply of seeds in tune with demand need to be evaluated before utilizing this methodology. By contrast, seed gardens can be viewed as a recombination tool for addressing the improvement of wild Amazonian populations. With this view, 50 Amazonian parents were analyzed in Côte d'Ivoire using microsatellite markers (Blanc et al. 2001). Most of the paternal contribution to the progenies was due to a restricted number of male parents with substantial flowering; hence, they were very far from a panmictic status. It is implicit that each seed garden needs to be evaluated with DNA fingerprinting.

Derivation and Evaluation of Recombinants

Recombination breeding starts with production of full-sib families, followed by a seedling evaluation trial (SET), a small-scale clonal trial (SSCT), and a large-scale clonal trial (LSCT) with selection practiced at every level. This process is cyclical, with the best clones becoming candidates for recombination in the next cycle. Yield improvement from 500 kg/ha in primary clones to 2,500 kg/ha in the current clones could be attained through recombination breeding and selection in RRIM and Prang Besar. The RRIC 100 series released in Sri Lanka during the 1970s is yet another example. Much of the hybridization work in Malaysia, Indonesia, India, Côte d'Ivoire, Brazil, Thailand, and Vietnam further strengthened the array of hybrid clones with differential genetic setup, obviously owing to selection pressure applied under varied conditions (Table 2). At least 16 primary clones are considered prime progenitors of many modern clones (Fig. 3). However, many valuable recombinants must have been lost during the course of assortative mating of primary and hybrid clones followed by subsequent directional selection for yield under varied geocli-

mates (Priyadarshan 2003a). The crossing of "the best with the best" (generation-wise assortative mating, GAM) with strong emphasis on selection for precocious yield within Wickham material (Wycherley 1976) has been practiced in all these recombination breeding programs. Breeding for disease resistance has to take account of specific aspects related to host × pathogen interactions. But this exercise has to go a long way before it achieves clones combining resistance and higher yield.

Genetic Resources Management

Yield in rubber analyzed according to different types of mating designs was shown to have a large additive genetic variance. Heritability and GCAs for yield and growth have been investigated at the RRIM and are tend to be high, thus justifying GAM (Gilbert et al. 1973; Nga and Subramaniam 1974; Tan 1977, 1978, 1981; Simmonds 1989). The importance of low female fertility of many parents emerges here as a limiting factor for producing every full-sib progeny. The need for selecting highly heterozygous clones and that for reducing the risk of narrowing the genetic base are the two prime attributes that need attention (Simmonds 1989). One option is to involve Amazonian germplasm in breeding programs. Though such crosses appear as the best way to introgress the new germplasm into breeding populations, most of the Amazonian genotypes bear a large part of alleles unfavorable for yield (genetic burden in heterozygous plants). Prebreeding appears necessary within the Amazonian groups before using them as progenitors in crossing with Wickham (Baudouin et al. 1997; Priyadarshan and Clément-Demange 2004). Since a detailed evaluation of whole *Hevea* germplasm is quite impossible, a working population of 287 accessions was extracted at the IRRDB African Germplasm Center (Clément-Demange et al. 1998). It was proposed to combine the use of field experiments and molecular markers (microsatellites) for extracting a clonal population of reduced size (core collection) containing maximized genetic variability (Hamon et al. 1998; Brown 1989; Clément-Demange et al. 2000).

Selections

The breeding cycle in rubber extends to 20–30 years between pollination and yield assessment, distributed over three selection stages. This justifies standardization of early selection methods to optimize and shorten the cycle as much as possible. One compo-

Table 2. Profile of prominent clones

Clone	Parentage	Yield (kg/ha)	Girth increment during tapping	Resistance to						
				Wind damage	Panel dryness	Pink disease	Oidium	Colletotrichum	Corynespora	Phytophthora
RRII 105[I]	Tjir 1 × Gl 1	2,210	3	3	5	5	3	5	5	1
RRII 203[I]	PB 86 × Mil 3/2	1,618	4	3	2	3	3	NA	3	3
RRII 208[I]	Mil 3/2 × AVROS 255	1,587	3	3	3	NA	3	NA	NA	NA
RRIC 100[M]	RRIC 52 × PB 83	1,774	3	5	3	3	4	3	5	NA
RRIM 600[M]	Tjir 1 × PB 86	2,199	4	4	4	1	3	3	1	1
RRIM 623[M]	PB 49 × PB 84	1,622	4	3	2–3	1–2	3–4	4	4	1
RRIM 712[M]	RRIM 605 × RRIM 71	2,264	2	5	4	3	3	1	3	3
RRIM 936[M]	GT 1 × PR 107	2,146	3	4	3	4	3	4	4	2
RRIM 937[M]	PB 5/51 × RRIM 703	2,483	2	5	3	4	3	3	5	3
RRIM 2015[M]	PB 5/51 × IAN 873	2,760	4	NA	NA	NA	4	4	4	3
PB 217[M]	PB 5/51 × PB 6/9	1,778	4	4	4	2	2	3	4	1
PB 235[M]	PB 5/51 × PB S/78	2,485	3	2	2	3	2	2	4	3
PB 255[M]	PB 5/51 × PB 32/36	2,283	3	4	2	2	2	2	4	2
PB 28/59[M]	Primary clone	2,023	1	3	3	2	2	2	4	2
PR 255[M]	Tjir 1 × PR 107	2,018	3	4	3	3	3	3	4	3
PR 261[M]	Tjir 1 × PR 107	1,838	3	4	3	3	1–2	4	3	3
IRCA 111[CD]	PB 5/51 × RRIM 600	1,446	5	3	3–4	NA	NA	NA	NA	NA
IRCA 230[CD]	PB 5/51 × GT 1	1,807	5	3	3–4	NA	NA	NA	NA	NA
RRIT 163[I]	PB 5/51 × RRIM 501	2,086	2	NA	NA	NA	3	NA	3	NA
Haiken 1[C]	Primary clone	1,500	3	4	2	3	NA	NA	NA	NA
REYAN 8-333[C]	SCATC 88-13 × SCATC 217	2,187	3	3	3	3	NA	3	NA	NA
BPM 24[M]	GT 1 × AVROS 1734	1,394	2	3	3	3	2	3	3–4	4
IAN 873[B]	PB 86 × FA 1717	1,920	4–5	3	4	4	4	4	NA	NA
IAC 301[B]	RRIM 501 × AVROS 1511	2,320	4	4	4	4	4	4	NA	NA
IAC 300[B]	RRIM 605 × AVROS 353	1,887	3	2	2	3	2	2	NA	2

REYAN is the new name for South China Academy of Tropical Crops (SCATC). Tapping system s/2 d/2 6d/7 86%; no. of tapping days per year 158 ± 11 (with wide regional variation depending on weather); 327 ± 34 trees per hectare. IAN 873 exhibits good tolerance to South American leaf blight. Tapping under Vietnamese (southeast) conditions s/2 d/3 6d/7 1 poor, 2 below average, 3 average, 4 good, 5 very good, NA not available, since the disease is not prominent; M Malaysia, I India, C China, CD Côte d'Ivoire, B Brazil

Table 2. (continued)

Clone	Parentage	Yield (kg/ha)	Girth increment during tapping	Resistance to						
				Wind damage	Panel dryness	Pink disease	Oidium	Colletotrichum	Corynespora	Phytophthora
Fx 3864[B]	PB 86 × PB 38	1,755	4	3	3	NA	2	2	NA	3
IAN 4493[B]	EX 441 × Tjir 1	1,711	3	3	2	NA	2	2	NA	2
IAC 303[B]	RRIM 505 × AVROS 1511	2,190[6Y]	3	3	2	NA	2	2	NA	2
PB 260[VN]	PB 5/51 × PB 49	1,691[10Y]	4	3–4	3–4	3	4–5	4	NA	3–4
RRIC 12[VN]	PB 28/59 × IAN 873	1,654[10Y]	4–5	4	4	4	2	3–4	NA	4
GT 1[VN]	Primary clone	1,459[10Y]	3	5	5	3	2–3	3	NA	3–4
RRIV 4[VN]	RRIC 110 × PB 235	2,103[10Y]	2	2	4	3–4	2–3	2	NA	4

VN Vietnam; 6Y average over 6 years, 10Y average over 10 years

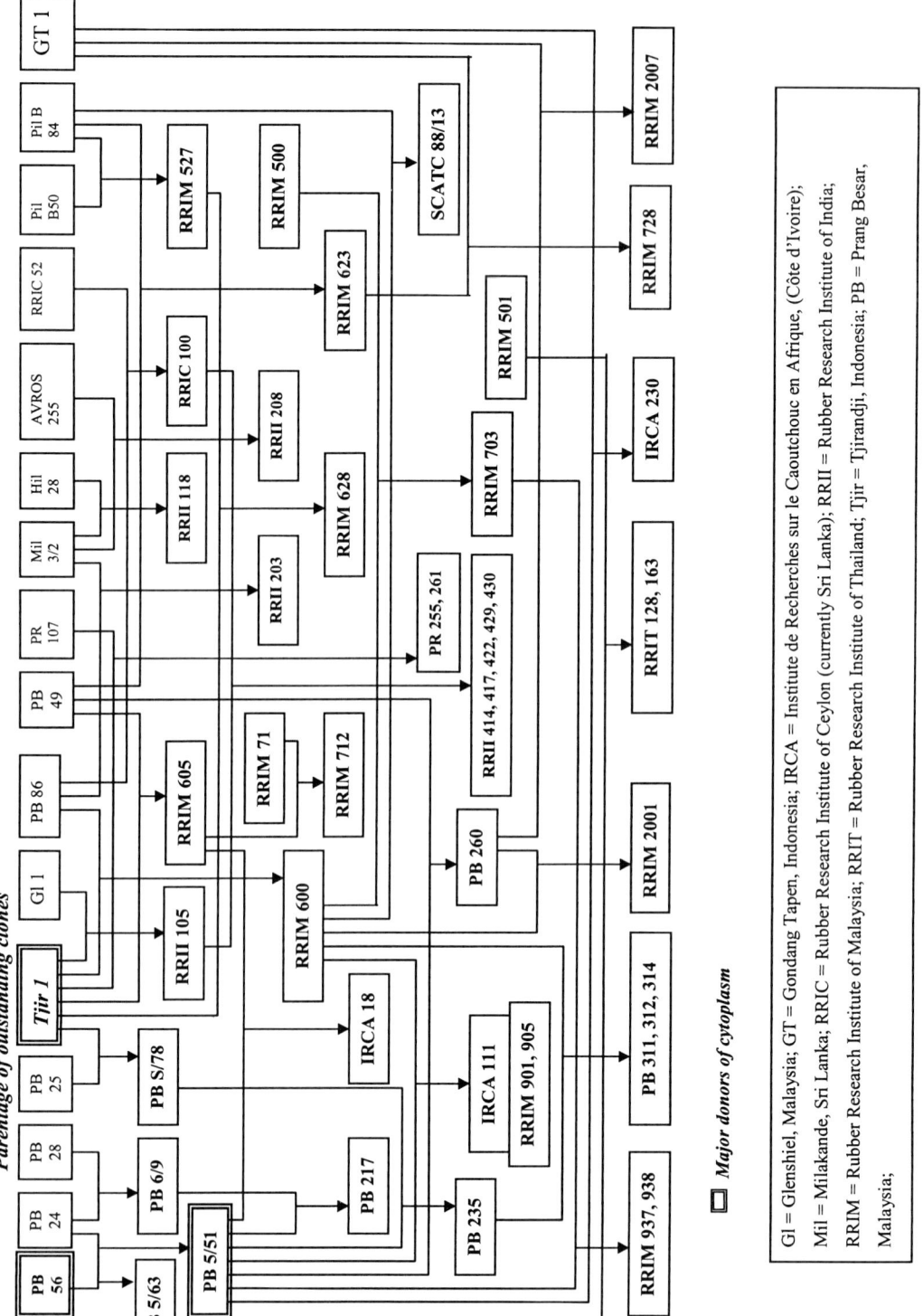

Fig. 3. Parentage of outstanding clones

nent of early selection is identification of traits at a young age that have correlated response with yield at maturity and the other is combined management of different selection stages to improve the accuracy of estimation of genetic value. Several parameters, viz., girth, height, bark thickness, latex vessel number, latex vessel and sieve tube diameters, and rubber hydrocarbon in bark and petiole, were inconsistent in having relations with yield both at seedling and at mature stages (Gunnery 1935). Also, parameters like quantity of latex oozing out of leaflets or petiolules, plugging index, photosynthetic rate, and number of stomata (Senanayake and Samaranayake 1970; Ho 1976; Zhou et al. 1982; Samsuddin et al. 1987) were studied but only plugging index and latex vessel number showed consistent and significant correlations with yield (Huang et al. 1981). Hénon and Nicolas (1989) justified the thickness of the bark not being considered as a reliable attribute to predict yield, but the number of latex vessel rings can help to differentiate poor yielders in both Amazonian and Wickham populations. The first stage of selection is applied to full-sib seedlings (SET) and information from this stage is used for selecting new clones to be evaluated as grafted trees in a SSCT (Fernando and De Silva 1971). Analysis of different procedures for assessing yield on young seedlings confirmed the use of only mild selection at the first nursery stage (Gnagne et al. 1998; Tan 1998).

For combining SET and SSCT, Gnagne et al. (1998) studied the relationships between the two stages of early selection. A combined family × individual selection was proposed in the form of a linear combination of family value and individual values. At the nursery stage, with only one seedling tree per genotype, it will almost be impossible to directly assess the environment effect, so limiting the predictive efficiency of this first stage. With this view, early selection does not aim at shortening the cycle but at improving the selection efficiency. Alternatively, molecular genetic markers are considered independent of the environment and using them as predictors can contribute to improving the accuracy of genetic value assessment according to the concept of marker-assisted selection (MAS; Lynch and Walsh 1998). But this technique needs further refinement. The third stage of selection is a LSCT, involving evaluation of individual genotypes in sets and on rather large plots over a long period at different locations. Biochemical markers like ATP can also be used as markers (Sreelatha et al. 2004). Further, Priyadarshan and Clément-Demange (2004) proposed an al-

ternative method where the seedlings raised at a moderate spacing will be evaluated at maturity and then cut back for budwood multiplication. The high yielders will be evaluated directly as a Clonal Block Trials before they are recommended for a specific location. This can reduce the experimental period from 34 to 20 years (Fig. 4).

Breeding Against Stresses

The increased global demand for rubber prompted the countries outside the hitherto traditional zone to focus their attention on the cultivation of rubber (Pushparajah 1983). Rubber was also extended to suboptimal environments of the countries coming under the traditional belt. This was mainly due to three reasons, viz., increasing demand for rubber, crop diversification in traditional areas, and efforts to upgrade the living standards of the people in the so-called suboptimal environments (Priyadarshan 2003a). Specific areas of China, Thailand, Vietnam, India, Côte d'Ivoire, and southern plateau of Brazil fall under suboptimal environment (nontraditional areas) that experiences one or more stress situations, viz., drought, low temperature, high altitude, diseases, and strong winds. Latitudinal range will be more than 10°N or 10°S of the equator. On the other hand, the traditional rubber growing tracts extend up to 10°N and 10°S of equator, and offer environmental conditions ideal for rubber cropping. They are (1) 2,000–4,000 of mm rainfall distributed over 100–150 days/year (Watson 1989), (b) mean annual temperature around $28 \pm 2\,°C$ with a diurnal variation of about $7\,°C$ (Barry and Chorley 1976), and (3) sunshine hours of about 2,000 h/year at the rate of 6 h/day in all months (Ong et al. 1998). In a study with hydrothermal index, Rao et al. (1993) rationalized Senai of Malaysia ($1°\,36'$ N; $103°\,39'$ E) to be the most suitable area for rubber cultivation and production.

Latitudinal increase will imply a fall in mean annual temperature and more prominent winter conditions during November to January. Northeastern states of India and the highlands and coastal areas of Vietnam and south China that lie between 18° and 24° N are regions well recognized as inhospitable for the crop, exhibiting stress situations like low temperatures and typhoons (Zongdao and Yanqing 1992; Priyadarshan and Gonçalves 2003). It may also be worthwhile to note that rubber areas of China and Tripura fall under the same latitude range, though climatic conditions in vivid pockets of China vary because its trop-

Fig. 4. Scheme for the production of clones

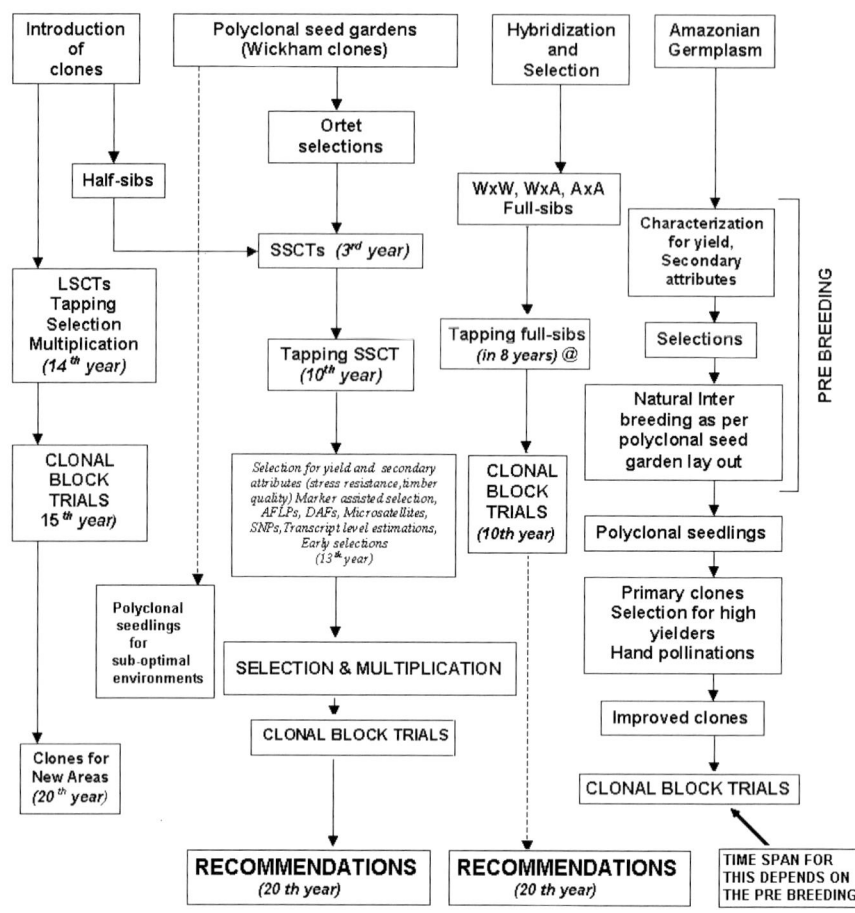

LSCTs= Large scale clone trials ; SSCTs = Small scale clone trials; W = Wickham ; A = Amazonian
@ Full sibs will be cut back and kept as budwood points after yield evaluation

ical and subtropical regions are undulating and diversified (Priyadarshan 2003a; Priyadarshan et al. 2005). The southern plateau of Brazil, especially São Paulo (20–22° S; 450–500 m above sea level) is a prominent rubber area. This move to grow rubber seasonally affected by dry and cold conditions is mostly motivated to escape from the climatic conditions congenial to SALB. These areas, apart from high altitude, offer high rainfall that often exceeds the basic requirements. North Côte d'Ivoire is also being experimented with rubber, where warm climatic conditions prevail (Dea et al. 1997). A geoclimatic comparison of various environments with Tripura, China, Brazil, Côte d'Ivoire, Indonesia, Vietnam, and Thailand would amply reveal a spectrum of climatic conditions over which rubber is grown (Table 3). In India, marginal areas delineated as nontraditional zones, spread over to the states of Maharashtra, Orissa, Tripura, Assam, West Bengal, Meghalaya, and Mizoram, pose a multitude of hazards, viz., moisture stress, low temperature, wind,

high altitude, and disease epidemics, apart from altered soil physical properties (Priyadarshan 2003a). Adaptation of existing clones to nontraditional environments with clone-specific/area-specific tapping schedules and fertilizer inputs are of prime importance to achieve latex yields compared with favorable zones.

Stress Factors Low temperature, wind, diseases, and climatic changes due to higher altitude and latitude are the stress factors influencing rubber culture. In China, two types of cold damage (chilling injury) have been identified, viz., radiative and advective (Zongdao and Xueqin 1983). In the radiative type, the night temperature falls sharply to 5°C, whereas the day temperature ranges between 15 and 20 °C or above; in the advective type, the daily mean temperature remains below 8–10 °C, with a daily minimum of 5 °C. In both types, under extreme circumstances, complete death of the plant is the ultimate outcome.

Table 3. Spectrum of weather variables under different geoclimates

Attributes	Bogor (Indonesia)[a]	Pindorama (São Paulo, Brazil)[b]	Kourou (French Guiana)[a]	Odienne (Côte d'Ivoire)[b]	Nong Khai (Thailand)[b]	Hainan (China)[b]	Agartala (Tripura, India)[b]	Senai (Malaysia)[a]	Dak Lak (Vietnam)[b]
Temperature (°C) (mean)	27.4	22.9	26.3	25.6	26.8	22.6	25.4	26.9	21.5
Daily temperature range (°C)	9.1	11.8	7.8	12.7	10.2	7.8	9.9	7.2	7.9
Relative humidity (%)	79	67	81.5	67	74	79.9	76.8	82.3	75.7
Sunshine (% h)	61	55.1	49.9	59.2	58.1	46.8	50.8	47.8	48.8
Wind run (m/s)	2.4	1.6	1.35	1.3	1.2	2.7	1.38	2.1	2.5
Rain fall (mm/year)	1,791.5	1,117.6	2,573.53	1,297.9	1,455.96	1,431.29	1,960.1	2,282.2	1,669.31
No. of rainy days	159	117	193	119	128	151	93	182	163
Moisture availability index	0.78	0.49	1.4	0.67	0.7	0.6	1.1	1.2	0.8
Penman ET_0 (mm/day)	4.4	3.87	3.78	4.3	3.97	3.48	3.39	3.9	3.57
Latitude	5° 9′ S	20° 25′ S	5° 7′ N	9° 30′ N	17° 51′ N	19° 2′ N	23° 49′ N	1° 36′ N	14° 55′ N
Longitude	106° 58′ E	49° 59′ W	52° 56′ W	7° 34′ W	102° 44′ E	109° 30′ E	91° 16′ E	103° 39′ E	108° 10′ E
Altitude (m)	16	505	48	451	164	671	31	13	655

Source: http://www.iwmi.org
Senai (Malaysia) is considered as the area offering the optimum environment.
[a]Traditional
[b]Nontraditional

An analogous situation prevails in northeastern states of India. Reports from China point out that while clones GT 1 and Haiken 1 can withstand temperatures as low as 0 °C for a short span, SCATC 93-114 can endure temperature as low as −1 °C. The cold-wave conditions in Tripura state (northeast India) can be conveniently classified as relating to the radiative type. Chinese clones like Haiken 1, SCATC 88-13, and SCATC 93-114 are being evaluated in Tripura. The yielding pattern shows Haiken 1 to be a high yielder among Chinese clones, compared with RRIM 600, which is used as a local check. Though SCATC 93-114 is known for its cold endurance, it never shows considerable yield potential under the conditions of Tripura, at least during the initial stages on B0-1 panel (Priyadarshan et al. 1998a, b). China has also developed Zhanshi 86, a clone borne out of a random cross between SCATC 93-114 and Wuxing I₃, and this has better cold endurance than SCATC 93-114 (Senyuan 1990). Further, clones like Zhanshi 306-15 (RRIM 600 × Guangxi 6-68) give around 10 kg of dry rubber per tree. But these contentions are to be proved at the block level. IAN 873, a SALB-resistant high-yielding clone developed in Brazil shows resistance to cold weather in China (Senyuan 1990).

In India, areas between 15° and 20° N of the western and the eastern side have been identified as nontraditional zones. For instance, the Konkan region of western India experiences long dry periods, high temperatures, low atmospheric humidity, and zero rainfall between September and May with daytime temperatures ranging between 38 and 41 °C during summer months (with a maximum of 47 °C). Though it gets rainfall of 2,430 mm, the distribution is uneven (Devakumar et al. 1998). The atmosphere during summer results in a high vapor pressure deficit. Almost an analogous situation prevails in the eastern part of India.

Wind is yet another abiotic stress influencing establishment and growth of rubber. Contributing to the drying effect of drought conditions, it induces regimes of long-lasting steady winds during the dry season in highlands of Vietnam. Wind speeds of 2.0–2.9 m/s retard rubber growth and latex flow, and speeds of 3.0 m/s or above severely inhibit normal growth. Wind over Beaufort force 10 (more than 24.5 m/s) plays havoc with branch breaks, trunk snaps, and uprooting of trees, mainly prevalent in China, during June to October (Watson 1989). Studies in China revealed that clones PR 107 and Haiken 1 can be wind-

enduring, and PB 5/51 is wind-enduring in Tripura (Priyadarshan et al. 1998a). Establishment of shelter-belts, consisting of fast-growing and wind-resistant species, is one remedial measure being followed in China (Zongdao and Xueqin 1983). But this exercise needs proof, taking into account their effects on the total stand as well as the economy of their implementation and land occupation. Alternatively, adoption of judicial pruning of branches and induction of branches at lower height can reduce wind damage from 25.3 to 13.7% (Zongdao and Xueqin 1983). In Côte d'Ivoire, rubber plantations often experience wind damage due to storms occurring at the onset of the rainy season (April to May) (Clement-Demange et al. 1995).

Diseases Despite having all-favorable climatic conditions, SALB prevents Latin America from developing rubber plantations and it represents a permanent major threat to rubber in Asia and Africa (Chee 1976; Dean 1987; Davies 1997). Breeding work mainly based on backcross technique was undertaken to incorporate resistance in high-yielding clones; however, the efforts were in vain owing to the unknown polygenic nature of the attributes, high variability of the pathogen, and multiple interactions between strains and clones (Rivano 1997a, b). Simmonds (1990, 1991) argues that the pathotype-specific resistance (vertical resistance) has resulted in catastrophic failures. Horizontal resistance should be more effective and durable (Rivano et al. 1989; Simmonds 1990). Amazonian germplasm with resistant sources is yet to be improved for yield. With these views, efforts have been reoriented toward the analysis of partial resistance components (Junqueira et al. 1990). Recently, the genetic determinism of the resistance source of *H. benthamiana* (F 4542), widely used in many former backcross programs, was characterized by a genetic map (Lespinasse et al. 2000b).

Resistance to other diseases was studied in some detail. Clones with an early refoliation like AVROS 2037, RRIC 100, RRIM 600, or PB 260 can develop their new leaves before the onset of the rainy season, and so are able to escape incidence of *C. gloeosporioides*. By contrast, the widely planted clone GT 1 with late defoliation has been seriously affected in many areas of Malaysia, Indonesia (Kalimantan), and Central Africa. This consequence of early defoliation on the resistance of some leaf diseases of rubber was successfully used for the development and implementation of artificial early defoliation by Ethephon® aerial spray in Cameroon and Gabon for escaping from *Colletotrichum* (Sénéchal 1986). *Corynespora* leaf fall disease has become a major threat for rubber cropping in Southeast Asia and West Africa. An escape strategy related to early defoliating clones or by the way of artificial defoliation is not operative. It was demonstrated that the fungus is acting by the emission of a toxin (*cassiicoline*) in the leaves (Onesirosan et al. 1975). Studies conducted under controlled conditions have not provided evidence of a significant interaction between clones and strains, but GT 1 is tolerant and PB 260 and RRIC 100 are highly susceptible (Breton et al. 2000). *Oidium heveae* seems to be favored by rather cold conditions prevalent toward the onset of refoliation (Rajalakshmy et al. 1997). In a comparative study with clones of various geographic origin, SCATC 93-114, Haiken 1, and RRIM 703 were adjudged as resistant to *Oidium* in the traditional areas of India (Rajalakshmy et al. 1997; Alice et al. 2000). While studying sensitivity relationships between clones, Alice et al. (2001) confirmed these results and noted SCATC 93-114, RRIM 703, Haiken 1, RRII 208, RRII 5, and PB 310 as stable sources of resistance over the years. Molecular markers for resistance to *Oidium* are to be developed to augment breeding programs with a cautious approach since the cost-effectiveness of this technique is yet to be proved.

Genotype × Environment Interactions Yielding trends in Tripura, Vietnam, and São Paulo indicate that there are low-yielding and high-yielding periods (Fig. 5). Under the hydrothermal situations of Tripura, in a study involving 15 clones of vivid geographical origin, almost all clones showed an increment in yield toward the onset of the cold season, i.e., during October to November. Onset of cold season renders a stimulatory effect to maximize yield and the trend continues till the temperature falls below 15 °C during January. The clones are classified under two categories: (1) one showing a slow escalation in yield from April onward, reaching the maximum during November, and receding sharply during December and January and (2) the other with a low-yield regime during April to October, and with the peak yield during November and December (high-yield regime), then receding during January. While PB 235 comes under the first category, all the other clones come under the second. The trend shows that the first category is appreciable since the clones give considerable yield during regime I, which ensures bet-

ter returns to the planter. The rationale is that a fall in temperature along with reduced evaporation and low wind speeds prevail upon the microenvironment to influence yield stimulation during October to December (Priyadarshan et al. 2000). The test of heterogeneity for the environmental index showed high significance, so indicating that the high stability values of a few clones ($s^2 i$) over the years were due to a linear effect of the climatic attributes (Priyadarshan 2003b). In São Paulo, RRIM 526 showed higher yield during the low-yield regime in comparison with RRIM 600 (Gonçalves, IAC, São Paulo, personal communication). These observations clearly rationalize the selection to be in favor of a consistent yielder (Priyadarshan et al. 2000).

Under Malaysian conditions, Tan (1995) accounted for genotype × environment interactions with a nonlinear effect of wind damage and disease. In fact, these hazards play a prominent role in differentiating the adaptation of clones to one or different locations. Grouping of clones with a high mean and a low coefficient of variation has been proved to be dependable in selecting better performers in a new environment (Tan 1995; Priyadarshan et al. 2002). Genotype × environment interactions were also significant for rubber production and girth increment under the conditions of São Paulo (Gonçalves et al. 1998; Costa et al. 2000). In an investigation with seven clones (GT 1, PR 255, PR 261, IAN 873, RRIM 701, PB 235, and RRIM 600), over five environments, IAN 873 was adjudged as the most stable clone over years and locations (Gonçalves et al. 2003). Though GT 1 and IAN 873 were stable for girth and yield, respectively, the change in rank of genotypes across the environments suggests that a breeding strategy of selecting specifically adapted clones in a megaenvironment will ensure the required productivity. Planters will also perceive yield stability as the most important socioeconomic aim to minimize crop failure, especially in suboptimal environments.

8.2
Application of in Vitro Culture

Long breeding cycle and larger size of the crop make the breeding process time-consuming. Attainment of yield plateau prompted researchers to employ biotechnology tools to induce, increase, and exploit new genetic variation. Biotechnology applied

Fig. 5. Contribution toward yield in GT 1 and PB 235 over months in Vietnam (highlands), India (Tripura), and Brazil (São Paulo) (São Paulo data provided by Dr. Goncalves)

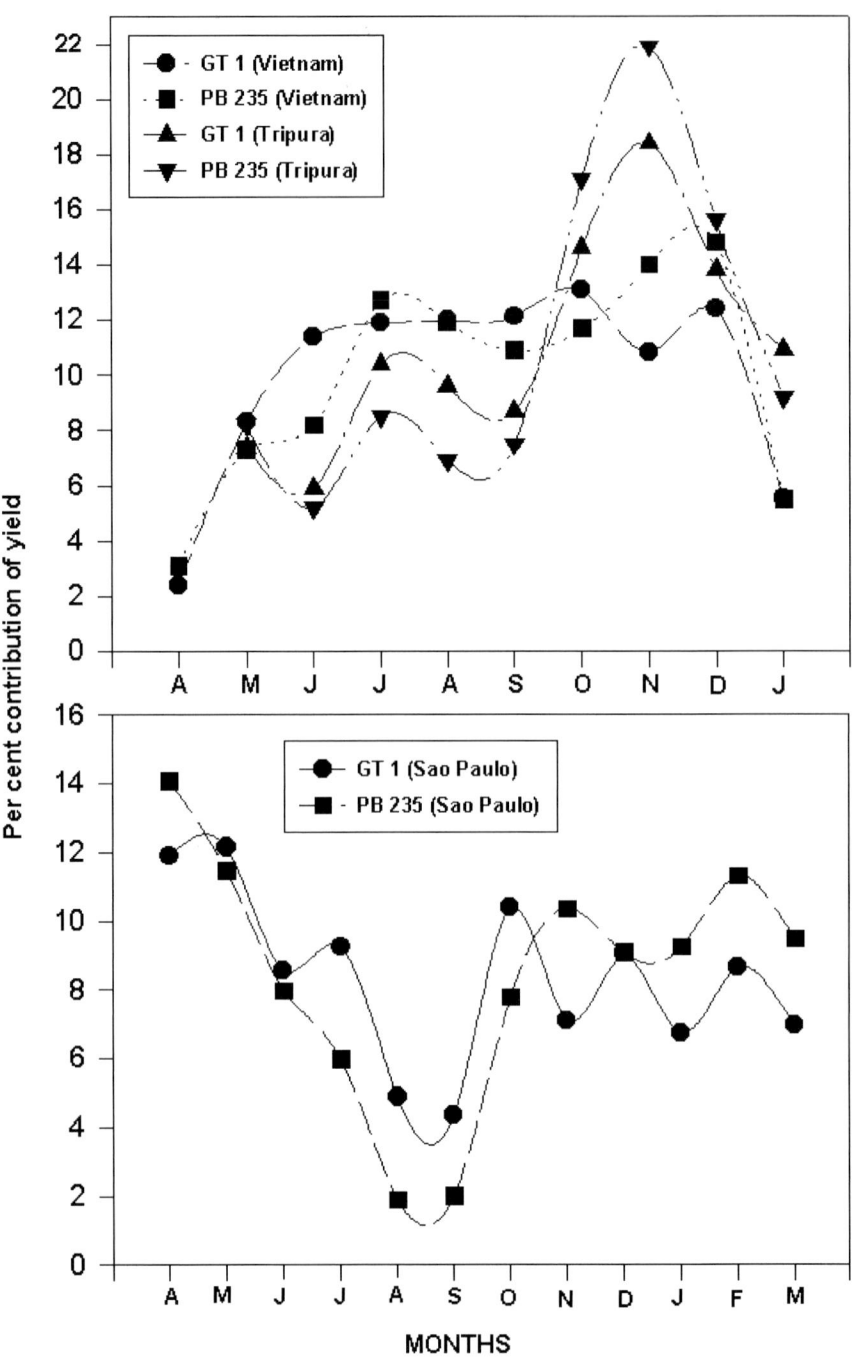

to *Hevea* can be discussed under two categories, viz., in vitro culture and molecular genetics. In vitro culture deals with regeneration and propagation, and molecular genetics involves identification, characterization, introduction, and expression of novel genes.

Chua (1966), attempting derivation of callus from plumule tissues of seedlings, was the first to attempt in vitro culture of rubber in 1960s. Further, the RRIM took the initiative of maintaining callus cultures from various explants that later expanded to somatic embryogenesis and micropropagation through stem explants (Paranjothy and Gandhimathi 1976). While anther culture was employed to achieve pure lines followed by exploitation of heterosis, micropropagation and somatic embryogeny were worked out to have homogeneous populations. Though research on in vitro culture commenced nearly 38 years ago, even after rigorous experimentations, due to shortcomings toward commercial applicability, these areas are still under experimentation.

8.2.1
Anther Culture

Plants from *Hevea* pollen were initially made available during 1977 at the Baoting Institute of Tropical Crops, Hainan, China (Chen et al. 1979). Since then, at least four laboratories in China have taken the lead in researching production of haploid plants in vitro (Carron et al. 1989). Carron et al. (1989) enumerated three phases for the production of haploids from anther culture, viz., production of embryos, maturity of embryos, and plant regeneration. Embryo production from callus takes nearly 50 days. The balance between callus development and initiation of embryos needs to be maintained though use of MB medium with the judicious addition of naphthaleneacetic acid (NAA), coconut water, nitrogen, potassium, and sugar, resulting in the production of calluses and embryos (Chen 1984). The somatic callus then degenerates and the embryos develop from microspores. Subculture must be carried out at this stage into the differentiation medium in order to ensure maturity of embryos (Chen et al. 1982). The cultures need 2–3 months for the apical bud to develop. Coconut water at this stage will be substituted with gibberellic acid (GA_3) for better development of cotyledons. For plant regeneration, progressive increment of GA_3, gradual withdrawal of other growth regulators, addition of 5-bromouracil, and reduction of sugar will result in the development of plants from embryos. Cytological investigations of callus, embryos, and plantlets showed mixoploidy (Qin et al. 1979); however, when the plants develop in vitro, there is a progressive tendency toward diploidy (Carron et al. 1989).

8.2.2
Somatic Embryogenesis and Meristem Culture

The first plants from somatic embryogeny were obtained simultaneously in China and Malaysia from anther wall (Carron et al. 1989). Later, inner integument that represents mother tissue was used to produce somatic embryos by the Centre de Coopération Internationale en Recherché Agronomique pour le Développement (CIRAD) in France (Carron and Enjalric 1982). The successive phases are callogenesis, differentiation, multiplication, and plant regeneration. The judicious combination of 2,4-dichlorophenoxyacetic acid (2,4-D), indoleacetic acid, and benzylaminopurine (BAP) and an increase in sucrose concentration promotes callogenesis in the dark. Cultures are then taken into the light with a changed macroelement composition to increase tissue proliferation (Carron et al. 1989). The differentiation medium is enriched with naphthoxyacetic acid and BAP with low sucrose concentration. It takes 5–6 months in this medium for the embryos to develop. Carron et al. (1989) claimed that nearly 3,000 globular or lanceolate embryoids could be achieved in 4 months. For plant regeneration, addition of indolebutyric acid (IBA) is crucial for promoting root and cotyledon formation. Successful plantlet formation and acclimatization have been achieved in Haiken 1, Haiken 2, and SCATC 88/13 (Wang et al. 1980). Anther wall requires 2,4-D, NAA and kinetin for both callogenesis and embryo induction. BAP and zeatin are essential in addition to NAA and 2,4-D. GA_3 is found to increase the number of embryoids. BAP and GA_3 together with lower sucrose level are shown to improve plant regeneration (Sushamakumari et al. 2000). Carron et al. (1995a, b) gave a detailed account of the procedure and media formulation for somatic embryogenesis in *Hevea*.

Significant genotype–medium interactions are experienced in the aforementioned procedures (El Hadrami et al. 1991; Montoro et al. 1993). Tissue–medium interactions are also very prominent. This is evident in integument culture, where a different additive of 234 mM sucrose, 9 mM BAP, and 2,4-D was needed for embryogenesis. Abscisic acid was essential for embryo development (Etienne et al. 1993; Veisseire et al. 1994 a, b). Plant regeneration takes 25 days. Low germination percentage and plant conversion are seen as setbacks in this procedure, since the mechanism involved in this technology is poorly understood (Cailloux et al. 1996; Linossier et al. 1997). For instance, initiation and germination of embryos are seen to progress at higher temperature of 24–27 °C (Wang and Chen 1995; Wang et al. 1998). Poly(ethylene glycol and high $CaCl_2$ concentration are seen to stimulate embryo production (Etienne et al. 1997a; Linossier et al. 1997). Thus, the clone tissue–media interactions prevail in this technology that necessitates extensive basic studies. More recently, Etienne et al. (1997b) standardized a pulsed-air

Fig. 6. a Embryogenic callus under immersion in an
autoclavable filtration unit
RITA^{TM}. b Somatic embryo
development from the callus.
c Rooted plantlet. d Trial
borne out of micropropa-
gated plants. (After Carron
et al. 1995b)

temporary immersion system for enhancing embryo
production, through culturing embryogenic callus
under immersion in an autoclavable filtration unit
RITA^{TM}. Somatic embryo production was 3–4 times
greater than that on a semisolid medium, to the
tune of 400 embryos per gram fresh weight with
a ower number of abnormal embryos (Fig. 8-6). The
Rubber Research Institute of Thailand (Bangkok) and
CNRA (Côte d'Ivoire) planted 13,000 embryo-derived
plants for field trials (Carron et al. 1995b). Clones
PR 107 and PB 260 were highly regenerative. This
is a leap toward regeneration of *Hevea* in vitro,
since higher regeneration should be ensured to have
homogeneous populations and a rapid gene transfer
system in *Hevea*.

Juvenile stem pieces are desirable for meristem
culture that follows three phases, viz., primary culture,
multiplication with rooting, and acclimatization. Pre-
treatments with a mixture of Gentamycin, Kanamycin,
Chlortetracycline, Chloramphenicol, Rifampicin and
the fungicide benomyl make the explants aseptic. Primary culture involves soaking explants in a solution of

growth regulators (IBA and BAP) for 2–3 h. Budding is
initiated in MB medium (Carron et al. 1989) without
growth regulators. Isolated buds are cultured in half-
strength Lepoivre medium with IBA and BAP. These
buds are subcultured to form microshoots that will
in turn be cultured as explants in the multiplication
phase. Soaking the base of the root in an IBA–NAA
mixture for 3–4 days induces roots. Rooted microcut-
tings can be transferred to soil in 4–5 weeks. A number of clones, such as RRII 105, PB 5/51, PB 235, IRCA
438, IRCA 440, IRCA 442, PR 107, and GT 1, have
been multiplied through micropropagation (Carron
et al. 1995a). However, the acclimatization of plants is
crucial, with a balance between relative humidity and
temperature governing the establishment of plants in
the soil.

Though gross experimentations were conducted
for standardizing in vitro technologies, there had
been many setbacks in commercializing these procedures (Carron et al. 1992). A number of aspects
inherent in the explant tissue, viz., release of phe-
nols, contamination of bacteria and fungi, recalci-

trant status, reduced axillary branches, lack of sufficient juvenility, and above all, increased sensitivity of in vitro raised plantlets toward environmental attributes, are responsible for the delay in commercialization. There are, however, remedial measures for these setbacks. Since the contamination of microorganisms is location-specific, newer chemicals are to be tried to raise aseptic cultures. Instead of treating the explants with antioxidants, the incorporation of them in the media decreased browning (Seneviratne and Wijesekara 1996). The use of support systems like cellulose plugs in liquid media reduced synthesis of polyphenols, and embryogenesis activity could be maintained for more than 200 days (Housti et al. 1992). On the other hand, the growth regulators used to induce axillary branches and somatic embryogenesis are more or less the same throughout. Judicious combination of new growth regulators that have shown positive results in other tree species can be tried in rubber. Also, metabolism of ethylene and polyamines during callus development must be controlled by appropriate adjustment of growth regulators (Carron et al. 1992). More prominently, water status of embryogenic callus is a governing factor to enhance embryogenesis (Etienne et al. 1991). Further, Lardet et al. (1999) demonstrated that protein and starch accumulation commenced from the 13th and 15th weeks, respectively, leading to development and maturity of zygotic embryos. However, the smaller size of somatic embryos that can accomplish a relatively small mass of starch and protein reserves can lead to lower vigor and conversion rates where vigor is directly related to acclimatization success. Hence, increasing the size of somatic embryos through nutrient supplies deserves priority. To increase juvenility, air layering and progression to three to four generations can be exercised and explants from such source plants should be used. If commercialized in the strict sense, these technologies can assist breeding programs and enhance productivity significantly.

8.3
Molecular Genetics and Breeding

Owing to long generation time and larger size of the crop, new tools could be developed in order to manage germplasm variability and assist breeders in their recombination strategies. Molecular markers were developed that can be classified into three categories, viz., first generation (restriction fragment length polymorphisms, RFLPs, randomly amplified polymorphic DNAs, RAPDs, and modifications), second generation, mainly based on targeted polymerase chain reaction (PCR) techniques with simple sequence repeats (SSRs) or microsatellites, amplified fragment length polymorphisms (AFLPs) and their modifications, and third-generation markers like expressed sequence tags (ESTs) and single nucleotide polymorphisms (SNPs) (Rudd et al. 2005). Though RFLPs are powerful for studying the genetic diversity and mapping, the technology is not preferred now since it is labor-intensive, requires large DNA samples, and often involves radioisotopes. Its marker index value is also low (expressed as the number of polymorphic products per sample) with only 0.10 compared with PCR-based marker systems like RAPDs (0.23), SSRs (0.60), and AFLPs (6.08) (Low et al. 1996). Ever since isozymes were utilized for clonal identification (Chevallier 1988; Yeang et al. 1998), tools like minisatellites (Besse et al. 1993a), RFLPs (Besse et al. 1993b, 1994), mitochrondial DNA (mtDNA) RFLPs (Luo et al. 1995), RAPDs and DNA amplification fingerprinting (DAF) (Low et al. 1996; Venkatachalam et al. 2001), AFLPs (Lespinasse et al. 2000a), and SSRs (Besse et al. 1993b; Atan et al. 1996; Low et al. 1996; Roy et al. 2004; Saha et al. 2005) were developed and used in detection and increment of molecular markers in *Hevea*. All marker systems, except SNPs, have been applied in *Hevea* so far. The following section deals with various aspects of the application of the aforementioned techniques in dealing with measurement of molecular diversity, formulation of gene linkage maps, detection of quantitative trait loci (QTL) and evaluation of laticifer specific gene expression.

8.3.1
Molecular Diversity

Initial studies on isozymic diversity showed the existence of three genetic groups and many new alleles could be found in the Amazonian populations (Chevallier 1988), which was later confirmed through molecular studies (Seguin et al. 1996b). These studies indicate that the genetic diversity available in Amazonian accessions that are yet to be utilized at the molecular level to enrich the Wickham population is immense. However, transfer of such diversity can only be accomplished through gene manipulations at the molecular level. Further, analysis of isozymes that are proteic genetic markers was developed at CIRAD through formulation of a diagnostic kit with 13 polymorphic isozymic systems for clonal identification along with a clonal identification database. This kit has proved to be able to differentiate a large set of cultivated clones (Leconte et al. 1994). However, the analyses have to be carried out near the field sites owing to fragility of isozymes to varied temperatures, or otherwise the samples need to be freeze-dried and transported to the laboratory. Hence, initiating molecular studies, Low and Bonner (1985) characterized the *Hevea* nuclear genome as containing 48% of most slowly annealing DNA (putative single copy) and 32% middle repetitive sequences, with the remaining amount being highly repetitive or palindromic DNA. The whole nuclear genome size was first estimated as 6×10^8 bp. Estimation with flux cytometry demonstrated 2×10^9 bp for *H. brasiliensis*, *H. benthamiana*, *H. guianensis*, *H. pauciflora*, and *H. spruceana* (Seguin et al. 2003).

Fingerprinting through RFLP minisatellite probes could be more powerful and identification of 73 Wickham clones was done with 13 probes associated with restriction enzyme *Eco*RI (Besse et al. 1993b). RFLPs were also used for identification of progeny with two common parents such as PR 255 and PR 261, RRIM 901 and RRIM 905, RRIM 937 and RRIM 938 (Low et al. 1996). Further, Besse et al. (1994) using 92 Amazonian and 73 Wickham clones did an assessment of RFLP profiles. RFLPs, as molecular genetic markers, were used with homologous probes from a CIRAD *Hevea* bank that showed genetic enrichment brought by Amazonian collections to *Hevea* germplasm, following a genetic structure based on geographical collection sites (Besse et al. 1993a; Seguin et al. 1996b). Exceptionally, a clone from Rondonia (RO/C/8/9), showed eight specific restriction fragments and a unique malate dehydrogenase allele, indicating its interspecific origin.

In a comparative analysis of SSRs of 20 clones of *H. brasiliensis* and six allied species of *Hevea*, Low et al. (1996) measured polymorphism in *H. pauciflora*, *H. guianensis*, *H. camargoana*, *H. benthamiana*, and *H. brasiliensis*. Three microsatellite sequences of the gene for hydroxylmethylglutarylcoenzyme A reductase 1 were polymorphic. Amplification of the $(GA)_9$ region with an appropriate primer converted these regions into sequence-tagged microsatellite sites (STMSs). Polymorphisms in STMSs were with regard to band number and band length. While intraspe-

cific polymorphism (in clones of *H. brasiliensis*) was mainly with the number of bands, both the number and the length of the bands contributed to interspecific polymorphism. The intraspecific polymorphism must be due to allelic differences arising from recombination. The interspecific polymorphism is the result of DNA insertion/deletion and point mutations. On the other hand, DAF profiles were very distinct for vivid species (Low et al. 1996). These polymorphisms must have played a role in delineating species during the course of evolution. Microsatellites that are tandem repeats of short (2–6-bp) DNA sequences are high-utility markers that are codominant, highly polymorphic, abundant, and uniformly dispersed in plant genomes. They allows precise discrimination of even closely related individuals (Mallet 1995). In a bid to select suitable parents for extending rubber breeding programs, Lekawipat et al. (2003) applied microsatellites in detecting diversity in 40 Wickham and 68 Amazonian accessions. This was accomplished with 170 alleles from 12 microsatellite markers spread among all genotypes. On average, 14 alleles were available per locus. Wickham clones were unambiguously less variable than Amazonian accessions. Also, microsatellites of wild accessions are more polymorphic than cultivated Wickham clones and could be divided into three clusters depending on the geographical origin of collections such as Acre, Rondonia, and Mato Grosso (Fig. 7). This conforms to the earlier studies on isozymes and RFLPs. Two clones (RO/OP/4 20/16 and RO/A/7 25/133) were unique as they do not fall under any cluster owing to a high level of specific alleles (Lekawipat et al. 2003). A microsatellite-enriched library was constructed in *H. brasiliensis* involving four types of simple sequence repeats: $(GACA)_n$ (10%), $(GATA)_n$ (9%), $(GA)_n$ (34%) and $(GC)_n$ (9%) (Atan et al. 1996). In cooperation with the French National Center for Sequencing, CIRAD developed different microsatellite-enriched libraries in order to identify a large collection of microsatellite markers. Two possible applications are (1) clonal identification with the advantage of leaf samples sent through normal mail from one site to a distant laboratory and (2) parental identification of seeds collected from an open-pollinated seed garden (Blanc et al. 2001).

The evolution of a cytoplasmic genome in *H. brasiliensis* was slower, owing to lack of genetic recombination through meiosis. The estimated mean molecular size of chloroplast DNA is 152 kb (Fong et al. 1994). mtDNA was also analyzed with heterologous probes from broad bean by CIRAD and

Fig. 7. Geographical origin of *Hevea* clones analyzed with isozymes or restriction fragment length polymorphism markers for genetic diversity assessment (modified from Priyadarshan and Goncalves (2003))

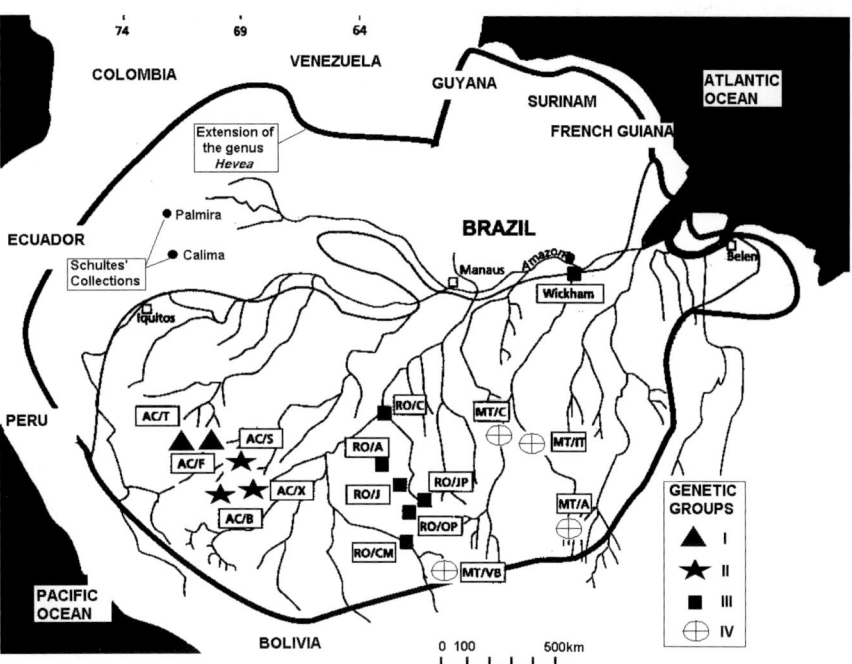

CNRA (Luo et al. 1995, Luo and Boutry 1995). A high mtDNA polymorphism was found in Amazonian accessions. The diversity of mtDNA of the Wickham population is almost nil as only GT 1, a male-sterile clone, exhibited a different type from that of 49 other Wickham clones analyzed. mtDNA appears to be a valuable tool for studies on classification and phylogeny in plants, resulting more from DNA rearrangements rather than nucleotide substitutions (Palmer and Herbon 1988). Sequencing of a highly polymorphic mtDNA fragment from 23 genotypes showed a real potential for phylogenetic analysis in *Hevea* (Luo and Boutry 1995). In chloroplast DNA analysis, much less polymorphism was found, indicating the high level of conservation of this genome.

As a synthesis of these diversity studies, the *Hevea* genetic structure clearly appears as geographically structured (Besse et al. 1994), in relationship with the hydrographic network of the Amazonian forest (Luo et al. 1995; Seguin et al. 1996b). Good relationships are found between the results from different genetic markers. Even if the contribution of isozymes is important by itself, molecular markers provided important clarifications for the distinction of different groups. There would be no barrier to migration of *Hevea* genes within the Amazonian basin. However, the wideness of the area and the limited dispersion of *Hevea* seeds allowed the preservation of the

current structure, which is assumed to have resulted from the fragmentation of the Amazonian forest during the Pleistocene, according to the refuge theory presented by Haffer (1982). The mtDNA of Wickham clones has less variation since their female progenitors are restricted to a very small set of primary clones. Cytoplasmic donors for most of the improved clones are either PB 56 or Tjir 1 (Fig. 8). Obviously, this is the reason for the mtDNA profile showing only two clusters (Priyadarshan and Gonçalves 2003). A possible explanation for greater polymorphism in mtDNA of wild accessions is that many must have evolved through interspecific hybridization. The mtDNA polymorphism in wild accessions needs to be exploited fully. A molecular survey of available Amazonian accessions and isolation of competent molecular variants in their progeny are the possible exercises that can give meaningful results.

8.3.2
Gene Linkage Maps and QTLs

The construction of a molecular gene linkage map in *Hevea* requires a specific methodology because of high heterozygosity. Unlike annual crops, a cross between two heterozygous parents in *Hevea* can yield information up to four alleles, which are segregated further. A comprehensive genetic linkage map of *H. brasiliensis* was formulated recently with the help

Fig. 8. Transmission of cytoplasm of PB 56 and Tjir 1 in the derivation of modern clones

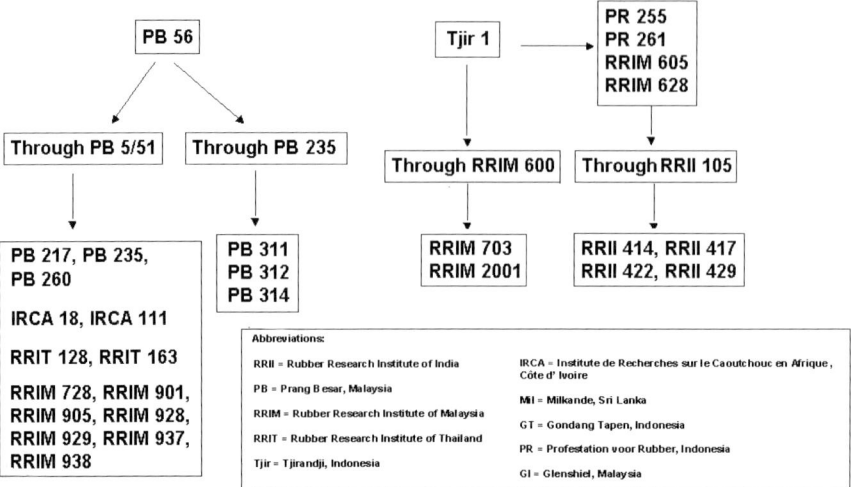

of RFLPs, AFLPs, microsatellites, and isozyme markers (Lespinasse et al. 2000a; Fig. 9). This was accomplished through a double pseudo-testcross as per the methodology of Grattapaglia and Sederoff (1994) and a map was constituted separately for each parent. Further, homologous markers segregating in both parents were ascertained and a consensus map was prepared. The parents used were PB 260 (PB5/51 × PB 49) and RO 38 (F4542 × AVROS 363). F4542 is a clone of *H. benthamiana*. The F_1 synthetic map of 717 markers was distributed in 18 linkage groups. These comprised 301 RFLP, 388 AFLP, 18 microsatellite, and ten isozyme markers. Identification of loci was based on mobility of electrophoresis bands, necessitating verification of the consistency of the location of alleles in both parental maps. The genetic length of 18 chromosomes was fairly homogeneous with an average map length per chromosome of 120 cM. Many AFLP markers were seen in clusters, which were attributed to reduced recombination frequency regions. Though the RFLP markers were well distributed all over the 18 linkage groups, these were insufficient to saturate the map. AFLPs and a few microsatellites together facilitated saturation of the map. However, these exercises are the initial steps for making a total genetic linkage map of *Hevea* in the future. The isozymes were found to inherit a following 1:1 ratio (Chevallier 1988). A partially nonrandom arrangement of duplicate loci observed in RFLP profiles indicates that they have homology descending from a common ancestor (Lespinasse et al. 2000a). The origin of such duplications is still unknown and *H. brasiliensis* continues to behave as a diploid.

As mentioned in the "Introduction," the upsurge of SALB looms over Asia and Africa as a potential threat for rubber plantations in the future. Complete resistance to SALB was believed to be monogenic (Simmonds 1990). However, QTLs for resistance to SALB (*M. ulei*) were mapped using 195 F_1 progeny derived from a cross between PB 260 (susceptible) and RO 38 (resistant) clones (Lespinasse et al. 2000b), which was done in continuation of a genetic analysis done earlier (Seguin et al. 1996 a). Eight QTLs were identified for resistance in the RO 38 map through a Kruskel–Wallis marker-by-marker test and an interval mapping method (Lander and Botstein 1989; van Oojen et al. 1992). The F_1 consensus map confirmed the results obtained in parental maps. It was further rationalized that the resistance (alleles) of RO 38 was inherited from a wild grandparent (*H. benthamiana*) and no favorable alleles came from AVROS 363, the Wickham parent. Specificity to resistance to different strains was persistent. Two distinct forms of resistance were identified, i.e., a complete resistance in the absence of sporulation lesions as in *H. benthamiana* and *H. pauciflora* and a partial resistance with a reduced rate of epidemic development (Le Guen et al. 2003). Investigations into the resistance mechanism were also conducted (Garcia et al. 1995). Le Guen et al. (2003) detected one major QTL (*M13-lbn*) located in g 13 in the RO 38 map responsible for 36–89% of phenotypic resistance. The effect of the QTL was large under natural conditions of French Guiana compared with that of controlled inoculation. This study should lead to MAS for identifying resistant genotypes with priority given to geographical extent of efficiency and predictable durability. More

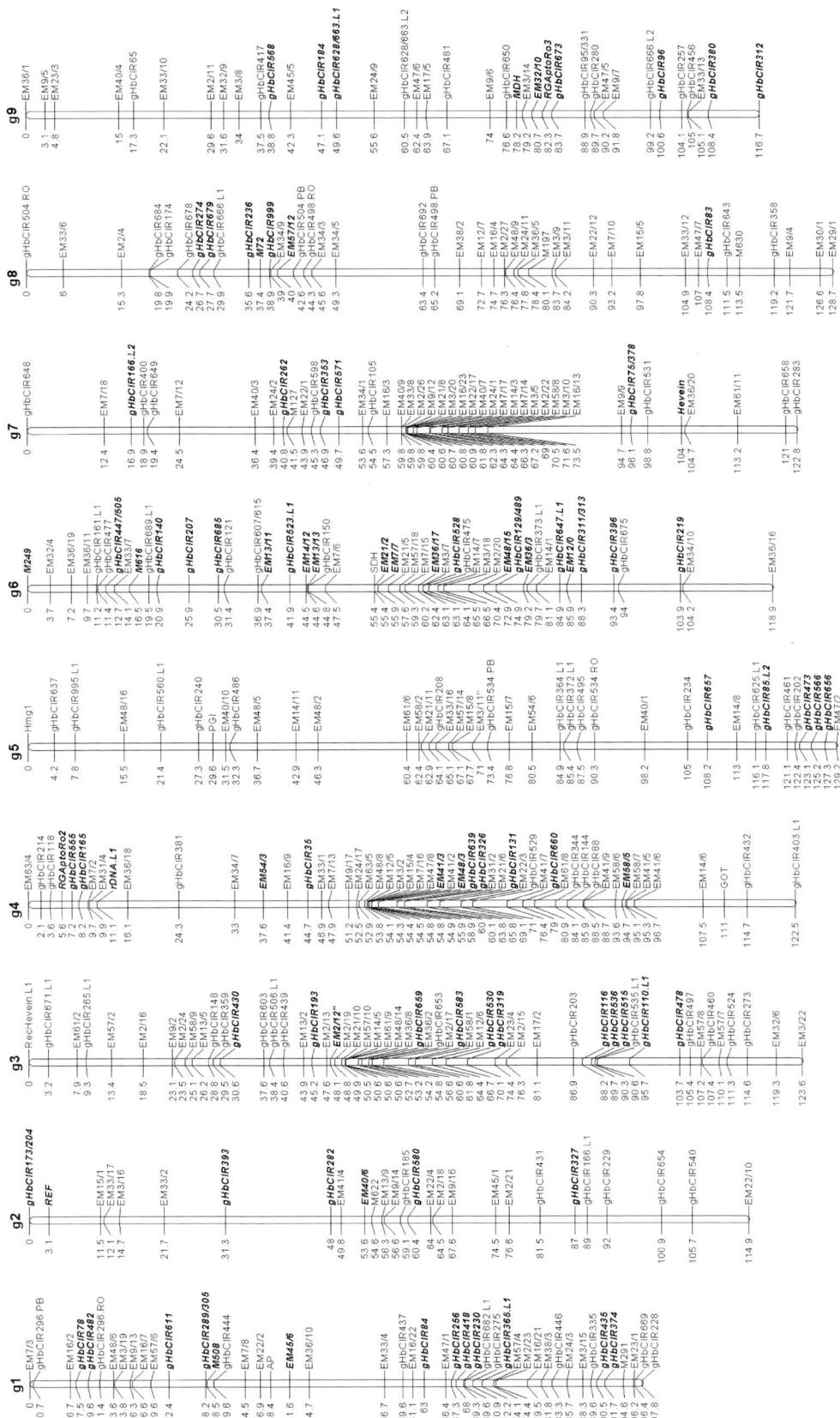

Fig. 9. F_1 synthetic map of 717 markers distributed in 18 linkage groups of *Hevea brasiliensis* (after Lespinasse et al. 2000a)

Fig. 9. (continued)

durable resistance will be available in other allied species and wild accessions of *Hevea* (Priyadarshan and Gonçalves 2003). However, the selection of clones having durable resistance with polygenic determinism is also important while undertaking such studies (Simmonds 1991). Darmono and Chee (1985) while studying lesion size on leaf discs identified SIAL 263, an illegitimate progeny of RRIM 501, as resistant to SALB.

8.3.3
Laticifer-Specific Gene Expression

Studies on rubber biosynthesis have gained momentum owing to inquisitiveness to synthesize artificial rubber. The gene responsible for the key enzyme for polymerization of polyisoprenes – the rubber transferase – is one of the most abundantly expressed genes in the latex. Genes expressed in the latex can be broadly categorized into three types, on the basis of their function: (1) defense genes, (2) genes for rubber synthesis, and (3) genes for allergenic proteins (Han et al. 2000). Hevein, a chitin-binding protein, is one of the defense proteins that plays a crucial role in the protection of wound sites from fungal attack. A complementary DNA (cDNA) clone (HEV 1) encoding *hevein* was isolated using PCR (Broekaert et al. 1990). HEV 1 is of size 1,018 bp and includes an open reading frame of 204 amino acids with a signal sequence of 17 amino acid residues followed by a 187 amino acid polypeptide. This polypeptide is found to contain striking features, like an amino terminal region (43 amino acids) with homology to other chitin-binding proteins and amino acid termini of wound-inducible proteins in potato and poplar. It was also seen that their genes were well expressed in leaves, stems, and latex (Broekaert et al. 1990). Nearly 12.6% of the proteins available in the latex are defense-related (Han et al. 2000).

Mainly three rubber synthesis related genes are expressed in the latex, viz., rubber elongation factor (REF; Dennis and Light 1989; Goyvaerts et al. 1991), hydroxylmethylglutarylcoenzyme A reductase (Chye et al. 1992), and small rubber particle protein (SRPP; Oh et al. 1999). They constitute the 200-odd distinct polypeptides (Posch et al. 1997). The most abundantly expressed gene is REF (6.1%) and then SRPP (3.7%) (Han et al. 2000). These expressed sequences (ESTs) were compared with public databases of identified genes. About 16% of the database matched ESTs encoding rubber biosynthesis related proteins. Analysis of ESTs revealed that rubber biosynthesis related genes are expressed most, followed by defense-related genes and protein-related genes (Han et al. 2000). Unlike photosynthetic genes, transcripts involved in rubber biosynthesis are 20–100 times greater in laticifers than in leaves (Kush et al. 1990). On the other hand, transcripts for chloroplastic and cytoplasmic forms of glutamine synthetase are restricted to leaves and laticifers, respectively (Kush et al. 1990), indicating thereby that the cytoplasmic form of glutamine synthetase plays a decisive role in amino acid metabolism of laticifers. Studies on laticifer-specific gene expression have important implications for selection and breeding. It would be worthwhile to use transcript levels as molecular markers for early selection (Kush et al. 1990). The transcript levels of hydrolytic enzymes, viz., polygalacturonase and cellulase, should be taken as indicators for better laticifer development. It is felt that extensive studies on expression of genes are mandatory to unravel the intricacy of latex production. Detection and evaluation of more molecular markers must also help to breed *Hevea* at the molecular level.

8.3.4
Direct Gene Transfer

While the in vitro plant regeneration system in rubber is being standardized in a few laboratories worldwide, efforts have been made to transform *Hevea* cells through *Agrobacterium tumefaciens* in order to complement plant-breeding efforts to increase genetic variation (Arokiaraj et al. 1990, 1994). The anther-derived calluses were transformed with *A. tumefaciens* harboring *gus* and *npt*II genes encoding β-glucuronidase and neomycin phosphotransferase, respectively. Fluorometric assay and enzyme-linked immunosorbent assay were performed to prove the expression of *gus* and *npt*II genes, respectively, in calluses and embryoids (Arokiaraj et al. 1996). The expression of foreign proteins in *Hevea* latex was demonstrated in 1998 (Arokiaraj et al. 1998). This transformation appeared stable even after three vegetative generations with no chimeras, indicating a single transformed plant is sufficient to achieve a population through budding. But this exercise would not be adequate enough to take care of the stock–scion interaction and the ensuing yield variation in a clonal population. Transformation of *Hevea* cells with genes for apomixis might be an alternative to circumvent stock–scion interaction. Lately, genes for human serum proteins have been expressed in rubber latex through genetic transformation (Arokiaraj et al. 2002; Yeang et al. 2002). However, the aforementioned studies have to go a long way to significantly assist breeding new clones with traits like resistance and capability to produce secondary proteins.

Another important achievement is with the expression of cDNA for farnesyl diphosphatesynthase from a rare rubber-producing mushroom, *Lactarius chrysorrheus*, in *Escherichia coli* (Mekkriengkrai et al. 2004). Such research has long way to go before rubber will be produced commercially in vitro or in vivo.

8.3.5
Conclusions and Future Outlook

It is quite evident that rubber breeding is time-consuming and labor-intensive. It is here that the biochemical and biotechnological tools come handy in assisting the rubber breeder in deriving and evaluating new recombinants/clones in as short a time span as possible. The foremost and essential way to shorten the period needed to derive a clone is to standardize and implement a routine biochemical/molecular MAS system to detect high-yielding accessions at the juvenile stage. Though many efforts have been made in this direction, a dependable method is yet to emerge from this research. Even if a real "MAS" applied to rubber is still to be developed and validated, so contributing to early selection, it is very probable that molecular markers, especially microsatellites, will be substantially used at different levels and will improve the efficiency of rubber breeding. However, this does not mean that the quality of field testing, associated with the methodology of quantitative genetics and modern statistics, must be overlooked.

Rubber is traditionally propagated through bud grafting. Variations among a bud-grafted population are significant and can influence the productivity levels. As in vitro techniques are yet to make a commercial impact in rubber, a propagation system that can circumvent the influence of stock–scion interactions needs to be achieved. One way is to derive somatic seeds that can produce true-to-mother plants. Introduction of genes for apomixis is the only way to have a homogeneous population.

Cryogenic preservation of endangered seedling trees is yet another important aspect to be looked into urgently.

Since the introduction of Wickham seedlings to Asia in 1876, rubber breeding in Southeast Asia has been based on Wickham material with focus on yield improvement, while research in Latin America has been devoted to create *Microcyclus*-tolerant and productive clones. This has led to two constant parallel ways of achieving clones suited to the needs of the respective regions. Since *Microcyclus* is a threat to Southeast Asian countries, international multilocation experiments must be given priority.

Because rubber is a perennial crop, rubber breeding has been influenced by the grafting technique, which permitted the development and multiplication of clones evolved from recombinants. In spite of the implementation of early selection techniques, rubber breeding is impeded by the length of the selection process, and the limited creation of full-sib families achieved through low success rates of hand pollination.

In order to broaden the genetic base, various attempts were made to introduce new germplasm to Asia, including species allied to *H. brasiliensis*, among which the international IRRDB collection was the most significant. But owing to the low yield of this germplasm and to the length of the breeding process, benefits will be distributed only over a long time period. Apart from the creation of new clones for development, this research requires specific germplasm prebreeding programs to produce new parents for recombination breeding prior to selection. The spectrum of useful genetic variation needs to be enlarged, especially through utilizing variable cytoplasmic donors, since most oriental clones received cytoplasm either from PB 56 (through PB 5/51) or from Tjir 1. Biotechnological tools should assist in finding useful variation among the wild germplasm and allied species, especially cytoplasmic genetic diversity and QTLs for resistance to diseases.

There must be an effort to split the Wickham population into different groups at the molecular level. This will pave the way to delimit the development of many related clones and of inbreeding depression. Also, with regard to rubber cropping in relation to overall economy(new locations, new objectives, new economic constraints), rubber breeding needs to address the derivation of a larger scope of clones adapted to various biotic or abiotic stresses, and to various specifications, including rubber quality. Consequently, it would be required not only to select elite clones but also to describe the behavior of a larger range of clones at the small-scale experimental level and in different environments (development of clonal databases). There are newer clones derived by fast developing economies like China, Vietnam, and Thailand. All these and the existing popular clones should be enlisted in the clonal database, probably under the umbrella of IRRDB. Such databases should provide details of not only clones but also judicious tapping schedules and discriminatory fertilizer doses for new locations. Such a description will also help to suggest better arguments for the diversification of recommended clones. More emphasis on ecophysiology research could provide the necessary results for achieving some or all of those goals.

Research devoted to SALB resistance involving recurrent backcrossing of Amazonian resistant clones (mainly the *H. benthamiana* F 4542 or derived clones) with Wickham high-yielding clones to evolve different resistance sources (clone or polyclonal seedling population) needs to be augmented. This strategy could also be applied to specific programs aimed at selecting clones resistant to *Corynespora* or *Oidium*, or other diseases within integrated approaches.

As rubber wood has become an ancillary source of income, rubber breeding now has to integrate new traits, especially traits based on architecture and on biomass production, in order to produce better-optimized "latex-timber" clones. Another ancillary income is rubber honey. Intensive multilocation trials will provide data on the optimum bee population and on the impact of seeding efficiency.

Although in vitro culture still meets different obstacles, somatic embryogenesis is the gateway for the implementation of targeted genetic transfers, so accelerating genetic progress on agricultural traits or widening the scope such as the possible production of proteins in the latex.

International cooperation and the interface between research institutions and the product-transforming private sector must be promoted in order to amplify the efficiency and efforts of rubber breeders. Even if the needs of smallholders are addressed with priority, industrial estates have provided a significant contribution in land facilities and logistics for large-scale and long-term testing of clones, and would continue to do this. With many projects in this direction, IRRDB can play a key-role in this field of cooperation in rubber breeding.

Acknowledgement. The authors are thankful to several researchers worldwide who provided valuable information on *Hevea* breeding in addition to their published papers. Special thanks are due to M.P. Carron, CIRAD, France, for permitting the use of photographs of in vitro culture. Prof. A. R. Leitch, School of Biological Sciences, Queen Mary, London has been kind enough to give useful comments.

References

Ahmad B (2001) Physiological and morphological characteristics of *Hevea* rootstock in response to water stress. J Rubb Res 4:177–198

Alice J, Joseph A, Meenakumari T, Saraswathyamma CK, Varghese YA (2000) Clonal variation in the intensity of powdery mildew (*Oidium heveae* Steinm.) disease of *Hevea*. Ind J Nat Rubb Res 13:64–68

Alice J, Nair RB, Rajalakshmy VR, Saraswathyamma CK, Varghese YA (2001) Sensitivity relationship of *Hevea* clones to the biotic stress of powdery mildew (*Oidium heveae* Steinm.). Ind J Nat Rubb Res 14:88–92

Arokiaraj P, Jones H, Jaafar H, Coomber S, Charlwood BV (1990) *Agrobacterium* mediated transformation of *Hevea* anther calli and their regeneration into plantlets. J Nat Rubb Res 11:77–87

Arokiaraj P, Jones H, Cheong KF, Coomber S, Charlwood BV (1994) Gene insertion into *Hevea brasiliensis*. Plant Cell Rep 13:425–431

Arokiaraj P, Jones H, Jaafar H, Coomber S, Charlwood BV (1996) *Agrobacterium* mediated transformation of *Hevea* anther calli and their regeneration into plants. J Nat Rubb Res 11:77–87

Arokiaraj P, Yeang H Y, Cheong KF, Hamzah S, Jones H, Coomber S, Charlwood BV (1998) CaMV 35S promoter directs ß-glucuronidase expression in the laticiferous system of transgenic *Hevea brasiliensis* (rubber tree). Plant Cell Rep 17:621–625

Arokiaraj P, Rueker F, Obermayr E, Shamsul Bahri AR, Jaafar H, Carter DC, Yeang HY (2002) Expression of human serum albumin in transgenic *Hevea brasiliensis*. J Nat Rubb Res 5:157–166

Arshad NL, Othman R, Yacob ARW (1995) *Hevea* wood availability in Peninsular Malaysia. RRIM Planters Bull 224–225:73-83

Atan S, Low FC, Saleh NM (1996) Construction of a microsatellite enriched library from *Hevea brasiliensis*. J Nat Rubb Res 11:247–255

Baldwin JJT (1947) *Hevea*: a first interpretation. A cytogenetic survey of a controversial genus, with a discussion of its implications to taxonomy and to rubber production. J Hered 38:54–64

Barry RG, Chorley RJ (1976) Atmosphere, Weather and Climate. Methuen, London, UK

Baudouin L, Baril C, Clément-Demange A, Leroy T, Paulin D (1997) Recurrent selection of tropical tree crops. Euphytica 96:101–114

Baulkwill WJ (1989) The history of natural rubber production. In: Webster CC, Baulkwill WJ (eds) Rubber. Longman Scientific and Technical, Essex, UK, pp 1–56

Besse P, Seguin M, Lebrun P, Lanaud C (1993a) Ribosomal DNA variations in wild and cultivated rubber tree (*Hevea brasiliensis*). Genome 36:1049–1057

Besse P, Lebrun P, Seguin M, Lanaud C (1993b) DNA fingerprints in *Hevea brasiliensis* (rubber tree) using human minisatellite probes. Heredity 70:237–244

Besse P, Seguin M, Lebrun P, Chevallier MH, Nicolas D, Lanaud C (1994) Genetic diversity among wild and cultivated populations of *Hevea brasiliensis* assessed by nuclear RFLP analysis. Theor Appl Genet 88:199–207

Blanc G, Rodier-Goud M, Lidah Y J, Clément-Demange A, Seguin M (2001) Study of open pollination in *Hevea* using microsatellites. Plantations, Recherché Développement 8:68–71

Brazil (1971) Ministério da Indústria e Comércio. Superintendência da Borracha. O gênero *Hevea*, descrição das espécies e distribuição geográfica. Rio de Janeiro, Sudhevea, 1971 (Plano Nacional da Borracha, anexo 7)

Breton F, Sanier C, d'Auzac J (2000) Role of cassiicolin, a host-selective toxin, in pathogenicity of *Corynespora cassiicola*, causal agent of a leaf fall disease of *Hevea*. J Nat Rubb Res 3:115–128

Broekaert N. Lee H, Kush A, Chua N H, Raikhel N (1990) Wound induced accumulation of mRNA containing a *Hevein* sequence in laticifer of rubber tree (*Hevea brasiliensis*). Proc Natl Acad Sci USA 87:7633–7637

Brown AHD (1989) Core collections: a practical approach to genetic resources management. Genome 31:818–824

Cailloux F, Julien-Guerrier J, Linosseer L, Coudret A (1996) Long term somatic embryogenesis and maturation of somatic embryos in *Hevea brasiliensis*. Plant Sci 120:185–196

Carron MP, and Enljalric F (1982) Studies on vegetative micropropagation of *Hevea brasiliensis* by somatic embryogenesis *and in vitro* micro-cutting. In: Fujiwara A (ed) Plant Tissue Culture. Maruzan, Tokyo, Japan, pp 751–752

Carron MP, Enjalric F, Lardet L, Deschamps A (1989) Rubber (*Hevea brasiliensis* Muell. Arg.) In: Bajaj YPS (ed) Biotechnology in Agriculture and Forestry Vol V. Springer, Berlin Heidelberg New York, pp 222–245

Carron MP, d'Auzac J, Etienne H, El Hadrami I, Housti F, Michaux-Ferriere N, Montoro P (1992) Biochemical and histological features of somatic embryogenesis in *Hevea brasiliensis*. Ind J Nat Rubb Res 5 7–17

Carron MP, Etienne H, Lardet L, Campagna S, Perrin Y, Leconte A, Chaine C (1995a) Somatic embryogenesis in rubber (*Hevea brasiliensis* Muell. Arg.). In: Jain S, Gupta P, Newton R (eds) Somatic Embryogenesis in Woody Plants: Angiosperms. Kluwer Academic Publ, The Netherlands, pp 117–136

Carron MP, Etienne H, Michaux-Ferriere N, and Montoro P (1995b) Somatic embryogenesis in rubber tree *Hevea brasiliensis* Muell. Arg. In: Bajaj YPS (ed) Biotechnology in Agriculture and Forestry: Somatic Embryogenesis and Synthetic Seeds, Vol 30. Springer, Berlin Heidelberg New York, pp 353–369

Chandrashekar TR, Mydin KK, Alice J, Varghese YA, Saraswathyamma CK (1997) Intraclonal variability for yield in rubber (*Hevea brasiliensis*). Ind J Nat Rubb Res 10:43–47

Chandrashekar TR, Gawai PP, Saraswathyamma CK (2002) Yield performance of trees grown from polycross seeds of rubber (*Hevea brasiliensis*) in a dry sub humid climate in India. Ind J Nat Rubb Res 15:19–27

Chee KH (1976) Assessing susceptibility of *Hevea* clones to *Microcyclus ulei*. Ann Appl Biol 84:135–145

Chen Z (1984) Rubber (*Hevea*). In: Sharp WR, Evans DA, Ammirato PV, Yamada Y (eds) Handbook of Plant Cell Culture – Crop Species, vol 2. Macmillan, UK, pp 546–571

Chen Z, Chen F, Chien C, Wang C, Chang S, Hsu H, Ou H, Ho Y, Lu T (1979) A process of obtaining pollen plants of *Hevea brasiliensis* Muell. Arg Sci Sin 22:81–90

Chen Z, Quian C, Qin M, Xu X, Xiro Y (1982) Recent advances in anther culture of *Hevea brasiliensis* (Muell. Arg.). Theor Appl Genet 2:103–108

Chevallier MH (1988) Genetic variability of *Hevea brasiliensis* germplasm using isozyme markers. J Nat Rubb Res 3:42–53

Chua SE (1966) Studies on tissue culture of *Hevea brasiliensis*. 1. Role of osmotic concentration, carbohydrates and pH values in induction of callus growth in plumule tissues from *Hevea* seedling. J Rubb Res Inst Malaya 19:272–276

Chye ML, Tan CT, Chua NH (1992) Three genes encode 3-hydroxy-3-methyl glutaryl-coenzyme A reductase in *Hevea brasiliensis*. hmg1 and hmg3 are differentially expressed. Plant Mol Biol 19:473–484

Clément-Demange A, Chapuset T, Legnaté H, Costes E, Doumbia A, Obouayeba S, Nicolas D (1995) Wind damage: the possibilities of an integrated research for improving the prevention of risks and the resistance of clones in Rubber tree. Proc IRRDB Symp on Physiological and Molecular Aspects of the Breeding of *Hevea brasiliensis*, 6–7 Nov 1995, Penang, Malaysia, pp 182–199

Clément-Demange A, Legnaté H, Chapuset T, Pinard F, Seguin M (1998) Characterization and use of the IRRDB germplasm in Ivory Coast and French Guyana: status in 1997. In: Cronin ME (ed) Proc IRRDB Symp Natural Rubber (*Hevea brasiliensis*) – General, Soils and Fertilization and Breeding and selection. IRRDB, UK, pp 71–88

Clément-Demange A, Legnate H, Seguin M, Carron MP, Le Guen V, Chapuset T, Nicolas D (2000) Rubber Tree. In: Charrier A, Jacquot M, Hamon S, Nicolas D (eds) Tropical Plant Breeding. Collection Repères, CIRAD-ORSTOM, Montpellier, France, pp 455–480

Cornish K, Siler DJ, Grosjean O, Goodman N (1993) Fundamental similarities in rubber particle architecture and function in three evolutionarily divergent plant species. J Nat Rubb Res 8:275–285

Costa RB, Resende MDV, Aranjo AJ, Gonçalves P de S, Martins ALM (2000) Genotype-environment interaction and the number of test sites for the genetic improvement of rubber trees (*Hevea*) in São Paulo State, Brazil. Genet Mol Biol 23:79–187

Darmono TW, Chee KH (1985) Reaction of Hevea clones to races of *Microcyclus ulei* in Brazil. J Rubb Res Inst Malaysia 33:1–8

Davies W (1997) The rubber industry's biological nightmare. Fortune 136:86

Dea GB, Keli R J, Eschbach J M, Omont H, and Tran-van-Canh (1997) Rubber tree (*Hevea brasiliensis*). In: Cronin ME (ed) Proc IRRDB Symp Agronomy Aspects of the Cultivation of Natural Rubber (*Hevea brasiliensis*). IRRDB, Hertford, UK, pp 44–53

Dean W (1987) Brazil and the struggle for rubber. Cambridge Univ Press, New York, USA

Dennis MS, Light DR (1989) Rubber elongation factor from *Hevea brasiliensis* Identification, characterization and role in rubber biosynthesis. J Biol Chem 264: 18608–18617

Devakumar AS, Sathik M, Jacob J, Annamalainathan K, Prakash GP, Vijayakumar KR (1998) Effects of atmospheric soil drought on growth and development of *Hevea brasiliensis*. J Rubb Res 1:190–198

Dijkman MJ (1951) *Hevea*- Thirty Years of Research in the Far East. Univ Miami Press, Coral Gables, Florida, USA

El Hadrami I, Carron MP, d' Auzac J (1991) Influence of exogenous hormones on somatic embryogenesis in *Hevea brasiliensis*. Ann Bot 67:511–515

Etienne H, Berger A, Carron MP (1991) Water status of callus from *Hevea brasiliensis* during induction of somatic embryogenesis. Physiol Plant 82:213–218

Etienne H, Sotta B, Montoro P, Miginiac E, Carron MP (1993) Comparison of endogenous ABA and IAA contents in somatic and zygotic embryos of *Hevea brasiliensis* Müll. Arg. during ontogenesis. Plant Sci 92:111–119

Etienne H, Lartaud M, Carron M P, Michaux-Ferriere N (1997a) Use of calcium to optimize long-term proliferation of friable embryogenic calluses and plant regeneration of *Hevea brasiliensis* (Muell. Arg.). J Exp Bot 48:129–137

Etienne H, Lartaud M, Michaux-Ferriere N, Carron MP, Berthouly M, Teisson C (1997b) Improvement of somatic embryogenesis in *Hevea brasiliensis* (Muell. Arg.) using the temporary immersion technique. In vitro Cell Dev Biol Plant 33:81–87

Fernando DM, De Silva MSC (1971) A new basis for the selection of *Hevea* seedlings. Quart J Rubb Res Inst Ceylon 48:19–30

Fong CK, Lek KC, Ping CN (1994) Isolation and restriction analysis of chloroplast DNA from *Hevea*. J Nat Rubb Res 9:278–288

Garcia D, Cazaux E, Rivano F, d'Auzac J (1995) Chemical and structural barriers in Microcyclus ulei, the agent of South American Leaf Blight in *Hevea* pp. Eur J For Pathol 25:282–292

Gilbert NE, Dodds KS, Subramaniam S (1973) Progress of breeding investigations with *Hevea brasiliensis*. V. Analysis of data from earlier crosses. J Rubb Res Inst Malaysia 23:365–380

Gnagne M, Clément-Demange A, Legnaté H, Chapuset T, Nicolas D (1998) Results of the rubber breeding programme in Ivory Coast. In: Cronin MJ (ed) Proc IRRDB Symp Natural Rubber, vol 1. IRRDB, UK, pp 101–113

Gonçalves P de S, Cardoso M, Ortolani AA (1990) Origin, variability and domestication of *Hevea*. Pesquisa Agropecuaria Brasileira 25:135–156

Gonçalves P de S, Segnini Jr I, Ortolani AA, Brioschi AP, Landell MG, Souza SR (1998) Components of variance and genotype × environment interaction for annual girth increment in rubber tree. Pesquisa Agropecuaria Brasileira 33:1329–1337

Gonçalves P de S, Bortoletto N, Martin LM, Costa RB, and Gallo PB (2003) Genotype-environment interaction and phenotypic stability for girth growth and rubber yield of *Hevea* clones in São Paulo state, Brazil. Genet Mol Biol 26:441–448

Goyvaeerts E, Dennis M, Light D, Chua NH (1991) Cloning and sequencing of cDNA encoding the Rubber Elongation Factor of *Hevea brasiliensis*. Plant Physiol 97:317–321

Grattapaglia D, Sederoff R (1994) Genetic linkage maps of *Eucalyptus grandis* and *Eucalyptus urophylla* using a pseudo-test cross mapping strategy and RAPD markers. Genetics 137:1121–1137

Gunnery H (1935) Yield prediction in *Hevea*. A study of sieve-tube structure and its relation to latex yield. J Rubb Res Inst Malaya 6:8–20

Haffer J (1982) General aspects of the refuge theory. In: Prance GT (ed) Biological Diversification in the Tropics. Columbia Univ Press, New York, USA, pp 6–26

Hamon S, Dussert S, Deu M, Hamon P, Seguin M, Glaszmann JC, Grivet L, Chantereau J, Chevallier MH, Flori A, Lashermes P, Legnate H, Noirot M (1998) Effects of quantitative and qualitative principal component score strategies on the structure of coffee, rubber tree, rice and sorghum core collections. Genet Sel Evol 30:S237–S258

Han K, Shin DO, Yang J, Kim IJ, Oh SK, and Chow KS (2000) Genes expressed in the latex of *Hevea brasiliensis*. Tree Physiol 20:503–510

Hénon JM, Nicolas D (1989) Relation between anatomical organization and the latex yield: search for early selection criteria. In: d'Auzac, JD, Jacob JL, Chrestin H (eds) Physiology of Rubber Tree Latex, CRC Press, Boca Raton, Florida, USA, pp 31–49

Ho CY (1976) Clonal characters determining the yield of *Hevea brasiliensis*. Proc Intl Rubb Conf, 1975, Kuala Lumpur, Malaysia 2:27–38

Ho CY, Khoo SK, Meiganaratnam K, Yoon PK (1979) Potential new clones from mother tree selection. Proc Rubb Res Inst Malaysia Planters' Conf, Kuala-Lumpur 1979, Kuala Lumpur, Malaysia, p 201

Housti F, Coupe M, d'Auzac J (1992) Browning mechanisms and factors of influence in *in vitro Hevea* calli cultures. Ind J Nat Rubb Res 5:86–99

Huang X, Wei L, Zhan S, Chen C, Zhou Z, Yuen X, Guo Q, Lin J (1981) A preliminary study of relations between latex vessel system of rubber leaf blade and yield prediction at nursery. Chinese J Trop Crops 2:16–20

Imle EP (1978) *Hevea* rubber: Past and future. Econ Bot 32:264–277

Junqueira NTV, Lieberei R, Kalil FAN, Lima LIPM (1990) Components of partial resistance in *Hevea* clones to rubber tree leaf blight, caused by *Microcyclus ulei*. Fitopatol Bras 15:211–214

Kush AE, Goyvaerts ML, Chye Chua NH (1990) Laticifer specific gene expression in *Hevea brasiliensis* (rubber tree). Proc Natl Acad Sci USA 87:1787–1790

Lander BS, Botstein D (1989) Mapping mendelian factors underlying quantitative traits using RFLP linkage maps. Genetics 121:185–199

Lardet R, Piombo G, Oriol F, Dechamp E, Carron MP (1999) Relations between biochemical characters and conversion ability in *Hevea brasiliensis* zygotic and somatic embryos. Can J Bot 77:1168–1177

Leconte A, Lebrun P, Nicolas D, Seguin M (1994) Electrophoresis: application to *Hevea* clone identification. Plantations, Recherché Developpement 1:28–36

Le Guen V, Lespinasse D, Oliver G, Rodier-Goud M, Pinard F, Seguin M (2003) Molecular mapping of genes conferring field resistance to south America leaf blight (*Microcyckus ulei*) in rubber tree. Theor Appl Genet 108:160–167

Leitch AR, Lim KY, Leitch IJ, O'Neill M, Chye ML, Low FC (1998) Molecular cytogenetic studies in rubber. *Hevea brasiliensis*. Müll. Arg. (Euphorbiaceae). Genome 41:64–467

Lekawipat N, Teerawatannasuk K, Rodier-Goud M, Seguin M, Vanavichit A, Toojinda T, Tragoonrung S (2003) Genetic diversity analysis of wild germplasm and cultivated clones of *Hevea brasiliensis* Muell. Arg. by using microsatellite markers. J Rubb Res 6:36–47

Lespinasse D, Rodier-Goud M, Grivet L, Leconte A, Legnaté H, Seguin M (2000a) A saturated genetic linkage map of rubber tree (*Hevea* spp.) based on RFLP, AFLP, microsatellite and isozyme markers. Theor Appl Genet 100:127–138

Lespinasse D, Grivet L, Troispoux V, Rodier-Goud M, Pinard F, Seguin M (2000b) Identification of QTLs involved in the resistance to South American Leaf Blight (*Microcyclus ulei*) in the rubber tree. Theor Appl Genet 100:975–984

Linossier L, Veisseire P, Cailloux F, Coudret A (1997) Effect of abscisic acid and high concentration of PEG on *Hevea brasiliensis* somatic embryos development. Plant Sci 124:183–191

Low FC, Bonner J (1985) Characterization of the nuclear genome of *Hevea brasiliensis*. Intl Rubb Conf, Kuala Lumpur, Malaysia, pp 1–9

Low FC, Atan S, Jaafar H, Tan H (1996) Recent advances in the development of molecular markers for *Hevea* studies. J Nat Rubb Res 11:32–44

Luo H, Van Coppenolle B, Seguin M, Boutry M (1995) Mitochondrial DNA polymorphism and phylogenetic relationships in *Hevea brasiliensis*. Mol Breed 1:51–63

Luo H, Boutry M (1995) Phylogenetic relationships within *Hevea brasiliensis* as deduced from a polymorphic mitochondrial DNA region. Theor Appl Genet 91:876–884

Lynch M, Walsh B (1998) Genetics and Analysis of Quantitative Traits. Sinauer Associates, Massachusetts, USA

Majumder SK (1964) Chromosome studies of some species of *Hevea*. J Rubb Res Inst Malysia 18:269–273

Mallet J (1995) The genetics of biological diversity: From varieties to species. In: Gaston KJ (ed) Biodiversity: A Biology of Numbers and Difference. Blackwell Science Publ, UK, pp 13–47

Mekkriengkrai D, Sando T, Hirooka K, Sakdapipanich J, Tanaka Y, Fukusaki E, Kobayashi A (2004) Cloning and characterization of farnesyl diphosphate synthase from the rubber-producing mushroom *Lactarius chrysorrheus*. Biosci Biotechnol Biochem 68:2360–2368

Montoro P, Etienne H, Carron MP (1993) Callus friability and somatic embryogenesis in *Hevea brasiliensis*. Plant Cell Tiss Org Cult 33:331–338

Mooibroek H, Cornish K (2000) Alternative sources of natural rubber. Appl Microbiol Biotechnol 53:355–365

MRB Annual Report (1999) Malaysian Rubber Board Annual Report 1999, p 27

Nga BH, Subramaniam S (1974) Variation in *Hevea brasiliensis*. 1. Yield and girth data of the 1937 hand pollinated seedlings. J Rubb Res Inst Malysia 24:69–74

Oh SK, Kang HS, Shin DS, Yang J, Chow KS, Yeang HY, Wagner B, Breiteneder H, Han KH (1999) Isolation, characterization and functional analysis of a novel cDNA clone encoding a small rubber particle protein (SRPP) from *Hevea brasiliensis*. J Biol Chem 274:17132–17138

Onesirosan P, Mabuni CT, Durbin, RD, Morin RB, Righ DH, Arny DC (1975) Toxin production by *Corynespora cassiicola*. Physiol Plant Pathol 5:289–295

Ong SH (1975) Chromosome morphology at the pachytene stage in *Hevea brasiliensis*: a preliminary report. Proc Intl Rubb Conf, vol 2. IRRDB, Kuala Lumpur, pp 3–12

Ong SH, Othman R, Benong M (1998) Breeding and selection of clonal genotypes for climatic stress condition. In: Cronin ME (ed) Proc IRRDB Symp Natural Rubber (*Hevea brasiliensis*) – General, Soils and Fertilization and Breeding and selection. IRRDB, UK, pp 149–154

Oojen van JW, Sandbrink H, Purimahua C, Vrielink R, Verkerk R, Zabel P, Lindhout D (1992) Mapping quantitative genes involved in a trait assessed on an ordinal scale: a case study with bacterial canker in *Lycopersicon peruvianum*. In: Yoder JI (ed) Molecular Biology of Tomato. Technomic, Lancaster, USA, pp 59–74

Palmer JD, Herbon LA (1988) Plant mitochondrial DNA evolves rapidly in structure, but slowly in sequence. J Mol Evol 28:87–97

Paranjothy K, Gandhimathi H (1976) Tissue and organ culture of *Hevea*. Proc Intl Rubb Conf 1975. Vol 2. Kuala Lumpur, Malaysia, pp 59–84

Pires JM (1973) Revisão do gênero *Hevea*: descrição da espécies e distribuição geográfica. Relatório Anual, 1972. Belém, Instituto de Pesquisa Agropecuária do Norte, 1973, (Projeto de Botânica – Subprojeto revisão do gênero *Hevea*. Sudhevea/Dnpea (Ipean), pp 6–66

Posch A, Chen Z, Wheeler C, Dunn MJ, Raulf-Heinsoth, Baur X (1997) Characterization and identification of latex allergens by two-dimensional electrophoresis and protein micro-sequencing. J Allerg Chem 99:385–395

Priyadarshan PM (2003a) Breeding *Hevea brasiliensis* for environmental constraints. Adv Agron 79:351–400

Priyadarshan PM (2003b) Contributions of weather variables for specific adaptation of rubber tree (*Hevea brasiliensis* Muell. – Arg.) clones. Genet Mol Biol 26:435–440

Priyadarshan PM, Clément-Demange A (2004) Breeding *Hevea* rubber: formal and molecular genetics. Adv Genet 52:51–115

Priyadarshan PM, Gonçalves P de S (2003) *Hevea* gene pool for breeding. Genet Resour Crop Evol 50:101–114

Priyadarshan PM, Vinod KK, Rajeswari M, Pothen J, Sowmyalatha MKS, Sasikumar S, Sethuraj MR (1998a) Breeding *Hevea brasiliensis* Muell. Arg. in Tripura (NE India): Performance of a few stress tolerant clones in the early phase. In: Mathew NM, Jacob CK (eds) Developments in Plantation Crops Research. Allied Publ, New Delhi, India, pp 63–65

Priyadarshan PM, Sowmyalatha MKS, Sasikumar S, Varghese YA, Dey SK (1998b) Relative performance of six *Hevea brasiliensis* clones during two yielding regimes in Tripura. Indian J Nat Rubb Res 11:67–72

Priyadarshan PM, Sowmyalatha MKS, Sasikumar S, Varghese YA, Dey SK (2000) Evaluation of *Hevea brasiliensis* clones for yielding trends in Tripura. Ind J Nat Rubb Res 13:56–63

Priyadarshan PM, Sasikumar S, Gonçalves P de S (2001) Phenological changes in *Hevea brasiliensis* under differential geo-climates. The Planter, Kuala Lumpur, Malaysia 77:447–459

Priyadarshan PM, Sasikumar S, Nair RB, Dey SK (2002) Long term stability in yielding potential in clones of *Hevea brasiliensis* in Tripura. In: Rethinam P, Khan HH, Mandal PK, Reddy VM, Suresh K (eds) Plantation Crops Research and Development in the New Millennium. Coconut Development Board, Kochi, India, pp 280–283

Priyadarshan PM, Hoa TTT, Huasun H, Gonçalves P de S (2005) Yielding potential of rubber (*Hevea brasiliensis*) in suboptimal environments. J Crop Improv 14:221–247

Pushparajah E (1983) Problems and potentials for establishing *Hevea* under difficult environmental conditions. The Planter, Kuala Lumpur, Malaysia 59:242–251

Pushparajah E (2001) Natural rubber. In: Last FT (ed) Tree Crop Ecosystems (Vol: 19 Ecosystems of the World Series). Elsevier Science, Amsterdam, The Netherlands, pp 379–407

Qin M, Qian C, Wang C, Chen Z, Xiao Y (1979) Investigation of ploidy in the process of anther culture of *Hevea brasiliensis*. Annu Rep Inst Genet, Acad Sin, pp 85–87

Raemer H (1935) Cytology of *Hevea*. Genetics 17:193

Rajalakshmy V K, Joseph A, Varghese YA, Kothandaraman R (1997) Evaluation of *Hevea* clones against powdery mildew caused by Oidium heveae Steinm. Ind J Nat Rubb Res 10:110–112

Rao PS, Jayaratnam K, Sethuraj MR (1993) An index to assess areas hydrothermally suitable for rubber cultivation. Ind J Nat Rubb Res 6:80–91

Rivano F (1997a) South Amercan Leaf Blight of *Hevea*. I. Variability of *Microcyclus ulei* pathogenicity. Plantations, Recherche, Développement 4:104–114

Rivano F (1997b) South American Leaf Blight of *Hevea* II. Early Evaluation of the resistance of clones. Plantations, Recherche, Développement 4:187–196

Rivano F, Nicolas D, Chevaugeon J (1989) Résistance de l'hévéa à la maladie sud-américaine des feuilles. Perspectives de lutte Caoutchoucs et Plastiques 690:199–206

Roy B, Nazeer MA, Saha T (2004) Identification of simple sequence repeats in rubber (*Hevea brasiliensis*). Curr Sci 87:807–811

RRIM Annual Report (1997) Rubber Research Institute of Malaysia, Annual Report 1997, p 13

Rudd S, Schoof H, Mayer K (2005) Plant Markers—a database of predicted molecular markers from plants. Nucl Acids Res (Database issue) 33:D628–D632

Saha T, Roy CB, Nazeer MA (2005) Microsatellite variability and its use in the characterization of cultivated clones of *Hevea brasiliensis*. Plant Breed 124:86–92

Samsuddin Z, Tan H, Yoon PK (1987) Correlation studies on photosynthetic rates on girth and yield of *Hevea brasiliensis*. J Nat Rubb Res 2:46–54

Sasikumar S, Priyadarshan PM, Dey SK, Varghese YA (2001) Evaluation of polyclonal seedling population of *Hevea brasiliensis* (Willd. Ex. Adr de Juss) Muell. Arg. in Tripura. Ind J Nat Rubb Res 14:125–130

Schultes RE (1977a) A new infrageneric classification of *Hevea*. Botanical Museum Leaflets of Harvard University 25:243–257

Schultes RE (1977b) Wild *Hevea*: an untapped source of germpplasm. J Rubb Res Inst Sri Lanka 54:227–257

Seguin M, Lespinasse D, Rodier-Goud M, Legnaté H, Troispoux V, Pinard F, Clément-Demange A (1996a) Genome mapping and genetic analysis of South American Leaf Blight resistance in rubber tree (*Hevea brasiliensis*). Proc Third ASAP Conf on Agri Biotech Hua-Hin, Thailand, vol I, pp 1–8

Seguin M, Besse P, Lespinasse, D, Lebrun P, Rodier-Goud M, Nicolas D (1996b) *Hevea* molecular genetics. Plantations, Recherché Développment 3:77–88

Seguin M, Flori A, Legnaté H, Clément-Demange A (2003) Rubber tree. In: Hamon P, Seguin M, Perrier X, Glaszmann JC (eds) Genetic Diversity of Tropical Crops. CIRAD, France; and Science Publ, Enfield, USA, pp 277–306

Sekhar BC (2004) Asia should tap new 'Tyre rubber'. Rubber Asia 18:49–52

Senanayake YDA, Samaranayake P (1970) Intraspecific variation of stomatal density in *Hevea brasiliensis*. Müll. Arg. Quat J Rubb Res Inst Ceylon 46:61–68

Seneviratne P, Wijesekara GAS (1996) The problem of phenolic exudates in vitrocultures of mature *Hevea brasiliensis*. J Plantation Crops 24:54–62

Senyuan G (1990) *Hevea* breeding and selection for cold resistance and high yield in China. Proc IRRDB Symp, Oct 5–6, Kunming, China, pp 154–164

Shepherd H (1969) Aspects of Hevea breeding and selection. Investigations undertaken on Prang Besar Estate. RRIM Planters' Bull 104:206–216

Simmonds NW (1985) Two-stage selection strategy in plant breeding. Heredity 55:393–399

Simmonds NW (1986) Theoretical aspects of synthetic/polycross populations of rubber seedlings. J Nat Rubb Res 1:1–15

Simmonds NW (1989) Rubber breeding. In: Webster CC, Baulkwill WJ (eds) Rubber. Longman Scientific and Technical, Essex, UK, pp 85–124

Simmonds NW (1990) Breeding horizontal resistance to South American Leaf Blight of rubber. J Nat Rubb Res 5:102–113

Simmonds NW (1991) Genetics of horizontal resistance to diseases of crops. Biol Rev 66:189–241

Sreelatha S, Simon SP, Jacob J (2004) On the possibility of using ATP concentration in latex as an indicator of high yield in *Hevea brasiliensis*. J Rubb Res 7:71–78

Sushamakumari S, Shobana S, Rekha K, Jayasree K, Asokan MP (2000) Influence of growth regulators and sucrose on somatic embryogenesis and plant regeneration from immature inflorescence of *Hevea brasiliensis*. Indian J Nat Rubb Res 13:19–29

Tan H (1977) Estimates of general combining abilities in *Hevea* breeding at the RRIM. 1. Phases 2 and 3A. Theor Appl Genet 50:29–34

Tan H (1978) Estimates of parental combining abilities in rubber based on young seedling progeny. Euphytica 27:817–823

Tan H (1981) Estimates of genetic parameters and their implications in *Hevea* breeding. In: Yap TC, Graham KM, Sukanu J (eds) Crop improvement Research. Society for the advancement of Breeding Research in Asia and Oceania (SABRAO), Kuala Lumpur, Malaysia, pp 439–446

Tan H (1987) Strategies in rubber tree breeding. In: Abbott AJ, Atkin RK (eds) Improving vegetatively propagated crops. Academic Press, London, UK, pp 28–54

Tan H (1995) Genotype × Environment interaction studies in rubber (*Hevea*) clones. J Nat Rubb Res 10:63–76

Tan H (1998) A study on nursery selection in *Hevea* breeding. In: Cronin ME (ed) Proc IRRDB Symp Natural Rubber (*Hevea brasiliensis*) – General, Soils and Fertilization and Breeding and selection. IRRDB, UK, pp 114–120

Tan H, Khoo SK, Ong S H (1996) Selection of advanced polycross progenies in *Hevea* improvement. J Nat Rubb Res 11:215–225

Thomas KK (2001) Role of Clement Robert Markham in the introduction of *Hevea* rubber into the British India. The Planter 77:287–292

Veisseire P, Cailloux F, Coudret A (1994a) Effect of conditioned media on the somatic embryogenesis of *Hevea brasiliensis*. Plant Physiol Biochem 32:571–576

Veisseire P, Linossier L, Coudret A (1994b) Effect of abscisic acid and cytokinins on the development of somatic embryos in *Hevea brasiliensis*. Plant Cell Tiss Org Cult 39:219–223

Venkatachalam P, Sailasree R, Priya P, Saraswathyamma CK, Thulaseedharan A (2001) Identification of a DNA marker associated with dwarf trait in *Hevea brasiliensis* (Muell.) Arg. through random amplified polymorphic DNA analysis. In: Sainte-Beuve J (ed) Proc Annu IRRDB Mtg. CIRAD, Montpellier, France, pp 73–81

Wang Z, Zeng X, Chen C, Wu H, Li Q, Fan G, Lu W (1980) Induction of rubber plantlets from anther of *Hevea brasiliensis* Muell. Arg. *in vitro*. Chinese J Trop Crops 1:25–26

Wang ZY, Chen XN (1995) Effect of temperature on stamen culture and somatic plant regeneration in rubber. Acta Agron Sin 21:723–726

Wang ZY, Wu HD, Chen XT (1998) Effects of altered temperatures on plant regeneration frequencies in stamen culture of rubber trees. J Trop Subtrop Bot 6:166–168

Watson GA (1989) Climate and Soil. In: Webster CC, Baulkwill WJ (eds) Rubber. Longman Scientific and Technical, Essex, UK, pp 124–164

Webster CC, Paardekooper EC (1989) Botany of the rubber tree. In: Webster CC, Baulkwill WJ (eds) Rubber. Longman Scientific and Technical, Essex, UK, pp 57–84

Wycherley PR (1969) Breeding of Hevea. J Rubb Res Inst Malaya 21:38–55

Wycherley PR (1976) Rubber. In: Simmonds NW (ed) Evolution of Crop Plants. Longman, London, UK, pp 77–80

Wycherley PR (1992) The genus *Hevea*: botanical aspects. In: Sethuraj MR, Mathew NM (eds) Natural Rubber: Biology, Cultivation and Technology. Elsevier, Amsterdam, The Netherlands, pp 50–66

Yeang HY, Sunderasan E, Wickneswari R, Napi D, Zamri ASM (1998) Genetic relatedness and identities of cultivated Hevea clones determined by isozymes. J Rubb Res 1:35–47

Yeang HY, Arokiaraj P, Jaafar H, Siti AM, Rajamanikam S, Chan JL, Jafri S, Leelavathy R, Hamzah S, Van Der Logt CPE (2002) Expression of a functional recombinant antibody fragment in the latex of transgenic *Hevea brasiliensis*. J Rubb Res 5:215–225

Zhou Z, Yuan X, Guo O, Huang X (1982) Studies on the method for predicting rubber yield at the nursery stage and its theoretical basis. Chinese J Trop Crops 3:1–18

Zongdao H, Xuequin Z (1983) Rubber cultivation in China. Proc Rubb Res Inst Malaysia Planters' Conf 1983, Kuala Lumpur, Malaysia, pp 31–43

Zongdao H, Yanqing P (1992) Rubber cultivation under climatic stresses in China. In: Sethuraj MR, Mathew NM (eds) Natural Rubber: Biology, Cultivation and Technology. Elsevier, Amsterdam, The Netherlands, pp 220–238

9 Medicinal and Aromatic Plants

Ajit K. Shasany, Ashutosh K. Shukla, and Suman P. S. Khanuja

Central Institute of Medicinal and Aromatic Plants, P.O. CIMAP, Lucknow 226015, India
e-mail: khanujazy@yahoo.com

9.1
Introduction

Many of the industrially and commercially used pharmaceuticals are products of secondary metabolism in microbial or plant systems. Out of the 350,000 plant species known so far, about 35,000 (some estimate up to 70,000) are used worldwide for medicinal purposes and less than about 0.5% of these have been chemically investigated. About 100 plant species are involved in 25% of all drugs prescribed in advanced countries (Comer and Debus 1996). The annual market value of herbal drugs used worldwide was estimated to be around US $45 billion in the late 1990s and the value varies in different reports (US $60 billion to US $100 billion) presently. Even then the values with lower limits also denote the importance of medicinal plants. However, the use of medicinal plants is faced with constraints of identification, inventory, and characterization. The status is similar for aromatic plants. The capability of plants to either cure a disease or generate aroma is due to the biosynthesis of secondary metabolites. These groups of secondary metabolites may be classified into alkaloids, terpenoids, phenylpropanoids, and the complexes of these metabolites (Croteau et al. 2000). Broadly, many plants even from diverse taxonomic groups share biosynthetic pathways for these secondary metabolites. Conversely, specific compounds are also synthesized by specific taxa. In the latter case, the plants become more valuable. Genetic maps of biosynthetic pathways of these compounds are still far from complete and the information on the regulation of these pathways is even less.

9.2
Metabolic Profiles and Significance of Some Prominent Medicinal and Aromatic Plants

9.2.1
Catharanthus roseus

Catharanthus roseus (L.) G.Don is one of the most extensively investigated medicinal plants. A tremendous amount of literature has been generated on this plant and the trend is steadily increasing (Verpoorte et al. 1997). Most research endeavor on this plant has focused on its pharmacologically important alkaloids. The roots of the plant accumulate ajmalicine and serpentine, which are important components of medicines for controlling high blood pressure and other types of cardiovascular maladies. The shoot of the plant is a source of bisindole alkaloids, vinblastine and vincristine, which are indispensable components of most anticancer chemotherapies. Low yields of the bisindole alkaloids coupled with their high market value due to huge demand has prompted funding of several research projects worldwide. The plant has historically been used to treat a wide range of diseases. The plant contains a treasure of useful alkaloids (more than 125). Some, such as catharanthine, leurosine sulfate, lochnerine, tetrahydroalstonine, vindoline, and vindolinine, lower blood sugar levels (thus easing the symptoms of diabetes). Others lower blood pressure, whereas a few act as hemostatics (arrest bleeding). However, the plant has received global recognition and attention owing to the presence of vincristine and vinblastine, which have anticancer properties. This plant also contains the alkaloids reserpine and serpentine, which are powerful tranquilizers. Since the alkaloids in this plant can have serious side effects, such as nausea, hair loss, bone marrow depression, and neurotoxicity, it is not recommended that people attempt to medicate themselves with periwinkles. The rosy periwinkle does not look like a plant that is worth US $200 million a year but it is one of the most important plants on the face of the earth.

C. roseus is known as the common or Madagascar periwinkle, though its name and classification may be contradictory in some literature because this plant was formerly classified as the species *Vinca rosea Lochnera rosea*, and *Ammocallis rosea*. Furthermore, lesser periwinkle (*Vinca minor*) may also be called common periwinkle. Both species are also known as myrtle. In 1835, George Don assigned the name *Catharanthus* to the genus typified by *V. rosea* in his *General System of Gardening and Botany* (Don 1838). The name *Catharanthus* is derived from the Latin words *katharos* (pure) and *anthos* (flower), referring to the neatness, beauty, and elegance of the flower.

Periwinkle is a perennial, evergreen herb in the dogbane family (Apocynaceae) that was originally native to the island of Madagascar. It has been widely cultivated for hundreds of years and can now be found growing wild in most warm regions of the world, including the southern USA. The plants grow 1–2 ft high, have glossy, dark green leaves (1–2 in. long) and flower all summer long. The blooms of the natural wild plants are pale pink with a purple "eye" in their centers, but horticulturists have developed varieties with colors ranging from white to hot pink to purple. DNA fingerprinting studies on the available *C. roseus* germplasm are rare and not many have been reported.

During the course of its spread and naturalization over the tropics and subtropics, *C. roseus* acquired many vernacular names. The genus *Catharanthus* comprises eight species of small shrubs and herbs, six of which are predominantly indigenous to Madagascar (Stearn 1975; Mishra and Kumar 2000).

Two mutants, one dwarf and one semidwarf with high alkaloid content in the roots, and one mutant with a wavy leaf margin and high alkaloid content in its leaves were obtained after induced chemical mutagenesis with ethyl methane sulfonate and *N*-nitroso-*N*-ethylurea in the variety 'Nirmal', which is resistant to dieback disease. These mutants were evaluated in the M_3 and M_4 generations. The dwarf and semidwarf mutants differed from the parental variety for many morphological characters, whereas the mutant

with the wavy leaf margin differed mainly in leaf size and leaf thickness. Although both dwarf and semidwarf mutants showed a significantly higher alkaloid content in their roots in both generations, only the semidwarf mutant gave a significantly higher (23%) root alkaloid yield than the parental variety. The mutant with the wavy leaf margin showed a significantly higher alkaloid content in the leaves in both M_3 and M_4 generations and also had a significantly higher (21%) leaf alkaloid yield than the parental variety. All these three mutants originated from recessive mutation in single locus; the genes for "dwarfness" and "semidwarfness" were allelic, with the allele for semidwarfness being dominant over the allele for dwarfness. The gene for the wavy leaf margin was inherited independently of the genes for dwarfness and semidwarfness (Kulkarni et al. 1999).

In a study at the Central Institute of Medicinal and Aromatic Plants (CIMAP; the authors' laboratory) more than 300 mutants of the *C. roseus* germplasm generated through mutagen treatment of selfed seeds of the 'Nirmal' variety and the controls (Nirmal and Dhawal) were taken up for chemical analysis of the bisindole alkaloids (vincristine and vinblastine) and their monomeric precursors (catharanthine and vindoline). Out of these, 13 mutants were selected as probable metabolic blocks on the basis of their distinct chemical profiles. Amplified fragment length polymorphism (AFLP) analysis (Fig. 1) of the selected 13 mutants and the two control accessions (Nirmal and Dhawal) was carried out using 12 *MseI*/*Eco*RI primer pairs identified to respond well with the *C. roseus* genome through an "explorer" gel. The plants under study were also analyzed for their morphological characters.

Many biosynthetic studies have been undertaken to elucidate the terpenoid indole alkaloid (TIA) biosynthetic apparatus of the plant. Most studies have involved in vitro cell suspension cultures. Studies involving intact mature plants are rare. Various types of biotic and abiotic stress factors, including wounding, pathogen (like the fungus *Pythium aphanidermatum*) attack, UV stress and temperature changes, have been found to elicit TIA biosynthesis. The commercially important bisindole alkaloids are derived from the monomeric alkaloids vindoline and catharanthine. Both the monomeric alkaloids are derived from strictosidine, the central intermediate in TIA biosynthesis, which in turn is derived from the coupling of secologanin (a secoiridoid) and tryptamine by strictosidine synthase. Tryptamine is

Fig. 1. Amplified fragment length polymorphism (*AFLP*) clustering of the mutants of *Catharanthus roseus* compared with the varieties Nirmal (*NIR*) and Dhawal (*DHW*)

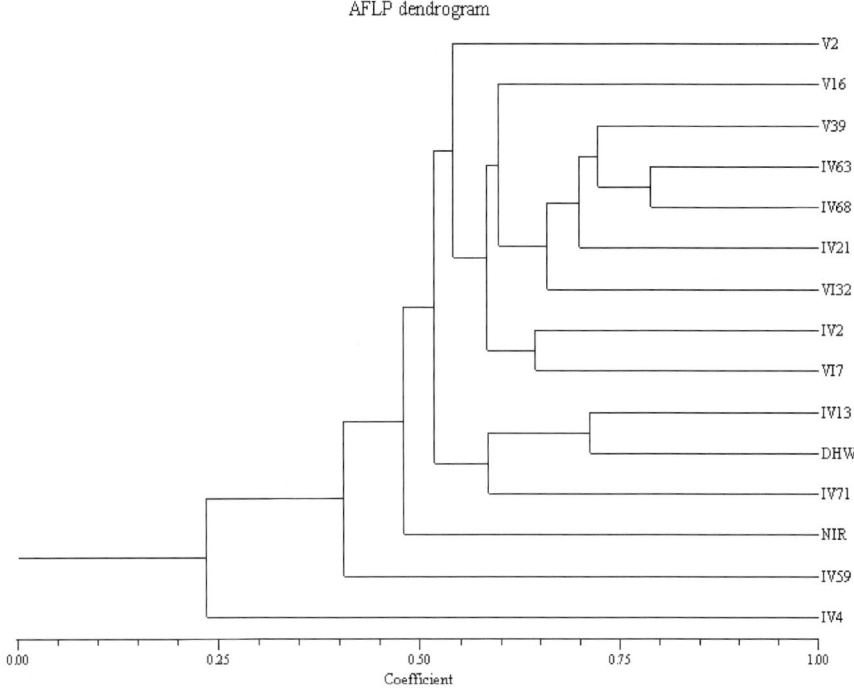

formed by the tryptophan decarboxylase mediated decarboxylation of tryptophan, and secologanin is obtained in a series of steps from isopentenyl pyrophosphate (IPP) via geraniol. It is now generally believed that IPP is formed via the non-mevalonate pathway in the plastids, but the metabolic crosstalk with the cytosolic mevalonate pathway route to IPP cannot be ruled out.

After strictosidine the pathway diverges to produce more than 3,000 TIAs and the enzyme responsible for creating this diversity is strictosidine β-D glucosidase. The pathway of catharanthine biosynthesis has not been clearly elucidated in the literature. In contrast, the vindoline biosynthetic pathway is nearly completely characterized with five out of six post-tabersonine steps elucidated. Vindoline biosynthesis is under developmental and light regulation. One of the steps catalyzed by *N*-methyltransferase (NMT) is confined to the green tissue of the plant owing to localization of this enzyme in the thylakoid membrane. The penultimate and the last steps in vindoline biosynthesis are catalyzed by desacetoxyvindoline 4-hydroxylase and deacetylvindoline 4-*O*-acetyltransferase, respectively, and both are light-dependent and possibly mediated by phytochrome. The coupling of the monomeric alkaloids to produce the bisindoles is not clearly understood. Although peroxidases (like horseradish peroxidase) have been reported to catalyze this reaction, possibilities of a nonenzymatic coupling cannot be ruled out.

Genes encoding many of the enzymes involved in TIA biosynthesis have been isolated/cloned as shown in Table 1. Efforts are ongoing to characterize the structural as well as regulatory genes, which have still not been cloned.

Future research effort on this plant demands use of modern genomic and proteomic approaches to unravel the regulatory mechanisms operating in the biosynthesis of TIAs, which will elucidate the biochemical diversity of the plant under different stress conditions. Although the major interest of researchers lies in the post-strictosidine pathway leading toward vindoline via tabersonine, many pathways like that of catharanthine biosynthesis are still not properly elucidated in the literature and require further research efforts.

9.2.2
Mentha spp.

The genus *Mentha* belongs to the family Lamiaceae. It consists of around 32 species, of which 16 are taxonomically accepted and these include *Mentha aquatica* L., *M. arvensis* L., *M. pulegium* L., *M. requienii* Benth., *M. spicata* L., *M. suaveolens* Ehrh., *M.* × *gen-*

Table 1. Cloned genes encoding enzymes involved in the terpenoid indole alkaloid biosynthetic pathway of *Catharanthus roseus*

Genes	References
hmgr	Maldonado-Mendoza et al. (1992)
dxps	Chahed et al. (20009
dxr	Veau et al. (2000)
mecs	Veau et al. (2000)
g10h	Collu et al. (2001)
cpr	Meijer et al. (1993)
as	Unpublished
tdc	De Luca et al. (1989)
sls	Irmler et al. (2000), Vetter et al. (1992)
ss	McKnight et al. (1990), Pasquali et al. (1992)
sgd	Geerlings et al. (2000)
t16h	Schroder et al. (1999)
d4h	Vazquez-Flota et al. (1997)
dat	St-Pierre et al. (1998)

tilis L., *Mentha* × *gracilis* Sole. (pro sp.), *Mentha* × *piperita* L. (pro sp.), *Mentha* × *piperita* var. *citrata* (Ehrh.) Briq., *Mentha* × *piperita* var. *piperita* L., *Mentha* × *rotundifolia* (L.) Huds. (pro sp.), *Mentha* × *smithiana* Graham, *Mentha* × *verticillata* L., and *Mentha* × *villosa* Huds. (pro sp.) (ITIS). Peppermint (*Mentha* × *piperita*), a sterile cross between *M. spicata* and *M. aquatica* (Murray et al. 1972; Harley et al. 1977), spearmint (*M. spicata*), scotch spearmint (*Mentha* × *gentilis*), and menthol mint (*M. arvensis*), yield essential oils of commercial importance. The oil principally comprises C-3 oxygenated *p*-menthone monoterpenes, the biosynthesis of which occurs in highly specialized secretory cells of epidermal oil glands (Mc Caskill et al. 1992). The commercial importance of the peppermint oil and its relatively simple vegetative propagation coupled with the easy isolation of the specialized oil glands, the site for monoterpene biosynthesis, for in vitro studies of the relevant enzymes and transcripts (Lange et al. 2000) have led to the development of peppermint as a model plant for the study of monoterpene metabolism.

Genetic diversity analysis of *Mentha* has been carried out both at interspecific as well as intraspecific level. Khanuja et al. (2000) assessed the relationships among *Mentha* spp. using randomly amplified polymorphic DNA (RAPD) technique. In this study, a set of 60 random primers was used to analyze 11 accessions from six taxa of *Mentha* developed at CIMAP. These accessions are maintained in the national gene bank for medicinal and aromatic plants at CIMAP. A total of 630 bands could be detected as amplified products upon polymerase chain reaction (PCR) amplification, of which 589 were polymorphic (93.5%). Further analysis of these RAPD profiles for band similarity indices clearly differentiated five of the *M. arvensis* accessions from the rest. Between two accessions of *M. spicata*, CIMAP/C33 could be distinguished from CIMAP/C32. *Mentha* × *gracilis* Sole'Cardiaca' showed a much higher similarity with *M. spicata* as well as *M. arvensis*, which amongst themselves showed rather a greater distance, indicating thereby that *Mentha* × *gracilis* Sole 'Cardiaca' might have evolved as a natural hybrid between *M arvensis* and *M spicata*. In terms of uniqueness of amplified bands for developing RAPD markers, it was observed that at taxa level 298 bands were unique to one of the six taxa, singly amounting to 47.3% of total amplified fragments. Primers MAP 10 and MAP 17 produced polymorphism only in the case of *M. spicata* and *M. spicata* 'Viridis', whereas MAP 08 produced polymorphic bands in all four other species but not in these two. Similarly unique patterns were observed differentiating all six species and can be used as RAPD markers for differentiating *Mentha* species.

Similarly, Shasany et al. (2002b) carried out the genetic diversity assessment of *M. spicata* germplasm through RAPD analysis. Fifteen elite accessions in the national gene bank germplasm of *M. spicata* at CIMAP were investigated for the level of their diversity in their morphological, oil quality characteristics and polymorphism in DNA. Using these morphotypic and chemotypic variation a graphic phenogram of the morphochemical relatedness on one hand and variation in RAPD profiles on the other, among these 15 accessions were generated by unweighted pair group method with arithmetic mean (UPGMA) cluster analysis. These genetic and morphochemical clusterings were compared for relatedness and differences. It was observed that RAPD analysis for the phylogenetic relationship was a better indicator of descendancy and origin among the germplasm accessions

Recently Shasany et al. (2005a) used RAPD and AFLP markers to identify interspecific and intraspecific hybrids of *Mentha*. Three controlled crosses were carried out involving *M. arvensis* and *M. spicata* [*M. spicata* CIMAP/C30 × *M. spicata* CIMAP/C33 ('Neera'); *M. arvensis* CIMAP/C18 × CIMAP/C17 ('Kalka'); and *M. arvensis* CIMAP/C17 × *M. spicata* CIMAP/C33]. The parents were subjected to

Fig. 2. Principal component analysis of *Mentha* accessions (AFLP) showing association of different species. *HYBRID* is a cross between *M. arvensis* and *M. spicata*

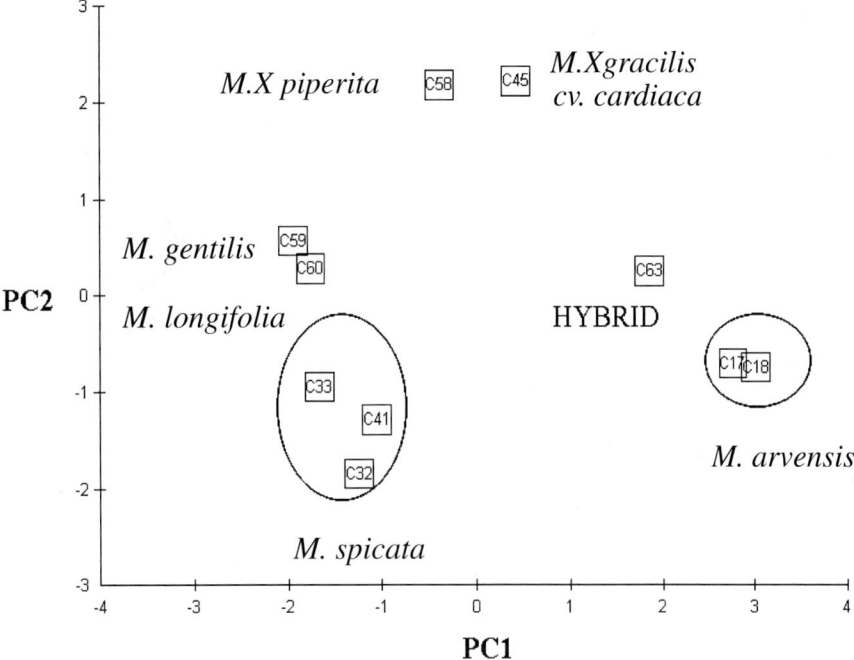

RAPD analysis with 80 primers and polymorphic primers were tested to detect coinherited RAPD profiles among the progeny of these crosses. Of 50 seedlings tested from each intraspecific cross, all resembled the profiles of female parents with the selected RAPD primers, except the detected hybrid from respective crosses. Coinherited markers could be detected with the primers OPJ 01, MAP 06, OPT 08, and OPO 20 for *M. arvensis*; OPJ 05, OPJ 14, OPO 19, and OPT 09 for *M. spicata*; and OPJ 07, OPJ 10, OPJ 11, OPJ 14, and OPO 02 for the cross *M. arvensis* × *M. spicata*. In the AFLP analysis, 40 coinherited marker fragments were identified for the cross involving *M. arvensis*, 32 for the cross involving *M. spicata*, and 41 for the interspecific cross between *M. arvensis* and *M. spicata*. In all crosses, similarity values between the parents were less than those between the parents and the hybrids. Though RAPD markers are generally considered as dominant, it is possible to identify a few codominant markers, which behave like RFLP markers. This molecular marker system may be helpful in screening out hybrids rapidly in crops where cross-pollination is a problem.

Patra et al. (2001) developed a unique interspecific hybrid spearmint clone with growth properties of *M. arvensis* and oil qualities of *M. spicata*. Later on, seven taxa of *Mentha* along with an interspecific hybrid were assessed through AFLP and it was found that the natural hybrid *Mentha* × *gracilis* Sole 'Car-

diaca' was nearly equidistant from *M. arvensis* and *M. spicata* clusters, whereas the interspecific hybrid 'Neerkalka' showed more proximity toward *M. arvensis* accessions (Fig. 2) upon cluster analysis (Shasany et al. 2005b).

A combination of time-course studies and direct feeding experiments at the cell free enzyme level has helped to elucidate the major pathway for the conversion of a universal acyclic precursor, geranyl diphosphate, to the various menthol isomers in peppermint as shown in Wise and Croteau (1999). The secondary metabolic pathway begins with the conversion of isopentenyl diphosphate and dimethylallyl diphosphate via the action of plastidial prenyltransferase and geranyl diphosphate synthase to geranyl diphosphate (Burke et al. 1999), which undergoes cyclizaton by plastidial limonene synthase to (-)-limonene (Alonso et al. 1992). (-)-Limonene serves as the common olefinic precursor of the essential oil terpenes of peppermint and spearmint (Kjonaas et al. 1983) by way of a series of secondary, largely redox, transformations (Croteau 1994). In peppermint, a microsomal cytochrome P450 limonene-3-hydoxylase introduces an oxygen atom at an allylic position to produce (-)-*trans*-isopiperitenol and thereby establishes the oxygenation pattern of all subsequent derivatives (Karp et al. 1990).

The oxidation of α, β-unsaturated alcohol, (-)-*trans*-isopiperitenol, to the corresponding ketone, (-)-isopiperitenone, occurs by a soluble NAD^+-dependent dehydrogenase. Further, a soluble NADPH-dependent endocyclic double-bond reductase reduces (-)-isopieritenone to (+)-*cis*-isopulegone. The remaining double bond is next moved into conjugation with the carbonyl by an isomerase to produce (+)-pulegone, and a second NADPH-dependent reductase reduces the conjugated double bond to yield both (-)-menthone and (+)-isomenthone. Two stereoselective, NADPH-dependent reductases convert (-)-menthone and (+)-isomenthone to (-)-menthol and (+)-neoisomenthol, respectively (i.e., 3R-specific), and (-)-menthone and (+)-isomenthone to (+)-neomenthol and (+)-neomenthol and (+)-isomenthol, respectively (i.e., 3S-specific). The allylic oxidation coupled to conjugate reduction is a common phenomenon of terpenoid metabolism (Croteau 1994; Wise and Croteau 1999), but a notable variation to this general rule is reported to occur in peppermint, in that the enzymatic reaction steps involve two conjugate reduction steps i.e., of (+)-isopiperitenone and of pulegone (Croteau 1986). In spearmint, following cyclization of geranyl diphosphate by limonene synthase (Alonso et al. 1992), the olefinic precursor, limonene, is oxygenated by (-)-4S-limonene-6-hydroxylase to form (-)-*trans*-carveol (Karp et al. 1990). This is oxidized to (-)-carvone, which is the principal component of spearmint oil (Lawrence 1981).

Several genes encoding early pathway enzymes have been cloned by reverse genetic methods and characterized, including geranyl diphosphate synthase (Burke et al. 1999), limonene synthase (Colby et al. 1993), and cytochrome P450 limonene-3-and limonene-6-hydroxylases (Lupien et al. 1999). The reverse genetic approach and homology-based cloning appeared nonplausible for elucidating the remaining genes of the pathway. An enriched trichome-specific complementary DNA (cDNA) library (Lange et al. 2000) together with a novel in situ screening assay, employing functional heterlogous expression in *E.coli* (Ringer et al. 2003), provided a source for the isolation of later-pathway genes.

The limonene synthase genes from peppermint and spearmint are very similar (share 93% identity at the amino acid level) (Colby et al. 1993). The hydroxylases, i.e., (-)-4S-limonene-3-hydroxylase in peppermint and (-)-4S-limonene-6-hydroxylase in spearmint which convert common precursor (-)-limonene to (-)-*trans*-isopiperitenol or (-)-trans-

carveol, respectively (Karp et al. 1990), share only 70% deduced identity with most of the sequence differences localized to the presumptive active sites (Lupien et al. 1999). (-)-Limonene-3-hydroxylase and menthofuran synthase represents the two cytochrome P-450 mediated steps of monoterpene metabolism in peppermint (Lupien et al. 1999; Bertea et al. 2001). At the amino acid level, (+)-menthofuran synthase exhibits 35% sequence identity to peppermint (-)-limonene-3-hydroxylase and 38% identity to spearmint limonene-6-hydroxylase (Bertea et al. 2001). The (-)-isopiperitenone dehydrogenase of peppermint is quite similar to (-)-carveol dehydrogenase of spearmint and shares more than 99% amino acid identity and both are capable of utilizing (-)-*trans*-isopiperitenol and (-)-*trans*-carveol. These are members of the short-chain dehydrogenase/reductase superfamily and are related to other short-chain dehydrogenases/reductases involved in secondary metabolism (lignan biosynthesis, stress responses, and phytosteroid biosynthesis), but are quite dissimilar (approximately 13% identity) to monoterpene reductases of mint involved in (-)-menthol biosynthesis (Ringer et al. 2005).

The two double-bond reductases of peppermint [(-)-isopiperitenone reductase and (+)-pulegone reductase] are highly substrate specific and show less than 12% identity. (-)-Isopiperitenone reductase is a member of the short-chain dehydrogenase/reductase superfamily and (+)-pulegone reductase is a member of the medium-chain dehydrogenase/reductase superfamily, implying these reductases did not arise by simple gene duplication and differentiation; rather these genes arose from very different ancestral sources. Further downstream among the enzymatic steps of the menthol biosynthesis pathway, the two NADPH-dependent keto reductases which convert (-)-menthone to (-)-(3R)-menthol and (+)-3S-neomenthol are menthone:(-)-(3R)-menthol reductase and menthone:(+)-(3S)-neomenthol reductase. These menthone reductases account for all the menthol isomers found in the essential oil of peppermint (Davis et al. 2005). These two reductases share 73% identity and belong to the short-chain dehydrogenase/reductase family of enzymes. The menthone reductases (carbonyl reductase activity) share 64–66% amino acid identity with isopiperitenone reductase ($\Delta^{1,2}$-double-bond reductase activity) and both are different from NAD^+-dependent isopiperitenol dehydrogenase (12–13% amino acid identity; Ringer et al. 2005). All

Table 2. Enzymes from the metabolic pathway operating in *Mentha* spp. for which the encoding genes have been characterized

Enzymes	References
Geranyl diphosphate synthase	Burke et al. (1999)
Limonene synthase	Colby et al. (1993)
Cytochrome P450 (-)-limonene-3-hydroxylase (in peppermint); cytochrome P450 limonene-6-hydroxylase (in spearmint)	Lupien et al. (1999)
(-)-*trans*-Isopiperitenone dehydrogenase	Ringer et al. (2005)
(-)-Carveol dehydrogenase (in spearmint)	Ringer et al. (2005)
(-)-Isopiperitenone reductase	Ringer et al. (2003)
(+)-Pulegone reductase	Ringer et al. (2003)
(-)-Menthone:(-)-menthol reductase	Davis et al. (2005)
(-)-Menthone:(-)-neomenthol reductase	Davis et al. (2005)
Menthofuran synthase	Bertea et al. (2001)

four of these peppermint redox enzymes belong to the short-chain dehydrogenase/reductase superfamily on the basis of the similar length (300±50 amino acids) and the presence of several highly conserved sequence motifs, including the N-terminal cofactor binding domain (GXXXGXG), a downstream structural domain [(N/C)NAG] of unidentified function, and a catalytic domain (YXXXK).

In contrast, pulegone reductase, mechanistically similar to isopieritenone reductase, belongs to the medium-chain dehydrogenase/reductase superfamily and shares little sequence identity with peppermint short-chain dehydrogenases/reductasess and contains none of the aforementioned characteristic short-chain dehydrogenase/reductase motifs. This leads to the suggestion that the origin of these enzymes cannot be related to the mechanistically similar reaction types they catalyze. They have most likely evolved from distant ancestral redox genes that encoded enzymes with shared binding determinants for these structurally similar monoterpenoid substrates (Davis et al. 2005).

Turner et al. (2004) studied the subcellular localization of some of the enzymes of the pathway by immunoctyochemical means. All these enzymes are localized to the secretory cells of peltate glandular trichomes with abundant labeling corresponding to the secretory phase of gland development (Turner et al. 2004). It was found that limonene synthase and geranyl diphosphate were localized to leucoplasts, limonene-6-hydroxylase (in spearmint) was localized to gland cell endoplasmic reticulum, (-)-*trans*-isopiperitenol dehydrogenase was localized to

secretory cell mitochondria, whereas (+)-pulegone reductase labeling was confined to secretory cell cytoplasm. The genes isolated from *Mentha* and responsible for monoterpene biosynthesis are described in Table 2.

Future research effort on *Mentha* will aim at production of designer crops with traits from different species incorporated into a single plant entity. Such research efforts will also stand the scrutiny of market forces as the global demand for tailor-made crops is going to rise in the future.

9.2.3
Papaver somniferum

The largest and most diverse group of alkaloids is the group of benzylisoquinolines with about 2,500 members. Some important members are present in the opium poppy *Papaver somniferum* L. (Papaveraceae). Several of them are pharmaceutically important, including the analgesics morphine, codeine, and thebaine, the muscle relaxant papaverine, the antitumorigenic drug noscapine, and the antimicrobial agent sanguinarine. This plant is a native of the western Mediterranean region. Poppy is an annual erect plant, 60–120-cm in height, rarely branching, with ovate-oblong leaves. It bears large showy flowers and nearly globose to spherical capsules, containing small white or black kidney-shaped seeds. It is mainly (90%) self-pollinated, since the dehiscence of anthers takes place before the opening of the flower.

Benzylisoquinoline alkaloid biosynthesis begins with the conversion of L-tyrosine to dopamine and 4-hydroxyphenylacetaldehyde by decarboxylation, *ortho*-hydroxylation, and deamination reactions. Tyrosine/dopa decarboxylase catalyzes the conversion of L-tyrosine and L-dopa to tyramine and dopamine, respectively. Dopamine and 4-hydroxyphenylacetaldehyde condense to form (S)-norcoclaurine, the central precursor to all benzylisoquinoline alkaloids in plants (Facchini 2001; Samanani and Facchini 2001). (S)-Coclaurine is produced from (S)-norcoclaurine by a 6-O-methyltransferase (Morishige et al. 2000), and is subsequently converted to (S)-N-methylcoclaurine by an NMT (coclaurine NMT; Choi et al. 2002). A P450 hydroxylase (CYP80B1; Pauli and Kutchan 1998) converts (S)-*N*-methylcoclaurine to (S)-3′-hydroxy-N-methylcoclaurine, from which (S)-reticuline is formed via a 4′-O-methyltransferase (Morishige et al. 2000). (S)-Reticuline is the key branch-point intermediate in the formation of morphine and sanguinarine. The first committed step in sanguinarine biosynthesis is the conversion of (S)-reticuline to (S)-scoulerine by the berberine bridge enzyme (Facchini et al. 1996). (S)-Scoulerine is then converted to (S)-stylopine by two P450-dependent oxidases, resulting in the formation of two methylenedioxy groups (Bauer and Zenk 1989, 1991). Sanguinarine is produced from (S)-stylopine by the successive action of an NMT (Rueffer et al. 1990), two additional P450-dependent enzymes, a spontaneous intramolecular rearrangement (Tanahashi and Zenk 1990), and an oxidase (Arakawa et al. 1992). In contrast, the conversion of (S)-reticuline to (R)-reticuline (De-Eknamkul and Zenk 1992) followed by intramolecular carbon–carbon phenol coupling to form salutaridine (Gerardy and Zenk 1993a) are the first steps in morphine biosynthesis. Salutaridine is reduced to (7S)-salutaridinol (Gerardy and Zenk 1993b), which is converted to salutaridinol-7-O-acetate by (7S)-salutaridinol-7-O-acetyltransferase via closure of an oxide bridge (Grothe et al. 2001). In the final steps, codeinone is produced from salutaridinol-7-O-acetate and is reduced to codeine by codeinone reductase (Unterlinner et al. 1999), which is demethylated to yield morphine. During the last few years several genes from the biosynthetic pathways for reticuline, sanguinarine, and morphine have been cloned. Although the biosynthesis is well understood at the enzymatic level, the molecular and biochemical mechanisms that regulate these pathways are not known. However, the biosynthesis of morphine and related alkaloids involves many more enzymes. Moreover, much remains to be learned about the control of alkaloid biosynthetic pathways, which are clearly under strict regulation in plants.

Tyrosine is converted to morphine after an estimated 17 steps but only seven genes have been identified and isolated. These are the genes that code for (1) tyrosine/dopa decarboxylase, (2) 6-O-methyltransferase, (3) coclaurine NMT, (4) CYP80B1, (5) 4′-O-methyltransferase, (6) (7S)-salutaridinol-7-O-acetyltransferase, and (7) codeinone reductase. The multistep biosynthesis of sanguinarine from reticuline has not been deciphered properly, with only one characterized gene, that for (berberine bridge enzyme). Still several minor alkaloid derivatives have not been studied properly.

9.2.4
Ocimum spp.

This plant belongs to the family Lamiaceae. The essential oil of most of the 65 species of *Ocimum* contains monterpene derivaties such as camphor, limonene, thymol, citral, geraniol, and linalool. The essential oil is produced in specialized tissue called trichomes. The most important species are *O. basilicum* L., *O. americanum* L., and their hybrid *O. citriodorum* Vis., which are used for essential oil production and as potherbs. In India ten species have been recognized, which are *O. sanctum*, *O. basilicum*, *O. americanum*, *O. canum*, *O. suave*, *O. viride*, *O. carnosum*, *O. grattissum*, *O. adscendens*, and *O. kilmandscharicum*. Some members of the genus, including sweet basil (*O. basilicum* L.), contain an essential oil based primarily on high proportions of phenolic derivatives, such as eugenol, methyl chavicol (estragole), and methyl cinnamate, often combined with various proportions of linalool, a monoterpenol. The genes and enzymes involved in biosynthetic pathways for these phenolic derivatives have been well worked out (Table 3). Some of the genes that code for these O-methyltransferase enzymes have been isolated from several plants, including from sweet basil (Wang et al. 1999). S-Adenosyl methionine allylphenol O-methyltransferase activities from *O. basilicum* are responsible for transfer of a methyl group from S-adenosyl methionine to the *p*-hydroxyl group of either chavicol or eugenol to generate either methyl chavicol (estragole) or methyl

Fig. 3. Species differentiation in *Ocimum* through
randomly amplified poly-
morphic DNA

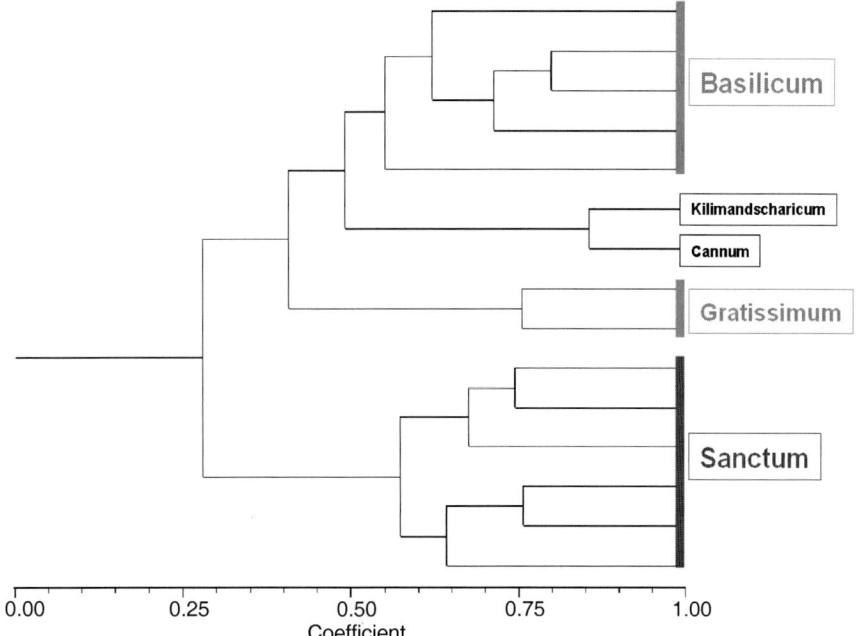

eugenol, respectively, along with *S*-adenosyl homo-
cysteine. The understanding of biosynthetic pathways of industrially important monoterpenes is incomplete, particularly in the well-known medicinal
plant *O. sanctum*. The enzymes and genes involved
in the synthesis of a few monoterpenes have been
worked out in other systems, which include copalyl
diphosphate synthetase, kurene synthetase, linalool
synthetase, limonene synthetase, myrcene synthetase,
pinene synthase, sabinenen synthase, cineole synthase, and bornyl diphosphate synthase. But these
are only the enzymes which are involved in the first
step of terpene synthesis after geranyl pyrophosphate.

Molecular genetics approaches to study genomic
architecture of *O. sanctum* species in relation to
biosynthesis of an array of industrially and medicinally important monoterpenes, such as α-pinene,
myrcene limonene, nerol, geraniol, ocimene, β-
caryophyllene, linalool, sabinene, and carvone, with
the ultimate objective of developing designer plants
using the array of biosynthetic genes and synthases of
these monoterpenes are the aims of current research
on this plant.

Five species of *Ocimum*, viz., *O. basilicum*, *O. sanctum*, *O. gratissimum*, *O. kilimandscharicum*, and *O.
cannum* were analyzed through RAPD for species differentiation at CIMAP. A total of 15 accessions were
analyzed. *O. kilimandscharicum* and *O. cannum* were
found to have a close relationship. Both species were
grouped along with *O. basilicum*. All the accessions of
O. sanctum clustered together but formed a separate
group. The accessions of *O. gratissimum* outgrouped
in the cluster of *O. basilicum* (Fig. 3).

Table 3. Genes encoding the intermediary enzymes of the
methyl eugenol and methyl chavicol pathway

Genes	Status
pal	–
c4h	–
4cl	–
ccr	–
cad	Isolated
cvomt	Isolated
c3h	–
cc3h	–
comt	–
ccomt	–
eomt	Isolated

9.2.5
Artemisia annua

Artemisia annua L. belongs to the family Asteraceae and is known for its sesquiterpene lactone
artemisinin, which is a known antimalarial com-

pound. Artemisinin or qinghaosu, an endoperoxide sesquiterpene lactone produced by aerial parts of *A. annua*, is effective even against multi-drug-resistant strains of the malarial parasite. The isolation and characterization of artemisinin from *A. annua* is considered as one of the most novel discoveries in recent medicinal plant research. It was isolated from the plant in 1972. Despite the use and importance of *A. annua*, little is known about the biosynthesis and developmental regulation of artemisinin. The genes and biochemical pathways for the production of different terpenes (specifically the sesquiterpenes) in the plant are poorly elucidated in the literature.

From the field population of *A. annua* L. growing at Pulwama (field station of CIMAP at Kashmir), ten seedlots possessing a concentration of 0.1% artemisinin were brought to Lucknow for evaluation and to utilize them as the parent genotypes for genetic improvement experiments and to analyze them chemotypically, morphologically, and lastly at the molecular level. On the basis of the morphological data, like plant height, branching capacity, and leaf morphology, three canopy types were observed, oval, spreading, and pyramidal, whereas in RAPD analysis 32–71% similarity was recorded in the parent genotypes. Through breeding and recurrent selection the amount of artemisinin was improved from 0.1% to above 1%. High and low artemisinin concentration plants were subjected to DNA isolation followed by PCR amplification using four kits of 20 primers each (MAP, OPO, OPJ, and OPT) to develop a sequence-characterized amplified region marker, in which MAP12, OPO03, and OPT13 showed clear-cut differences in RAPD profiles. MAP12 showed a fragment of 856 bp, which was present in all the high artemisinin concentration genotypes but was absent in the low artemisinin concentration genotypes; and this particular fragment was cloned in vector pBluescript SK+II at the *Hind*III restriction site. The clone obtained (pMAP12ART) was sequenced on the basis of which 22-mer forward and 25-mer reverse oligonucleotide primers were designed and synthesized and were further used to amplify the high artemisinin concentration genotypes. This particular fragment of 856-bp size was used as a probe for the identification of high artemisinin concentration genotypes, from which a genotype CIM-Arogya was released containing more than 1% artemisinin (Khanuja et al. 2005a).

Terpenoid biosynthesis in *A. annua* has received considerable attention because this plant (sweet wormwood; qinghao in traditional Chinese medicine) is the source of the endoperoxide sesquiterpene lactone antimalarial drug artemisinin (Klayman 1985). Several probable steps of the pathway have been described (Bouwmeester et al. 1999; Wallaart et al. 1999; Dhingra and Narasu 2001), and a number of groups have reported the molecular cloning of the sesquiterpene cyclase amorpha-4, 11-diene synthase responsible for catalyzing the committed step in the biosynthesis of artemisinin (Chang et al. 2000; Mercke et al. 2000; Wallaart et al. 2001). A number of cDNAs from *A. annua* that encode other terpenoid synthases (Jia et al. 1999; van Geldre et al. 2000), including that for epicedrol synthase (Hua and Matsuda 1999), which catalyzes a very complex sesquiterpene cyclization from the precursor farnesyl diphosphate, have been characterized. Another sesquiterpene cyclase cDNA from *A. annua* (AF472361) encodes a β-caryophyllene synthase. Two other members of the sesquiterpene cyclase gene family have been isolated from *A. annua*. Both full-length isolated genes are new potential candidates involved in the rate-limiting cyclization step of artemisinin biosynthesis, thereby enabling a molecular approach for improvement of artemisinin production (van Geldre et al. 2000).

9.2.6
Capsicum spp.

Capsicum, belonging to the Solanaceae family, is a genus distributed throughout the tropics, and represents 30 species. *C. annuum* L., *C. frutescens* L., *C. chinense* Jacq, *C. baccatum* L., and *C. pubescens* Ruiz & Pavon are widely cultivated throughout the world (Heiser and Pickersgill 1969). Since many of the *Capsicum* species are facultatively cross-pollinated, the genus exhibits considerable morphological variation (especially in flower and fruit color, shape, and size). Inflorescences vary from solitary to 14 flowers at one node (Tewari 1991). Seeds are cream colored except those of *C. pubescens*, which has black to dark brown seeds (D'Arcy et al. 1974). *Capsicum* spp. are diploids ($2n = 24$) with a few exceptions, including *C. compylopodium* ($2n = 26$) (Moscone et al. 1993).

The taxonomic identity of *Capsicum* species has always been difficult to establish since they displays variations especially in fruit morphology and capsaicinoid content, which is an economically major component of this species. Many investigators (Eshbaugh 1970; McLeod et al. 1982, 1983) attempted to

Fig. 4. AFLP principal component analysis of *Capsicum* accessions. *ca C. annuum, cb C. baccatum, cc C. chinence, ce C. eximium, cf C. frutescens, cl C. luteum)*

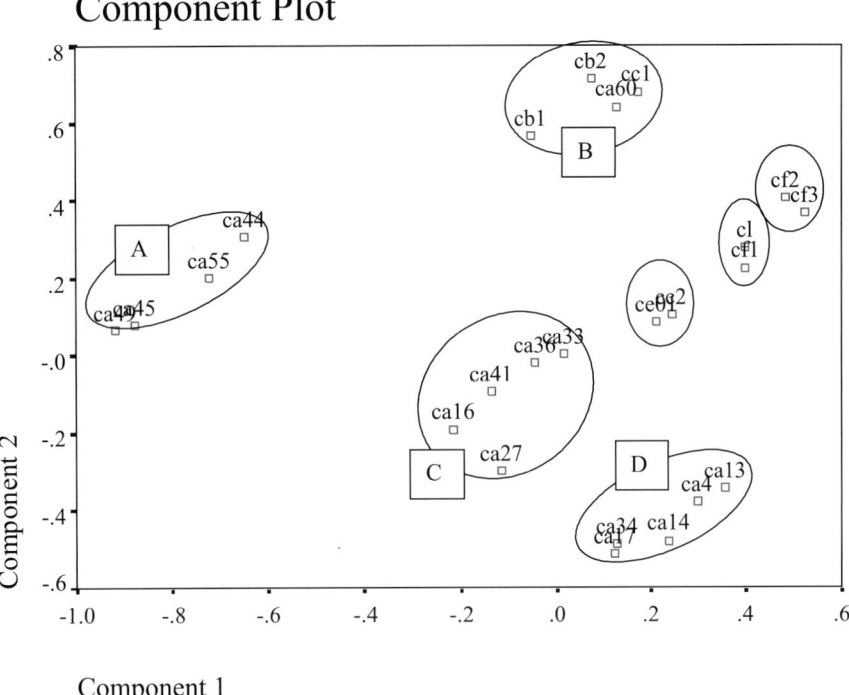

Component Plot

classify the *Capsicum* genus by using various parameters ranging from morphological characters, especially flower color, to molecular parameters.

The genus *Capsicum* was introduced to Europe from South and Central America by Columbus during his voyages and was soon incorporated into many cultures during the fifteenth century. Within 50 years it spread from Spain to England and then to India (Deb 1979). The center of origin is considered to be South–Central America as a lot of wild relatives are detected in this region. Since their introduction, the Indian chilies have undergone a lot of metamorphosis and diverse phenotypes in terms of flower, fruit, and the gross morphology of the plant. From time to time plant breeders have selected and developed varieties suited to the local needs and taste. In a recent investigation of the Indian chilies, the domesticated taxa of *C. annuum* widely grown in different parts of India were analyzed and compared with some of the introduced species to determine the extent of diversity and to predict the relationships among the species. Further, Intraspecific and interspecific genetic variation analysis was conducted using AFLP profiling in *Capsicum* accessions in the germplasm collected from different geographical locations in India. A total of 24 accessions were investigated belonging to six species, viz., *C. annuum, C. baccatum, C.*

chinence, C. eximium, C. frutescens, and *C. luteum.* The average similarity within the 15 accessions of *C. annuum* was the highest (100%) between accessions CIMAP/CA45 and CIMAP/CA49 obtained from IISR, Kerala (India) and was 43% among the species CIMAP/CC1 and CIMAP/CB2. In this analysis, accessions were clustered more pronouncedly according to their geographical locations than to their taxonomic labels (Fig. 4). A great degree of intermixing of present day domesticated chilies is evident from a current study (Thul et al. 2006). Earlier, the complex genus *Capsicum* was classified into different groups on the basis of flower color (Eshbaugh 1970; McLeod et al. 1982, 1983), enzymatic studies (McLeod et al. 1982; Loaiza-Figueroa et al. 1989), cytological analyses (Shapova 1966; Moscone et al. 1993), numerical taxonomy (Jensen et al. 1979), and molecular techniques like RAPD and AFLP (Paran et al. 1998; Hulya 2003). Variation among and within *Capsicum* was studied by Rodriguez et al. (1999) to characterize germplasm and develop diagnostic RAPDs to discriminate between the *Capsicum* species. Almost all the chili varieties cultivated on a field scale in India belong to *C. annuum* (Tewari 1991).

The burning sensation one gets from eating the *Capsicum* fruits is caused by the alkaloids known as capsaicinoids. Chief among them is capsaicin, which is chemically known as *trans*-8-methyl-*N*-vanillyl-6-nonenamide. All capsaicinoids consist of a common aromatic moiety, vanillylamine, and differ in length and degree of unsaturation of a fatty acid side chain. The most common capsaicinoids, capsaicin and dihydrocapsaicin, which together make up 80–90% of the total capsaicinoids, differ in the degree of unsaturation of a nine-carbon fatty acid side chain; five other naturally occurring capsaicinoids, nordihydrocapsaicin, homodihydrocapsaicin, homocapsaicin, norcapsaicin, and nornorcapsaicin differ in chain length ($n = 7$–10) as well as the degree of unsaturation. These compounds are uniquely synthesized and accumulated via the phenylpropanoid biosynthetic pathway in the epidermal tissues of the placenta of *Capsicum* fruits. Capsaicinoid biosynthetic activity occurs in the vacuoles and the capsaicinoids are secreted extracellularly into receptacles between the cuticles and epidermal layers of the placenta. These filled receptacles often appear as pale yellow to orange droplets on the placenta of most of the *Capsicum* fruits. Capsaicinoids are synthesized by the condensation of vanillylamine with a side short-chain branched fatty acid. Vanillylamine is synthesized from the aromatic amino acid phenylalanine via the phenylpropanoid pathway and the branched-chain fatty acid moieties are synthesized from a branched-chain amino acid, valine.

9.2.7
Vetiveria zizanioides

Vetiveria zizanioides is a useful aromatic grass known for its fragrant root and essential oil (Lemberg and Hale 1978) and is distributed in the warmer parts of the globe. Records indicate its origin to be in the region covering India to Vietnam and the usage of fragrant roots in these areas for centuries has also been reported (National Research Council 1993). In addition to its use as a source for perfumery, it is also used against central nervous system disorders and skin diseases. Besides, it has been used in water conservation and flow management since the adventitious root system of this plant is a very good soil binder and hence is referred as the "wonder grass". Adams and Dafforn (1997) advocated that the desirable feature of essential oil producing vetiver should be nonseed setting, which is hence propagated through vegetative means only. According to them, vetivers do not reproduce by seed as for centuries it has been relatively a uniform grass throughout the tropics and subtropics that

has not escaped cultivation as a weed. However, Lal et al. (2003) have used RAPD and AFLP techniques for the characterization of the vetiver germplasm and found that vetiver in the natural population does propagate through seeds. Genotyping on molecular basis through RAPD and AFLP analysis reveals the reproductive and dispersal nature of this plant species. In this study, detailed genetic diversity analysis of *V. zizanioides* germplasm available in the National Gene Bank of medicinal and aromatic plants at CIMAP was carried out using RAPD and AFLP approaches of DNA profiling and clustering. RAPD profiling of 51 accessions followed by AFLP profiles of a selected 14 representations from different locations of the country and also two introductions from South Africa were developed. Diversity to the extent of 40% was observed in RAPD analysis. This could be further resolved by an additional 25% divergence due to further polymorphism revealed by AFLP on selected accessions in nine primer combinations (Fig. 5). While genetic variation due to cross-pollination in nature is likely, a closer relationship of some geographically distant accessions is also likely owing to the free dispersal nature of vetiver in the wild. The combined analysis of phenotypic characters including chemomorphic analysis should provide variation and their mode of reproduction. Cultivated accessions through vegetative means as effected show limited variation falling in line with their phylogeny and locations.

9.2.8
Taxus wallichiana

The yew plant, which was once feared for its poisonous nature, is now a much valued and much studied plant among researchers throughout the world because it possess the compound paclitaxel (also known as taxol), which has proven anticancerous activity. The antitumor property of an extract of the bark of a *Taxus* species was first reported in 1964 against murine leukemia cells (Appendino 1993). The active principle of the extract, paclitaxel, was later found to be very effective against ovarian and breast cancers (Mc Guire et al. 1989; Holmes et al. 1991). There are about ten known species of the genus *Taxus* but many authors have regarded them as simple geographical varieties (Appendino 1993). Among these, the ones which have attracted the attention of researchers include European yew (*T. baccata* Linn), Himalayan yew (*T. wallichiana* Tucc.), Japanese yew (*T. cuspi-*

Fig. 5. AFLP-based phylogenetic tree showing relationship among *Vetiveria* accessions

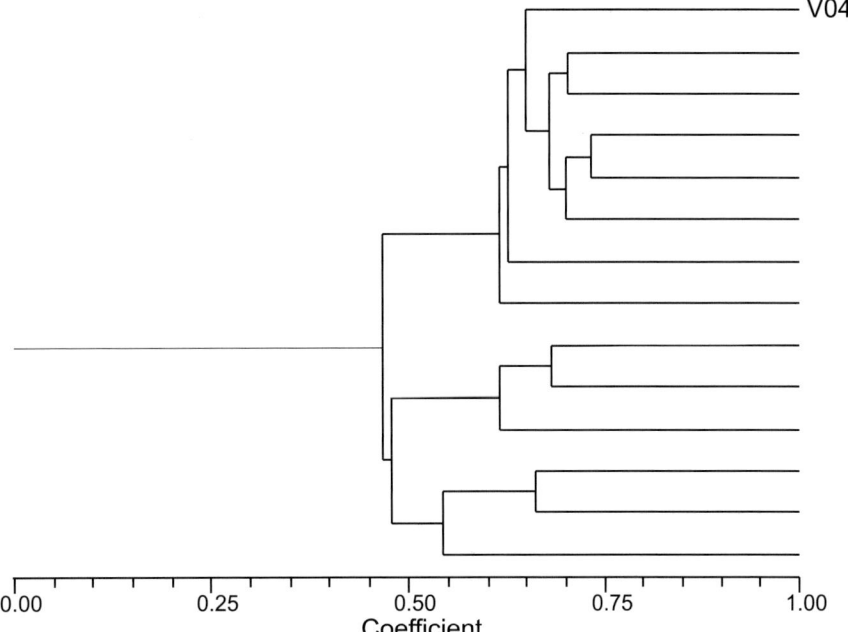

data), and Pacific yew (*T. brevifolia* Nutt.). Because of more resemblance than difference among them, various yews have also been considered as a single species descending from *PalaeoTaxus rediviva*, a fossil angiosperm that abounded in the Triassic (200 million years ago), and which later became confined to temperate zones of the northern hemisphere during the various glaciations (Appendino1993). The classification of the *Taxus* is further complicated by the existence of cultivated varieties. Over 100 cultivars of *T. baccata* alone are known which have been developed for certain gardening features.

The assessment of the genetic diversity in the *Taxus* germplasm has been carried out in various plant collections and a narrow genetic base in spite of a high level of polymorphism has been obtained (Iqbal et al. 1997). The unique bands/profiles have been pointed out for one or a few accessions in these cases but, in general, in spite of using a very large number of primers it was not possible to identify a single primer which could distinguish between all the accessions (Iqbal et al. 1997). RAPD analysis was carried out to determine the molecular relatedness among Himalayan yew (*T. wallichiana*) plants growing in the northeastern Himalayan region of India (Saikia et al. 2000). Twenty-four trees of *T. wallichiana* growing at three different locations in this region were randomly sampled for the analysis. Fifteen random primers for arbitrarily primed PCR generated 106 distinct bands as the amplification products. Among these bands,

only 48 (45%) were polymorphic, three primers did not produce any polymorphic bands at all, and nine of the 24 plants were distinguished by unique bands specific to the individuals. An UPGMA dendrogram, based on the similarity indices, showed two of the accessions, viz., AP 151 and WB 214, to be more distinct as separate clusters than others. The low level of polymorphism and close clustering with small molecular distances indicated a narrow genetic base among the trees growing in this region. However, the applicability of RAPD analysis by the generation of a few individual plant-specific unique bands is quite evident.

9.2.9
Phyllanthus spp.

The genus *Phyllanthus* L. of the family Euphorbiaceae consists of about 800 species, of which 200 are American, 100 African, 70 Madagascan, and the remainder Asian and Australasian (Webster 1994). *P. amarus* is a highly important medicinal plant species owing to its antiviral properties useful against hepatitis infection. The species is also used in stomach troubles like dyspepsia, colic, diarrhea, and dysentery; dropsy and urinogenital problems; and also as an external application for edematous swelling and inflammation. It is also used as an ingredient in many *Ayurvedic* preparations, especially those used in the treatment of jaundice. The taxonomic revision on this genus by Webster included closely related genera. *P. amarus* comes under the subsection *Swartiziani* of the section *Phyllanthus*. The information regarding the nomenclature, distinctness, and the close relatives was addressed in detail on the basis of morphology and geographical distribution (Webster 1957; Mitra and Jain 1987). *P. amarus* is said to be related to *P. abnormis*, which is endemic to sandy areas in Texas and Florida of the southern USA (Webster 1957). It is, therefore, most likely that *P. amraus* originated in the Caribbean area as a vicarious species of *P. abnormis* of the southern USA and was spread around the tropics by trading vessels (Webster 1957).

This species is distributed all over India and is considered as most widely occurring species of *Phyllanthus* in India (Chowdhury and Rao 2002). The presence of dioceous cymules (Mitra and Jain 1987) at the end of the branches is considered to be a unique character, though it resembles in many respects its close relatives *P. debilis* and *P. fraternus* of the same subsection *Swartiziani* (Webster 1994). This is the only subsection in the section *Phyllanthus*, which consists of herbaceous species that are all widespread throughout the tropics.

Jain et al. (2003) assessed molecular diversity in *P. amarus* germplasm through RAPD analysis and found that there was up to 65% variation among the accessions. Their study indicated random hybridization across the populations falling within the range of possible cross-pollination in terms of physical distance. The growth habit also displays marked variation within the species and leaves scope for measuring genetic diversity of this important medicinal plant.

9.2.10
Pelargonium graveolens

Cultivated *Pelargonium graveolens* (Geraniaceae), generally known as "rose-scented gerenium" popular for their aromatic oil qualities and ornamental value have been bred for approximately 200 years. The genus *Pelargonium* comprises over 250 species in 13–17 sections (Renou et al. 1997). The oil of *P. graveolens* possesses antibacterial, antifungal, nematicidal, and insecticidal properties and hence offers scope for medicinal and agrichemical applicability. The possibility of using its essential oil or its constituents geraniol and citronellol in the postharvest preservation of fruits against anthracnose has been reported (Chandravandana and Nidiry

1994). Besides its powerful roselike odor, the oil of *P. graveolens* is one of the most important essential oils used for isolation of rhodinol and its esters, which form part of most high-grade perfumes (Rao et al. 1993). The three genotypes of *P. graveolens*, Algerian, Kelkar, and Bourbon, are difficult to distinguish morphologically, particularly at the nursery stage for planting although these are distinct in their aroma notes.

Shasany et al. (2002a) have described detectable differences among these cultivars of *P. graveolens* through the RAPD approach, which is useful for maintaining genetic purity of the available germplasm and quality control in commercial fields. Comparable oil profiles were studied by gel–liquid chromatography for these genotypes. Algerian, Bourbon, and Kelkar genotypes showed the distinct citronellol-to-geraniol ratio in their oils. For RAPD analysis, of the 120 primers used in this study to differentiate the accessions, OPA03, OPA06, OPA08, OPA15, OPA18, OPA19, OPB02, OPB03, OPB05, OPB16, OPO02, OPO15, OPO16, OPO18, OPQ06, and OPQ07 were more informative to differentiate the three cultivars Algerian, Bourbon, and Kelkar. Among these primers, OPA18, OPA19, OPB02, OPB03, OPB05, OPB16, OPO02, OPO15, OPO16, OPO18, OPQ06, and OPQ07 generated unique fragments at least for one genotype. The primers OPB03 and OPB05 differentiated all three genotypes when used individually for amplification. In this investigation, the three genotypes Algerian, Kelkar, and Bourbon could be distinguished at the genetic level at the early growth stage itself through the genotype-specific RAPD profiles.

9.2.11
Cymbopogon spp.

The genus *Cymbopogon* (Poaceae) is known to include about 140 species. Among these, more than 52 have been reported to occur in Africa, 45 in India, six each in Australia and South America, four in Europe, two in North America, and the remainder are distributed in South Asia (Jagdish Chandra 1975). Most of these species produce characteristic aromatic essential oils that have commercial importance in perfumery, cosmetic, and pharmaceutical applications. The *Cymbopogon* essential oils are characterized by monoterpene constituents like citral, citronellol, citronellal, linalool, elemol, 1,8-cineole, limonene, geraniol, β-

carophyllene, methyl heptenone, geranyl acetate, and geranyl formate . Citral is one of the important components of the oils present in several species of *Cymbopogon*, with wide industrial uses as a raw material for perfumery, confectionery, vitamin A, etc. Taxonomically, the species of *Cymbopogon* have been divided into three series, viz., 'Shoenanthi', 'Rusae,' and 'Citrati' (Stapf 1906).

The morphological variation and oil characteristics of various species and varieties of *Cymbopogon* were reported earlier (Husain 1994), but such information is not able to precisely define the relatedness among the morphotypes and chemotypes. For instance, *C. martini* var. *sofia* and *C. martini* var. *motia* are morphologically almost indistinguishable, but show distinct chemotypic characteristics in terms of oil constituents (Guenther 1950). Conversely, phenotypically and taxonomically well distinguishable species produce oils of almost identical chemical compositions, such as lemongrass oils from *C. citratus* and *C. flexuosus* (Anon 1988). Such phenotypic traits, whether morphological or chemotypic, are basically the phenotypic expression of the genotype, while DNA markers are independent of environment, age, and tissue and are expected to reveal the genetic variation more conclusively in assessing such variations. Introgression of various traits, intermittent mutations, and selection through human intervention may lead to variation in chemotypic characters across geographical distributions (Kuriakose 1995). Although natural hybridization may lead to the formation of morphological or chemotypic intermediates, defining taxa purely on this basis may not be appropriate. The earlier works on phytochemical (Patra et al. 1990; Dhar et al. 1993), biochemical, and physiological parameters (Nandi and Chatterjee 1987), development of agrotechnology (Nair et al. 1979; Rao et al. 1985), morphoeconomic features and performance (Kole et al. 1981a, b), genetic improvement (Ganguly et al. 1979; Kole et al. 1983; Kole 1985, 1989; Kole and Sen 1985; Maheshwari and Sethi 1987), and breeding approaches (Kole et al. 1980; Kole 1986; Kole and Sen 1986; Kulkarni and Rajgopal 1986; Rao and Sobti 1987; Shyalaraj and Thomas 1993) of different *Cymbopogon* species and varieties appear scattered, and do not address the question of relationships among different taxa.

An attempt was made using RAPD analysis to trace the ancestors of cultivar Java II within *C. winterianus* (Shasany et al. 2000). This study indicated the extent of change the *C. nardus* (L.) Rendle geno-type has undergone in the process of evolution to different varieties, differentiation of *C. winterianus* (Jowitt) as a species, and further evolution of cultivars through selection occurring naturally and that imposed by human intervention and selections. But at the interspecific level, information on molecular characterization for establishing the phylogenetic relationships among *Cymbopogon* spp. is scarce. Khanuja et al. (2005b) have estimated the extent of genetic diversity using RAPD analysis in relation to chemovariation observed in oil constituents among different taxa of *Cymbopogon* in India. Nineteen *Cymbopogon* taxa belonging to 11 species, two varieties, one hybrid taxon, and four unidentified species were analyzed for their essential oil constituents and RAPD profiles to determine the extent of genetic similarity and thereby the phylogenetic relationships among them. Remarkable variation was observed in the essential oil yield, ranging from 0.3% in *C. travancorensis* Bor. to 1.2% in *C. martinii* (Roxb.) Wats var. *motia*. Citral, a major essential oil constituent, was employed as the base marker for chemotypic clustering. On the basis of genetic analysis, elevation of *C. flexuosus* var. *microstachys* (Hook.F.) Soenarko to species status and separate species status for *C. travancorensis* Bor., which has been merged under *C. flexuosus* (Steud.) Wats, were suggested in an attempt to resolve some of the taxonomic complexes in *Cymbopogon* .The separate species status for the earlier proposed varieties of *C. martinii* (*motia* and *sofia*) was further substantiated by these analyses. The unidentified species of *Cymbopogon* have been observed as intermediate forms in the development of new taxa.

9.2.12
Bacopa monnieri

Bacopa monnieri (L.) Penn. ($2n = 64$) commonly known as Neera-Brahmi, a member of the family Scrophulariaceae, is a small prostrate herb that grows wild in marshy and damp places near water logs throughout India. The plant has been used since ancient time in folklore and is used in the traditional system of medicine as a nervine and cardiotonic and diuretic (Chopra et al. 1956) and is prescribed against epilepsy and snakebite. It is a very common amphibious plant of the "Pan-Tropics" where it occupies the banks of slow flowing rivers and lakes and accepts soft to hard, even brackish water. The plant has been extensively worked upon for its chemical constituents, especially bacosides, which are the active agents responsible for memory enhancement (http://www.dalmiaindustries.com/ safety.htm). The crude extract of *B. monnieri* has been shown to improve the performance of rats in several learning tests as evidenced by better acquisition, improved retention, and delayed extinction of the information (Singh and Dhawan 1978, 1982). In experimental studies, the saponin-rich highly potent nootropic extracts have not shown any endocrine, metabolic, gastrointestinal, anabolic, or behavioral side effects (http://www.dalmiaindustries.com/safety.htm).

Despite the major and ancient position of *B. monnieri* as an important medicinal plant, little is known about its phylogeny and genetics. Darokar et al. (2001) generated a series of RAPD profiles in order to assess genetic diversity in the germplasm comprising several accessions of *B. monnieri* collected from different geographical regions of India. All the accessions shared a similarity range of 80–100%. The observed low level of genetic variation in their study was attributed to the interplay of sexual and vegetative modes of reproduction and the similarity of local environments in habitats of *B. monnieri*. It was possible to differentiate individual accessions, showing differences in morphological and growth properties at the DNA level.

9.2.13
Aloe

Aloe, a member of the Liliaceae family, is a tropical to subtropical plant with more than 300 species known, mostly indigenous to Africa, for which systematic information in relation to chemotaxonomy has also been attempted (Reynolds 1966). It is one of the most popular natural sources for a range of home remedies in use today. *Aloe vera* introduced in India long ago as the type specimen of this species is quoted as coming from an Indian garden (Wood 1982). The major component of the leaf mass is the gel from the parenchymatous cells consisting mainly of polysaccharides (Femenia et al. 1999), whereas the acetylated mannose sugar is the major bioactive component (Manna and McAnalley 1993). The species of *Aloe* have been used for centuries for their laxative, anti-inflammatory, immunostimulant, antiseptic (Capasso et al. 1998) wound and burn healing (Heggers et al. 1995; Chithra et al. 1998), antiulcer (Saito 1993; Koo 1994), and antitumor (Winters et al. 1981) ac-

tivities. There are reports on the antidiabetic activity of *Aloe* extracts also (Ghannam et al. 1986; Ajabnoor 1990). Variation in some economically important *Aloe* species based on phenolic constitution of leaves has been reported (Van Der Bank et al. 1995; Viljoen et al. 1995, 1999).

Since morphochemical characters are dependent on age and environment, it is essential to characterize this medicinally and economically important plant at the genetic level. Darokar et al. (2003) systematically characterized the *Aloe* species at the molecular level in relation to morphological variation for estimation of genetic diversity in the germplasm collection from different geographical regions of India using RAPD as well as AFLP techniques. The pattern of phylogeny was visibly parallel in AFLP analysis compared with the RAPD pattern. But the degree of diversity revealed by AFLP did show a zooming effect over RAPD analysis. The profiles in relation to growth and morphological traits will be useful in future genetic improvement and authentication of species and genotypes of this medicinally and economically important genus.

9.3
Concluding Remarks

This chapter illustrates the significant advances made in the field of identification of genes involved in biosynthetic pathways of some medicinally and aromatically important chemical agents and genetic diversity analysis of some prominent medicinal and aromatic plants. Conventional taxonomic approaches for diversity estimation and phylogenetic relationship mainly rely upon the ability to differentiate morphological characters at macroscopic and microscopic levels. Molecular analysis, on the other hand, can be used not only to reveal cryptic differences at the genetic and the functional level, but also to precisely quantify the same. Similarly the diversity in secondary metabolites results from the differential expression of genes in the biochemical pathway. Though some of the advances are being made in a few plants, we are still far from deciphering the mystery of their biosynthesis.

Molecular markers have been looked upon as tools for a large number of applications ranging from localization of a gene to improvement of plant varieties by marker-assisted selection. They are also becoming extremely popular markers for phylogenetic analysis, adding new dimensions to evolutionary theories. With the use of molecular markers, it is now routine to trace the valuable alleles in a segregating population and map them mostly in crop plants. These markers, once mapped, will enable more precisely the dissection of complex traits into component genetic units (Hayes et al. 1993), thus providing breeders with new tools to manage these complex units more efficiently. Plant improvement, either by natural selection or through different breeding techniques, has always relied upon creating, evaluating, and selecting the right combination of alleles. Manipulation of a large number of genes is often required for the improvement of even the simplest of characteristics. Tagging of useful genes, not only like the ones responsible for conferring resistance to plant pathogens, for the synthesis of plant hormones, for drought tolerance, and developmentally regulated or pathway genes for biosynthesis of a complex array of secondary metabolites, is a major challenge. Such tagged markers can also be used for detecting the presence of useful genes in the new genotypes in a hybrid program or in transgenomics, etc. As RAPD markers invariably behave as dominant markers, consequently a simple cosegregation analysis does not always reveal a tight linkage between the gene of interest and the RAPD marker. This limitation has been overcome in a number of cases, specifically by carrying out RAPD analysis for bulk plant representing and segregating for a specific character (bulked segregant analysis; Michelmore et al. 1991).

Like in all other plants, in medicinal and aromatic plants, it is not only the general traits but also the spatial and temporal expression of genes responsible for biosynthesis of secondary metabolites that hold the key for the yield and quality. Again the combination of compounds at a particular time in a specific genotype is preferable compared with other genotypes. A similar pathway operating in diverse plants and different pathways occurring in the same plants are first attributed to the presence of the genes, followed by their differential expression. Diverse plants have similar activities and similar plants have diverse activities. So it is necessary that the genes for secondary metabolites are marked and mapped to the genome. Comprehensive molecular maps for alkaloid, terpenoid, and phenylpropanoid biosynthetic genes and their regulatory counterparts are the need of the hour considering the synteny in secondary metabolites and the possibility of generating designer plants with specific combination of molecules.

Acknowledgement. The authors thankfully acknowledge the Council of Scientific and Industrial Research (CSIR), and the Department of Biotechnology, Government of India, for support leading to works on diversity analysis. The authors also acknowledge the direct and indirect help of "Team CIMAP."

References

Adams RP, Dafforn MR (1997) DNA fingerprints (RAPDS) of the pantropical grass vetiver, *Vetiveria zizanioides* (L.) Nash (Gramineae), reveal a single clone, sunshine, is widely utilized for erosion control. Vetiver Newsl 18:27–33

Ajabnoor MA (1990) Effects of *Aloe* on blood glucose level in normal and alloxan diabetic mice. J Ethanopharmacol 28:215–220

Alonso WR, Rajaonarivony JIM, Gershenzon J, Croteau R (1992) Purification of 4−S limonene synthase, a monoterpene cyclase from glandular trichomes of peppermint (*Mentha × piperia*) and spearmint (*M. spicata*). J Biol Chem 267:7582–7587

Anonymous (1988) Wealth of India, vol 2 (ed Ambasta SP). Publication and Information Directorate, CSIR, Hillside Road, New Delhi, India

Appendino G (1993) Taxol (Paclitaxel): Historical and ecological aspects. Fitoterapia LXIV:5–25

Arakawa H, Clark WG, Psenak M, Coscia CJ (1992) Purification and characterization of dihydrobenzophenanthridine oxidase from *Sanguinaria canadensis* cell cultures. Arch Biochem Biophys 299:1–7

Bauer W, Zenk MH (1989) Formation of both methylenedioxy groups in the alkaloid (S)-stylopine is catalyzed by cytochrome P-450 enzymes. Tetrahedron Lett 30:5257–5260

Bauer W, Zenk MH (1991) Two methylenedioxy bridge forming cytochrome P-450 dependent enzymes are involved in (S)-stylopine biosynthesis. Phytochemistry 30:2953–2961

Bertea CM, Schalk M, Karp F, Maffei M, Croteau R (2001) Demonstration that menthofuran synthase of mint (*Mentha*) is a cytochrome P450 monooxygenase: cloning, functional expression, and characterization of the responsible gene. Arch Biochem Biophys 390:279–286

Bouwmeester HJ, Wallaart TE, Janssen MH, van Loo B, Jansen BJ, Posthumus MA, Schmidt CO, De Kraker JW, König WA, Franssen MC (1999) Amorpha-4,11-diene synthase catalyzes the first probable step in artemisinin biosynthesis. Phytochemistry 52:843–854

Burke CC, Wildung MR, Croteau R (1999) Geranyl diphosphate synthase: cloning, expression and chareacterization of this prenyl transferase as heterodimer. Proc Natl Acad Sci USA 96:13062–13067

Capasso F, Borrelli F, Capasso R, Di Carlo G, Izzo AA, Pinto L, Mascolo N, Castaldo S, Longo R (1998) Aloe and its therapeutic use. Phytother Res 12:S124–S127

Chahed K, Oudin A, Guivarch N, Hamdi S, Chenieux J-C, Rideau M, Clastre M (2000) 1-Deoxy-D-xylulose 5-phosphate synthase from periwinkle: cDNA identification and induced gene expression in terpenoid indole alkaloid-producing cells. Plant Physiol Biochem 38:559–566

Chandravandana MV, Nidiry ESJ (1994) Antifungal activity of essential oil of *Pelargonium graveolens* and its constituents against *Colletotrichum gloesporoides*. Intl J Environ Biol 32:908–909

Chang YJ, Song SH, Park SH, Kim SU (2000) Amorpha-4,11-diene synthase of *Artemisia annua*: cDNA isolation and bacterial expression of a terpene synthase involved in artemisinin biosynthesis. Arch Biochem Biophys 383:178–184

Chithra P, Sajithlal GB, Chandrakasan G (1998) Influence of *Aloe vera* on glycosaminoglycans in the matrix of healing dermal wounds in rats. J Ethanopharmacol 59:179–186

Choi K-B, Morishige T, Shitani N, Yazaki K, Sato F (2002) Molecular cloning and characterization of coclaurine N-methyltransferase from cultured cells of *Coptis japonica*. J Biol Chem 277:830–835

Chopra RN, Nayar SL, Chopra IC (1956) Glossary of Indian Medicinal Plants. Council of Scientific and Industrial Research, New Delhi, India

Chowdhury LB, Rao RR (2002) Taxonomic study of herbaceous species of *Phyllanthus* L. (Euphorbiaceae) in India. Phytotaxonomy 2:143–162

Colby SM, Alonso WR, Katahira E, McGarvey DJ, Croteau R (1993) 4S-Limonene synthase from the oil glands of spearmint (*Mentha spicata*): cDNA isolation, characterization and bacterial expression of the catalytically active monoterpene cyclase. J Biol Chem 268:23016–23024

Collu G, Unver N, Peltenburg-Looman AMG, van der Heijden R, Verpoorte R, Memelink J (2001) Geraniol 10-hydroxylase, a cytochrome P450 enzyme involved in terpenoid indole alkaloid biosynthesis. FEBS Lett 508:215–220

Comer M, Debus E (1996) A partnership: Biotechnology, biopharmaceuticals and biodiversity. In: di Castri F, Younnes T (eds) Biodiversity. Science and Development. CABI Publ, Oxford, UK, pp 488–499

Croteau R, Gershenzon J (1994) Genetic control of monoterpene biosynthesis in mints (Mentha: lamiaceae). Genetic engineering of plant secondary metabolism. Rec Adv Phytochem 28:193–229

Croteau R, Venkatachalam KV (1986) Metabolism of monoterpenes: demonstration that (+)-cis-isopulegone, not piperitenone, is the key intermediate in the conversion of (-)-isopiperitenone to (+)-pulegone in peppermint (*Mentha piperita*). Arch Biochem Biophys 249:306–315

Croteau R, Kutchan TM, Lewis NG (2000) Natural products (secondary metabolites). In: Buchanan B, Gruissem W, Jones R (eds) Biochemistry and Molecular Biology of Plants. American Society of Plant Physiologists, Rockville, MD, USA, pp 1250–1318

D'Arcy WG, Eshbaugh WH (1974) New World Peppers [*Capsicum*—Solanaceae] North of Colombia: Baileya 19:93–103

Darokar MP, Khanuja SPS, Shasany AK, Kumar S (2001) Low levels of genetic diversity detected by RAPD analysis in geographically distinct accessions of *Bacopa monnieri*. Genet Resour Crop Evol 48:555–558

Darokar MP, Rai R, Gupta AK, Shasany AK, Rajkumar S, Sundaresan V, Khanuja SPS (2003) Molecular assessment of germplasm diversity in *Aloe* species using RAPD and AFLP analysis. J Med Arom Plant Sci 25:354–361

Davis EM, Ringer KL, McConkey ME, Croteau R (2005) Monoterpene metabolism. Cloning, expression, and characterization of menthone reductases from peppermint. Plant Physiol 137:873–881

Deb DB (1979) Solanaceae in India. In: Hawkes JG, Lester RN, Skelding AD (eds) The Biology and Taxonomy of the Solanaceae. Academic Press, London, UK, pp 87–112

De-Eknamkul W, Zenk MH (1992) Purification and properties of 1,2-dehydroreticuline reductase from *Papaver somniferum* seedlings. Phytochemistry 31:813–821

De Luca V, Marineau C, Brisson N (1989) Molecular cloning and analysis of cDNA encoding a plant tryptophan decarboxylase: comparison with animal DOPA decarboxylases. Proc Natl Acad Sci USA 86:2582–2586

Dhar AK, Sapru R, Lattoo KS (1993) Changes in oil concentration and its constituents, herbage and oil yield in five genotypes of *Cymbopogon jwarancusa* following foliar application of nitrogen. Ind Perfumer 37:303–310

Dhingra V, Narasu ML (2001) Purification and characterization of an enzyme involved in biochemical transformation of arteannuin B to artemisinin from *Artemisia annua*. Biochem Biophys Res Comm 281:558–561

Dobelis Inge N (ed) (1989) Magic and Medicine of Plants. Reader's Digest Books, Pleasantville, NY, USA

Don G (1838) General System of Gardening and Botany, vol 4. Rivington, London, p 95

Eshbaugh WH (1970) A biosystematic and evolutionary study of *Capsicum baccatum* (Solanaceae). Brittonia 22:31–43

Facchini PJ (2001) Alkaloid biosynthesis in plants: biochemistry, cell biology, molecular regulation, and metabolic engineering applications. Annu Rev Plant Physiol Plant Mol Biol 52:29–66

Facchini PJ, Penzes C, Johnson AG, Bull D (1996) Molecular characterization of berberine bridge enzyme genes from opium poppy. Plant Physiol 112:1669–1677

Femenia A, Sanchez ES, Simal S, Rossello C (1999) Compositional features of polysaccharides from *Aloe vera* (*Aloe barbadensis* Miller). Carbohydr Polym 39:109–117

Ganguly D, Singh KK, Bhagat SD, Upadhyay DN, Chauhan YS, Gupta NK, Singh HS (1979) "RRLJOR-3-1970" an improved strain of Java Citronella (*Cymbopogon winterianus* Jowitt.). Ind Perfumer 23:107–111

Geerlings A, Ibanez MM-L, Memelink J, van der Heijden R, Verpoorte R (2000) Molecular cloning and analysis of strictosidine β-D-glucosidase, an enzyme in terpenoid indole alkaloid biosynthesis in *Catharanthus roseus*. J Biol Chem 275:3051–3056

Gerardy R, Zenk MH (1993a) Formation of salutaridine from (R)-reticuline by a membrane-bound cytochrome P-450 enzyme from *Papaver somniferum*. Phytochemistry 32:79–86

Gerardy R, Zenk MH (1993b) Purification and characterization of salutaridine: NADPH 7-oxidoreductase from *Papaver somniferum*. Phytochemistry 34:125–132

Ghannam, N, Kingston M, Al-Meshaal IA, Tariq M, Parman NS, Woodhouse N (1986) The antidiabetic activity of *Aloes*: preliminary clinical and experimental observation. Hormone Res 24:288

Grothe T, Lenz R, Kutchan TM (2001) Molecular characterization of the salutaridinol 7-O-acetyltransferase involved in morphine biosynthesis in opium poppy *Papaver somniferum*. J Biol Chem 276:30717–30723

Guenther E (1950) The Essential Oils, vol IV. Van Nostrand, New York, USA

Harley RM, Brighton CA (1977) Chromosome numbers in the genus *Mentha*. Bot J Linn Soc 74:71–96

Hayes PM, Liu BH, Knapp SJ, Chen F, Jones B, Blake T, Franckowiak J, Rasmusson D, Sorrells M, Ullrich SE, Wesenberg D, Kleinhofs A (1993) Quantitative trait locus effects and environmental interaction in a sample of North American barley germplasm. Theor Appl Genet 87:392–401

Heggers JP, Kuchukcelebi A, Stabenare CJ (1995) Wound healing effects of Aloe gel and other tropical antibacterial agents on rat skin. Phytother Res 9:455–457

Heiser BC Jr, Pickersgill B (1969) Names for the cultivated *Capsicum* species (Solanaceae). Taxon 18:277–283

Heywood VH (ed) (1993) Flowering Plants of the World. Oxford Univ Press, New York, USA

Holmes FA, Walters RS, Theriault RL (1991) Phase II trial of taxol, an active drug in the treatment of metastatic breast cancer. J Natl Cancer Inst 83:1795–1805

Hua L, Matsuda SP (1999) The molecular cloning of 8-epicedrol synthase from *Artemisia annua*. Arch Biochem Biophy 369:208–212

Hulya I (2003) RAPD marker assisted varietal identification and genetic purity test in pepper, *Capsicum annuum*. Sci Hort 97:211–218

Husain A (1994) Chapter V. Family gramineae: 1. Citronella (*Cymbopogon winterianus* and *Cymbopogon nardus*). 2. Lemongrass (*Cymbopogon flexuosus* and *Cymbopogon citratus*). 3. Palmarosa (*Cymbopogon martini*). 4. Vetiver (*Vetiveria zizanioides*). In: Essential Oil Plants and their Cultivation. Central Institute of Medicinal and Aromatic Plants, Lucknow, India, pp 39–48

Iqbal MJ, Aziz N, Saeed NA, Zafar Y, Malik KA (1997) Genetic diversity evaluation of some elite cotton varieties by RAPD analysis. Theor Appl Genet 94:139–144

Irmler S, Schroder G, St-Pierre B, Crouch NP, Hotze M, Schmidt J, Strack D, Matern U, Schroder J (2000) Indole alkaloid

biosynthesis in *Catharanthus roseus*: new enzyme activities and identification of cytochrome P450 CYP72A1 as secologanin synthase. Plant J 24:797–804

Jagadish Chandra KS (1975) Cytogenetical evolution in some species of *Cymbopogon*. In: Kachroo P (ed) Advancing Frontiers in Cytogenetics. Hindustan Publ Co, New Delhi, India

Jain N, Shasany AK, Sundaresan V, Rajkumar S, Darokar MP, Bagchi GD, Gupta AK, Kumar S, Khanuja SPS (2003) Molecular diversity in *Phyllanthus amarus* assessed through RAPD analysis. Curr Sci 85:1454–1458

Jensen RJ, McLeod MJ, Eshbaugh WH, Guttman SI (1979) Numerical taxonomic analyses of allozymic variation in *Capsicum* (Solanaceae). Taxon 28:315–327

Jia JW, Crock J, Lu S, Croteau R, Chen XY (1999) (3R)- Linalool synthase from *Artemisia annua* L.: cDNA isolation, characterization and wound-induction. Arch Biochem Biophys 372:143–149

Karp F, Mihaliak CA, Harris JL, Croteau R (1990) Monoterpene bisynthesis: specificity and hydroxylations of (-)-limonene by enzyme preparations from peppermint (*Mentha* × *piperiata*), spearmint (*Mentha spicata*) and perilla (*Perilla frutescens*) leaves. Arch Biochem Biophys 220:219–226

Khanuja SPS, Shasany AK, Srivastava A, Kumar S (2000) Assessment of genetic relationships in *Mentha* species. Euphytica 111:121–125

Khanuja SPS, Paul S, Shasany AK, Gupta AK, Darokar MP, Gupta MM, Verma RK, Ram G, Kumar A, Lal RK, Bansal RP, Singh AK, Bhakuni RS, Tandon S (2005a) Genetically tagged improved variety 'CIM-Arogya' of *Artemisia annua* for high artemisinin yield. J Med Arom Plant Sci 27:520–524

Khanuja SPS, Shasany AK, Pawar A, Lal RK, Darokar MP, Naqvi AA, Rajkumar S, Sundaresan V, Lal N, Kumar S (2005b) Essential oil constituents and RAPD markers to establish species relationship in *Cymbopogon* Spreng. (Poaceae). Biochem Syst Ecol 33:171–186

Kjonaas R, Croteau R (1983) Demonstration that limonene is the first cyclic intermediate in the biosynthesis of oxygenated *p*-menthane monoterpenes in *Mentha piperita* and other *Mentha* species. Arch Biochem Biophys 220:79–89

Klayman DL (1985) Qinghaosu (artemisinin): an antimalarial drug from China. Science 228:1049–1055

Kole C (1985) Improvement of *Cymbopogon winterianus* Jowitt, through mutagenesis. Ind Perfumer 29(1&2):129–138

Kole C (1986) Path-coefficient analysis in citronella. Ind J Agric Sci 56(4):241–244

Kole C (1989) Isolation of hardy and high yielding mutants in citronella (*Cymbopogon winterianus*). EUCARPIA Congr, Heidelberg, Germany, Vortrage fur pflanzenzuchtg 15:20–6

Kole C, Sen S (1985) Role of gamma-irradiation in improvement of *cymbopogon pendulus*. In: Proc 5th ISHS Intl Symp, Darjelling, India, p 42

Kole C, Sen S (1986) Selection strategy for improvement of oil yield in citronella, *Cymbopogon winterianus*. Environ Ecol 4(4):613–618

Kole C, Patra NK, Sen S (1980) Variation of some important traits in citronella (*Cymbopogon winterianus* Jowitt). Ind Perfumer 24(4):185–191

Kole C, Biswas S, Sen S (1981a) Growth and performance of *Cymbopogon citratus* Staf, the West Indian lemongrass and *Cymbopogon pendulus* (Nees ex Steud.) Wats., the Jammu lemongrass in West Bengal. Ind Perfumer 25(1):56–60

Kole C, Biswas S, Sen S (1981b) Morpho-economic features of Burma citronella (*Cymbopogon winterianus* Jowitt). Ind Perfumer 25(1):61–65

Kole C, Sarkar K, Sen S (1983) Improvement of oil of citronella, *Cymbopogon winterianus* jowitt, by x-irradiation. In: Abstr of Contributed Papers, Part I, Intl Congr Genet, New Delhi, p 475

Koo MWL (1994) *Aloe vera*: Antiulcer and antidiabetic effects. Phytother Res 8:455–457

Kulkarni RN, Rajgopal K (1986) Broad and narrow sense heritability estimates of leaf yield, leaf width, tiller number and oil content in east Indian lemongrass. J Plant Breed Berlin 96:135–139

Kulkarni RN, Baskaran K, Chandrashekara RS, Kumar S (1999) Inheritance of morphological traits of periwinkle mutants with modified contents and yields of leaf and root alkaloids. Plant Breed 118:71–74

Kuriakose KP (1995) Genetic variability in East Indian lemongrass (*Cymbopogon flexuosus* Stapf). Ind Perfumer 39:76–83

Lal N, Shasany AK, Lal RK, Darokar MP, Rajkumar S, Sundaresan V, Khanuja SPS (2003) Diversity analysis of vetiver (*Vetiveria zizanioides*) gene bank accessions using RAPD and AFLP analysis. J Med Arom Plant Sci 25:25–32

Lange BM, Wildung MR, Stauber EJ, Sanchez C, Pouchnik D, Croteau R (2000) Probing essential oil biosynthesis and secretion by functional evaluation of expressed sequence tags from mint glandular trichomes. Proc Natl Acad Sci USA 97:2934–2939

Lawrence BM (1981) Monoterpene interrelationships in *Mentha* genus: a biosynthetic discussion. In: Mookherjee BD, Mussinan CJ (eds) Essential Oils. Allured publ, Wheaton, IL, USA, pp 1–81

Lemberg S, Hale RB (1978) Vetiver oil of different geographical origins. Perfum Flav 3:23–27

Loaiza-Figueroa F, Ritland K, Laborde Cancino JA, Tanksley SD (1989) Patterns of genetic variation of the genus *Capsicum* (Solanaceae) in Mexico. Plant Syst Evol 165:159–188

Lupien S, Karp F, Wildung M, Croteau R (1999) Regiospecific cytochrome P450 limonene hydroxylases from mint (*Mentha*) species: cDNA isolation, characterization, and functional expression of (-)-4S-limonene-3-hydroxylase and (-)-4S-limonene-6-hydroxylase. Arch Biochem Biophys 368:181–192

Maheshwari ML, Sethi KL (1987) Selection and improvement in palmarosa. Ind Perfumer 31:17–32

Maldonado-Mendoza IE, Burnett RJ, Nessler CL (1992) Nucleotide sequence of a cDNA encoding 3-hydroxy-3-methylglutaryl coenzyme A reductase from *Catharanthus roseus*. Plant Physiol 100:1613–1614

Manna S, McAnalley BH (1993) Determination of the position of O-acetyl group in a b(1-4) mannan (acemannan) from *Aloe barbadensis* Miller. Carbohyd Res 241:317–319

McCaskill D, Gershenzon J, Croteau R (1992) Morphology and monoterpene biosynthetic capabilities of secretory cell clusters isolated from glandular tichomes of peppermint (*Mentha piperita* L.). Planta 187:445–454

Mc Guire WP, Rowinsky EK, Rosenshein NB, Grumbine FC, Ettinger DS, Armstrong DK (1989) Taxol: a unique antineoplastic agent significant activity in advanced ovarian epithelial neoplasmas. Ann Intern Med 111:273–279

McKnight TD, Roessner CA, Devagupta R, Scott AI, Nessler CL (1990) Nucleotide sequence of a cDNA encoding the vacuolar protein strictosidine synthase from *Catharanthus roseus*. Nucl Acids Res 18:4939

McLeod MJ, Guttman SI, Eshbaugh WH (1982) Early evolution of chilli peppers (*Capsicum*). Econ Bot 36:361–368

McLeod MJ, Guttman SI, Eshbaugh WH, Rayle RE (1983) An electrophoretic study of the evolution in *Capsicum* (Solanaceae). Evolution 37:562–574

Meijer AH, Cardoso MIL, Voskuilen JT, de Waal A, Verpoorte R, Hoge JHC (1993) Isolation and characterization of a cDNA clone from *Catharanthus roseus* encoding NADPH: cytochrome P-450 reductase, an enzyme essential for reactions catalysed by cytochrome P-450 mono-oxygenases in plants. Plant J 4:47–60

Mercke P, Bengtsson M, Bouwmeester HJ, Posthumus MA, Brodelius PE (2000) Molecular cloning, expression, and characterization of amorpha-4,11-diene synthase, a key enzyme of artemisinin biosynthesis of *Artemisia annua* L. Arch Biochem Biophys 381:173–180

Michelmore RW, Paran I, Kesseli RV (1991) Identification of markers linked to disease resistance genes by bulked segregant analysis: a rapid method to detect markers in specific genomic region using segregating populations. Proc Natl Acad Sci USA 88:9828–9832

Mishra P, Kumar S (2000) Emergence of Periwinkle *Catharanthus roseus* as a model system for molecular biology of alkaloids: Phytochemistry, pharmacology, plant biology and *in vivo* and *in vitro* cultivation. J Med Arom Plant Sci 22:306–337

Mitra RL, Jain SK (1987) Concept of *Phyllanthus niruri* (Euphorbiaceae) in Indian Floras. Bull Bot Surv India 27:161–176

Morishige T, Tsujita T, Yamada Y, Sato F (2000) Molecular characterization of the S-adenosyl-L-methionine: 3′-hydroxy-N-methylcoclaurine-4′-O-methyltransferase of isoquinoline alkaloid biosynthesis in *Coptis japonica*. J Biol Chem 275:23398–23405

Moscone EA, Lambrou M, Hunziker AT (1993) Giesma C-banded karyotypes in *Capsicum* (Solanaceae). Plant Syst Evol 186:213–229

Murray MJ, Lincoln DE, Marble PM (1972) Oil composition of *Mentha aquatica* x *Mentha spicata* F_1 hybrids in relation to the origin of *M. X piperita*. Can J Genet Cytol 14:13–29

Nair EGV, Nair KC, Chinnamma NP (1979) Field experiments with micronutrients on yield of grass and oil and citral content of east Indian lemongrass, *Cymbopogon flexuosus* variety OD-9. Ind Perfumer 23:55–58

Nandi RP, Chatterjee SK (1987) Effect of gibberlic acid on growth, development and essential oil formation in *Cymbopogon winterianus* jowitt. Ind Perfumer 31:72–77

National Research Council (1993) Vetiver Grass. A Thin Green Line Against Erosion. National Academy Press, Washington, DC, USA

Paran I, Aftergoot E, Shifriss C (1998) Variation in *Capsicum annuum* revealed by RAPD and AFLP markers. Euphytica 99:167–173

Pasquali G, Goddijn OJM, de Waal A, Verpoorte R, Schilperoort RA, Hoge JHC, Memelink J (1992) Coordinated regulation of two indole alkaloid biosynthetic genes from *Catharanthus roseus* by auxin and elicitors. Plant Mol Biol 18:1121–1131

Patra NK, Srivastava RK, Chauhan SP, Ahmed A, Mishra LN (1990) Chemical features and productivity of a geraniol rich variety (GRL 1) of *Cymbopogon flexuosus*. Planta Medica 56:239–240

Patra N, Tanveer H, Khanuja SPS, Shasany AK, Singh HP, Singh VR, Kumar S (2001) A unique interspecific hybrid spearmint clone with growth properties of *Mentha arvensis* L. and oil qualities of *Mentha spicata* L. Theor Appl Genet 102:471–476

Pauli HH, Kutchan TM (1998) Molecular cloning and functional heterologous expression of two alleles encoding (S)-N-methylcoclaurine 3'-hydroxylase (CYP80B1), a new methyl jasmonate-inducible cytochrome P-450-dependent mono-oxygenase of benzylisoquinoline alkaloid biosynthesis. Plant J 13:793–801

Rao BRR, Bhattacharya AK, Kaul PN, Chand S, Ramesh S (1993) Changes in profiles of essential oil of rose scented geranium (*Pelargonium* species) during leaf ontogeny. J Essen Oil Res 5:301–304

Rao EVSP, Singh M, Rao RSG (1985) Effect of NPK fertilizers on yield and nutrient uptake in lemongrass. Ind J Trop Agric 3:123–127

Rao BL, Sobti SN (1987) Breeding of a high oil yielding lemongrass for flavour industry. Ind Perfumer 31:32

Renou JP, Aubry C, Seveau M, Jalouzot P (1997) Evolution of the genetic variability in the genus *Pelargonium* using RAPD markers. J Hort Sci 72:229–237

Reynolds GW (1966) The Aloes of Tropical Africa and Madagascar. The Aloes Book Fund, Mbabane, Swaziland

Ringer KL, Davis EM, Croteau R (2005) Monoterpene metabolism. Cloning, expression, and characterization of (-)-isopiperitenol/(-)-carveol dehygrogenase of peppermint and spearmint. Plant Physiol 137:863–872

Ringer KL, McConkey ME, Davis EM, Rushing GW, Croteau R (2003) Monoterpene double-bond reductases of the (-)-menthol biosynthetic pathway: isolation and characterization of cDNAs encoding (-)-isopiperitenone reductase and (+)-pulegone reductase of peppermint. Arch Biochem Biophys 418:80–92

Rodriguez JM, Berke T, Engle L, Nienhuis J (1999) Variation among and within *Capsicum* species revealed by RAPD markers. Theor Appl Genet 99:147–156

Rueffer M, Zumstein G, Zenk MH (1990) Partial purification of S-adenosyl-L-methionine:(S)-tetrahydroprotoberberine-cis-N-methyltransferase from suspension cultured cells of *Eschscholtzia* and *Corydalis*. Phytochemistry 29:3727–3733

Saikia D, Khanuja SPS, Shasany AK, Darokar MP, Kukreja AK, Kumar S (2000) Assessment of diversity among *Taxus wallichiana* accessions from North East India using RAPD analysis. Plant Genet Resour Newsl 121:27–31

Saito H (1993) Purification o factive substances of *Aloe arborescens* Miller and their biological ad pharmacological activity. Phytother Res 7:S14–S19

Samanani N, Facchini PJ (2001) Isolation and partial characterization of norcoclaurine synthase, the first committed step in benzylisoquinoline alkaloid biosynthesis, from opium poppy. Planta 213:898–906

Schroder G, Unterbusch E, Kaltenbach M, Schmidt J, Strack D, De Luca V, Schroder J (1999) Light-induced cytochrome P450-dependant enzyme in indole alkaloid biosynthesis: tabersonine 16-hydroxylase. FEBS Lett 458:97–102

Shapova M (1966) Studies in the genus *Capsicum*. Chromosoma (Berl) 19:340–348

Shasany AK, Lal RK, Darokar MP, Patra NK, Garg A, Kumar S, Khanuja SPS (2000) Phenotypic and RAPD diversity among *Cymbopogon winterianus* Jowitt accessions in relation to *Cymbopogon nardus* Rendle. Genet Resour Crop Evol 47:553–559

Shasany AK, Aruna V, Darokar MP, Kalra A, Bahl JR, Bansal RP, Khanuja SPS (2002a) RAPD marking of three *Pelargonium graveolens* genotypes with chemotypic differences in oil quality. J Med Arom Plant Sci 24:729–732

Shasany AK, Srivastava A, Bahl JR, Sharma S, Kumar S, Khanuja SPS (2002b) Genetic diversity assessment of *Mentha spicata* L. germplasm through RAPD analysis. Plant Genet Resour Newsl 130:1–5

Shasany AK, Darokar MP, Dhawan S, Gupta AK, Gupta S, Shukla AK, Patra NK, Khanuja SPS (2005a) Use of RAPD and AFLP markers to identify inter- and intraspecific hybrids of *Mentha*. J Hered 96:542–549

Shasany AK, Shukla AK, Gupta S, Rajkumar S, Khanuja SPS (2005b) AFLP analysis for genetic relationships among *Mentha* species. Plant Genet Resour Newsl 144:14–19

Shyalaraj KS, Thomas J (1993) Ideal plant type concept of palmarosa (*Cymbopogon martinii* var. *motia*). Ind Perfumer 37:213–217

Simpson BB, Conner-Ogorzaly M (1986) Economic Botany: Plants in Our World. McGraw-Hill Publ, New York, USA

Singh HK, Dhawan BN (1978) The effect of *B. monniera* on the learning ability of rats. Ind J Pharmacol 10:72

Singh HK, Dhawan BN (1982) Effect of *B. monniera* Linn. (Brahmi) extracts on avoidance response in rats. J Ethnopharmacol 5:205–214

St-Pierre B, Laflamme P, Alarco AM, De Luca V (1998) The terminal O-acetyltransferase involved in vindoline biosynthesis defines a new class of proteins responsible for coenzyme A-dependant acyl transfer. Plant J 14:703–713

Stapf O (1906) The oil grasses of India and Ceylon. Kew Bull 8:297–463

Stearn WT (1975) A synopsis of the genus *Catharanthus* (Apocynaceae). In: Taylor WI, Farnsworth NR (eds) The *Catharanthus* Alkaloids – Botany, Chemistry, Pharmacology, and Clinical Use. Dekker, New York, USA, pp 9–44

Tanahashi T, Zenk MH (1990) Elicitor induction and characterization of microsomal protopine-6-hydroxylase, the central enzyme in benzophenanthridine alkaloid biosynthesis. Phytochemistry 29:1113–1122

Tewari VP (1991) A multipurpose perennial chilli 'Pusa Sadabahar'. Ind Hort Jan-Mar 30–31

Thul ST, Shasany AK, Darokar MP, Khanuja SPS (2006) AFLP analysis for genetic diversity in *Capsicum annuum* and related species. Nat Prod Commun 1:223–228

Turner GW, Croteau R (2004) Organization of monoterpene biosynthesis in *Mentha*. Immunochemical localizations of geranyl diphosphate synthase, limonene-6-hydroxylase, isopiperitenol dehydrogenase, and pulegone reductase. Plant Physiol 136:4215–4227

Unterlinner B, Lenz R, Kutchan TM (1999) Molecular cloning and functional expression of codeinone reductase: the penultimate enzyme in morphine biosynthesis in the opium poppy *Papaver somniferum*. Plant J 18:465–475

Van Der Bank H, Van Wyk BE, Van Der Bank M (1995) Genetic variation in two economically important *Aloe* species (Aloaceae). Biochem Syst Ecol 23:251–256

van Geldre E, de Pauw I, Inze D, van Montagu M, van den Eeckhout E (2000) Cloning and molecular analysis of two new sesquiterpene cyclases from *Artemisia annua* L. Plant Sci 158:163–171

Vazquez-Flota F, De Carolis E, Alarco A-M, De Luca V (1997) Molecular cloning and characterization of desacetoxyvindoline-4-hydroxylase, a 2-oxoglutarate dependant-dioxygenase involved in the biosynthesis of vin-

doline in *Catharanthus roseus* (L.) G. Don. Plant Mol Biol 34:935–948

Veau B, Courtois M, Oudin A, Chenieux JC, Rideau M, Clastre M (2000) Cloning and expression of cDNAs encoding two enzymes of the MEP pathway in *Catharanthus roseus*. Biochim Biophys Acta 1517:159–163

Verpoorte R, van der Heijden R, Moreno PRH (1997) Biosynthesis of terpenoid indole alkaloids in *Catharanthus roseus* cells. In: Cordell GA (ed) The Alkaloids, vol 49. Academic Press, San Diego, CA, USA, pp 221–299

Vetter H-P, Mangold U, Schröder G, Marner F-J, Werck-Reichhart D, Schröder J (1992) Molecular analysis and heterologous expression of an inducible cytochrome P-450 protein from periwinkle (*Catharanthus roseus* L.). Plant Physiol 100:998–1007

Viljoen AM, Van Wyk BE, Dagne E (1995) The chemotaxonomic value of 10-hydroxyaloin B and its derivatives in Aloe series Asperifoliae Berger. Kew Bull 51(1):159–168

Viljoen AM, van Wyk BE, Van Heerden FR (1999) The chemotaxonomic value of two cinnamoyl chromones, aloeresin E and F in *Aloe* (Aloaceae). Taxon 48:747–754

Wallaart TE, Bouwmeester HJ, Hille J, Poppinga L, Maijers NC (2001) Amorpha-4,11-diene synthase: cloning and functional expression of a key enzyme in the biosynthetic pathway of the novel antimalarial drug artemisinin. Planta 212:460–465

Wallaart TE, van Uden W, Lubberink HG, Woerdenbag HJ, Pras N, Quax WJ (1999) Isolation and identification of dihydroartemisinic acid from *Artemisia annua* and its possible role in the biosynthesis of artemisinin. J Nat Prod 62:430–433

Wang J, Dudareva N, Kish CM, Simon JE, Lewinsohn E, Pichersky E (1999) Nucleotide sequences of two cDNAs encoding caffeic acid O-methyltransferases (Accession Nos. AF154917 and AF154918) from sweet basil (*Ocimum basilicum*). Plant Physiol 120:1205

Webster GL (1957) A monographic study of the West Indian species of *Phyllanthus*. J Arnold Arboric Harv Univ 39:49–100

Webster GL (1994) Synopsis of the genus and suprageneric taxa of Euphorbiaceae. Annu Missouri Bot Gar 81:33–144

Winters WD, Benavides R, Clouse WJ (1981) Effects of *Aloe* extracts on human normal and tumor cells in vitro. Econ Bot 35:89–95

Wise ML, Croteau R (1999) Monoterpene biosynthesis. In: Cane DE (ed) Comprehensive Natural Products Chemistry: Isoprenoids including Carotenoids and Steroids, vol 2. Elsevier, Oxford, UK, pp 97–153

Wood JRI (1982) The Aloes of the Yemen Arab Republic.Kew Bull 38:13–31

Subject Index